NATO ASI Series

Advanced Science Institutes Series

A series presenting the results of activities sponsored by the NATO Science Committee, which aims at the dissemination of advanced scientific and technological knowledge, with a view to strengthening links between scientific communities.

The Series is published by an international board of publishers in conjunction with the NATO Scientific Affairs Division

A	Life Sciences	Plenum Publishing Corporation
B	Physics	London and New York
C	Mathematical and Physical Sciences	Kluwer Academic Publishers
D	Behavioural and Social Sciences	Dordrecht, Boston and London
E	Applied Sciences	
F	Computer and Systems Sciences	Springer-Verlag
G	Ecological Sciences	Berlin Heidelberg New York
H	Cell Biology	London Paris Tokyo Hong Kong
I	Global Environmental Change	Barcelona Budapest

PARTNERSHIP SUB-SERIES

1. Disarmament Technologies	Kluwer Academic Publishers
2. Environment	Springer-Verlag/Kluwer Acad. Publishers
3. High Technology	Kluwer Academic Publishers
4. Science and Technology Policy	Kluwer Academic Publishers
5. Computer Networking	Kluwer Academic Publishers

The Partnership Sub-Series incorporates activities undertaken in collaboration with NATO's Cooperation Partners, the countries of the CIS and Central and Eastern Europe, in Priority Areas of concern to those countries.

NATO-PCO DATABASE

The electronic index to the NATO ASI Series provides full bibliographical references (with keywords and/or abstracts) to about 50 000 contributions from international scientists published in all sections of the NATO ASI Series. Access to the NATO-PCO DATABASE compiled by the NATO Publication Coordination Office is possible in two ways:

- via online FILE 128 (NATO-PCO DATABASE) hosted by ESRIN,
  Via Galileo Galilei, I-00044 Frascati, Italy.

- via CD-ROM "NATO Science & Technology Disk" with user-friendly retrieval software in English, French and German (© WTV GmbH and DATAWARE Technologies Inc. 1992).

The CD-ROM can be ordered through any member of the Board of Publishers or through NATO-PCO, Overijse, Belgium.

Series I: Global Environmental Change, Vol. 47

Springer
Berlin
Heidelberg
New York
Barcelona
Budapest
Hong Kong
London
Milan
Paris
Santa Clara
Singapore
Tokyo

Past and Future Rapid Environmental Changes:

The Spatial and Evolutionary Responses of Terrestrial Biota

Edited by

Brian Huntley

University of Durham
Environmental Research Centre
Department of Biological Sciences
South Road, Durham DH1 3LE, U.K.

Wolfgang Cramer

Potsdam Institute for Climate Impact Research
Telegrafenberg, P.O. Box 6012 03
D-14412 Potsdam, Germany

Alan V. Morgan

Quaternary Sciences Institute
University of Waterloo
Waterloo, Ontario N2L 3G1, Canada

Honor C. Prentice

Lund University
Department of Systematic Botany
Östra Vallgatan 18–20
S-223 61 Lund, Sweden

Judy R. M. Allen

University of Durham
Environmental Research Centre
Department of Biological Sciences
South Road, Durham DH1 3LE, U.K.

With 84 Figures and 5 Colour Plates

Springer

Published in cooperation with NATO Scientific Affairs Division

Proceedings of the NATO Advanced Research Workshop "Past and Future
Rapid Environmental Changes: The Spatial and Evolutionary Responses of
Terrestrial Biota", held at Crieff, Scotland, June 26–30, 1995

Library of Congress Cataloging-in-Publication Data

Past and future rapid environmental changes : the spatial and
  evolutionary responses of terrestrial biota / edited by Brian
  Huntley ... [et al.].
      p.   cm. -- (NATO ASI series. Series I, Global environment
  change ; vol. 47)
    "Proceedings of the NATO Advanced Research Workshop "Past and
  Future Rapid Environmental Changes--the Spatial and Evolutionary
  Responses of Terrrestrial Biota", held at Crieff, Scotland, June
  26-30, 1995"--T.p. verso.
    Includes bibliographical references and index.
    ISBN-13:978-3-642-64471-9
    1. Paleoecology--Holocene--Congresses. 2. Climatic changes-
  -Congresses.   I. Huntley, Brian, 1952-   . II. NATO Advanced
  Research Workshop "Past and Future Rapid Environmental Changes: the
  Spatial and Evolutionary Responses of Terrestrial Biota" (1995 :
  Crieff, Scotland) III. Series.
  QE720.P38   1997
  560'.4522--dc21
                                                        96-45079
                                                            CIP

ISBN-13:978-3-642-64471-9   e-ISBN-13: 978-3-642-60599-4
DOI: 10.1007/978-3-642-60599-4

© Springer-Verlag Berlin Heidelberg 1997
Softcover reprint of the hardcover 1st edition 1997

Typesetting: Camera ready by authors/editors
Printed on acid-free paper
SPIN: 10502907   31/3137 - 5 4 3 2 1 0

PREFACE

This volume contains the Proceedings of the NATO Advanced Research Workshop on '*Past and future rapid environmental changes: The spatial and evolutionary response of terrestrial biota*', held at Crieff, Scotland, June 26 – 30, 1995. A total of 40 participants took part in the workshop; in addition a member of the NATO '*Science of Global Environmental Change*' Programme Committee, Mr Max Beran, attended for one day.

As well as 33 papers submitted by workshop participants, 2 papers are included that were written by intending participants who were, at the last moment, prevented from attending (Dr Denis-Didier Rousseau and Prof. Allen M. Solomon). An Introduction and a final section reporting the discussions and conclusions from the working group sessions also are included. These have been written by the editors, although the latter draws heavily upon the reports of the working group rapporteurs.

The workshop brought together a diverse group of scientists with disparate backgrounds; inevitably, as a result, the papers in the present volume also deal with a wide range of topics that includes, for example, population genetics of plants, Quaternary evolutionary history of vertebrates, mechanisms of invasion by plants and large-scale range displacements of beetles. The unifying theme was the response of terrestrial biota to rapid environmental changes and in this context papers on modelling past and future environmental changes and the response of organisms to these changes also are included.

The overwhelming and unanimous conclusion of the workshop participants was that forecast global environmental changes pose a severe threat to the integrity of global ecosystems and to the survival of at least some species. The papers here presented provide some of the evidence upon which this conclusion is based. It is the hope of all who participated in this workshop that policy makers worldwide will heed the warnings of scientists before severe damage is caused to the biosphere.

Brian Huntley (Workshop Director)
Durham, 25 July, 1996

Photographs by John R.G. Daniell

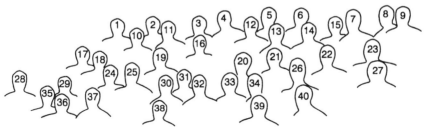

1 Alan J. Gray;   2 Philip Grime;   3 Richard N. Mack;   4 Dale S. Solomon;
5 Patrick J. Bartlein; 6 Håkan Hytteborn; 7 Richard C. Preece; 8 Rob Hengeveld;
9 John C. Avise; 10 Thompson Webb III; 11 Martin T. Sykes; 12 Matt. S. McGlone;
13 Adrian M. Lister; 14 Andrei V. Sher; 15 Margaret B. Davis; 16 Honor C. Prentice;
17 John C. Coulson; 18 Max Beran; 19 Russell W. Graham; 20 Nathalie de Noblet;
21 Allan C. Ashworth;   22 Philippe Ponel;   23 Herman H. Shugart;
24 Yvonne C. Collingham;   25 Wolfgang Cramer;   26 Paul N. Dolman;
27 Alan V. Morgan;   28 Judy R. M. Allen;   29 Jennifer E. L. Butterfield;
30 George A. King; 31 Delphine Texier 32 Harald Bugmann; 33 Vera Markgraf;
34 Alberte Fischer   35 Annika Hofgaard;   36 Brian Huntley;   37 Peter Coxon;
38 Elisabeth S. Vrba; 39 Csaba Mátyás; 40 Peter R. Evans.

ACKNOWLEDGEMENTS

The workshop whose proceedings comprise the present volume could not have taken place without the encouragement of the Global Environmental Change Committee of NATO, and Max Beran in particular, and financial support from the Scientific and Environmental Affairs Division of NATO. The staff of the Crieff Hydro Hotel, Crieff, Perthshire, Scotland, especially Sheonagh McLeod, are thanked for all of their help in ensuring the smooth running of the workshop and for making our visit to Crieff comfortable and enjoyable.

John R.G. Daniell and Yvonne C. Collingham provided invaluable assistance to the organisers and delegates with respect to transport arrangements and other practical matters during the workshop. James and Alexina Huntley packed and labelled many envelopes and provided various other organisational assistance during the period of preparation for the workshop; without their help the Workshop Director's task would have been much more onerous.

The Workshop Director and the other members of the Organising Committee must express their thanks to Allen M. Solomon for his contribution to the planning of the workshop; his advice and recommendations for participants helped make the workshop as successful as it was. We only are sorry that circumstances prevented him from joining us in Crieff and from playing a rôle in the editing of this volume.

The authors of the papers that comprise this volume must all be thanked for their prompt delivery of manuscripts, for keeping to the deadlines set by the editors and for their co-operation in the preparation of the final camera-ready typescript. All of the papers were subjected to peer review by two referees; the editors are grateful to those workshop participants who acted in this capacity and especially to those referees who were not workshop participants but who willingly agreed to undertake this task, namely:

Dr Stefan Andersson, Department of Systematic Botany, Lund University, Sweden

Dr Robert Baxter, Environmental Research Centre, University of Durham, UK

Dr Martin Claußen, Potsdam Institut für Klimafolgenforschung, Potsdam, Germany

Prof. Eric A. Colhoun, Department of Geography, The University of Newcastle, New South Wales, Australia

Dr H. Basil S. Cooke, White Rock, British Columbia, Canada

Dr Richard A. Ennos, Institute of Ecology and Resource Management, University of
Edinburgh, UK

Dr Bruno J. Ens, Institute for Forestry and Nature Research, Den Burg,
The Netherlands

Dr Daniel C. Fisher, University of Michigan, Anne Arbor, Michigan, USA

Dr Richard C. Harington, Canadian Museum of Nature, Ottawa, Ontario, Canada

Dr Mark O. Hill, Institute of Terrestrial Ecology, Monks Wood, UK

Dr Philip E. Hulme, Department of Biological Sciences, University of Durham, UK

Dr Felix Kienast, Botanisches Institut, Universität Bern, Switzerland

Prof. Gundolf Kohlmaier, Institut für Physikalische und Theoretische Chemie, Johann-
Wolfgang Goethe Universität, Frankfurt am Main, Germany

Dr John V. Matthews Jr., Geological Survey of Canada, Ottawa, Ontario, Canada

Dr Barry B. Miller, Kent State University, Kent, Ohio, USA

Dr R.F. Miller, New Brunswick Museum, St John, New Brunswick, Canada

Dr Anne Morgan, University of Waterloo, Waterloo, Ontario, Canada

Prof. Ramakrishna Nemani, School of Forestry, University of Montana, Missoula,
Montana, USA

Mr Ralf Otto, Institut für Physikalische und Theoretische Chemie, Johann-Wolfgang
Goethe Universität, Frankfurt am Main, Germany

Prof. I. Colin Prentice, Department of Ecology, Lund University, Sweden

Prof. Outi Savolainen, Department of Genetics, University of Oulu, Finland

Dr Håkan Tegelström, Department of Genetics, Uppsala University, Sweden

Prof. W.A. Watts, School of Botany, Trinity College, Dublin, Ireland

In preparing the concluding discussion section of this volume the editors have drawn
heavily upon the notes and reports of the rapporteurs to the various workshop
discussion groups. We would like to thank Margaret B. Davis, Russell W. Graham,
Adrian M. Lister, Herman H. Shugart and Elisabeth S. Vrba for acting as rapporteurs.

Last, but by no means least, the Workshop Director must express his gratitude to
Judy R.M. Allen for her unstinting assistance with organisational and practical details
during the workshop and, on behalf of himself and the other editors, also must thank
her for the huge effort that she has put into the preparation of the final camera-ready
typescript.

TABLE OF CONTENTS

Introduction

Brian Huntley

Wolfgang Cramer

Alan V. Morgan

Honor C. Prentice

Judy R.M. Allen

During the Quaternary ice-age that has occupied the recent geological past (the last
ca. 2·4 Ma) the terrestrial biota has experienced a series of rapid environmental
changes. In order to survive such changes, organisms either must change their
spatial distribution ('migrate') so as to track the geographical location of suitable
conditions, or else must evolve so as to adapt to the new environmental conditions;
some may exhibit a combination of the two responses. Although the most visible
response of many organisms to the rapid environmental changes during the
Quaternary has been spatial, large changes in their geographical distributions having
taken place, major morphological evolution also has occurred in some groups. Many
other species probably have undergone cryptic evolution of their physiological
characteristics and tolerances. Few species are likely to have survived Quaternary
climatic changes without at least some adaptive evolutionary response to their
changing environment. It can be predicted that spatial and evolutionary elements
also will be combined in the response of organisms to future rapid environmental
changes, including those forecast for the next one or two centuries that will result
from human additions of greenhouse gases to the global atmosphere.

The principal aim of the workshop was to assess the relative importance of spatial
and evolutionary mechanisms in the response of terrestrial biota to changes in their
environment both in the past and in the future. It was intended that some general
conclusions should emerge from the workshop as to maximum rates and magnitudes
of environmental change to which organisms of different groups may attain a
sufficient response, by each of these two mechanisms, to enable them to avoid
substantial population reduction and/or potential extinction.

NATO ASI Series, Vol. I 47
Past and Future Rapid Environmental Changes:
The Spatial and Evolutionary Responses of
Terrestrial Biota
Edited by Brian Huntley et al.
© Springer-Verlag Berlin Heidelberg 1997

In order to address this overall aim, four broad topics were discussed:

1. The Quaternary palaeoecological and other evidence documenting the migration of biota in response to the rapid environmental changes associated with alternating glacial and interglacial stages.

2. The Quaternary palaeontological evidence of evolutionary changes seen in some elements of the terrestrial biota — changes that may, at least in part, have been responses to the rapid environmental fluctuations during this geological period.

3. Ecological studies of those dispersal and population dynamic attributes of organisms that contribute to their capacity to achieve rapid spatial responses to a changing environment.

4. Contemporary studies of micro-evolution and of population genetics that document adaptive responses to environmental heterogeneity and/or change, as well as the mechanisms whereby these responses are achieved.

An overall emphasis was placed upon quantifying both the rates of the responses attained by means of the two alternate mechanisms, and the magnitudes of change to which each mechanism might provide a sufficient response.

The 35 invited papers written by participants and intending participants are organised below into six sections that reflect the six themes around which the plenary sessions of the workshop were organised. The sections are outlined below.

Past environmental changes — the late-Quaternary

The two papers in this section serve a scene-setting rôle. The first, by Pat Bartlein, provides general background on the nature of the climatic changes that have characterised the Quaternary, focusing especially upon placing in context the climate of the present and the changes of the geologically recent past, as well as predicted future changes. In the second paper Nathalie de Noblet discusses recent advances in the modelling of palaeoclimates using atmospheric general circulation models (GCMs); she focuses particularly upon the feedbacks between the biosphere and climate.

Spatial responses to past changes

This section is comprised of ten papers that present evidence from studies of fossil remains of terrestrial organisms showing that both plant and animal taxa have in general migrated in response to Quaternary environmental changes. The first five papers consider the palaeovegetation record; the first three dealing with the responses of vegetation in three contrasting geographical areas. In eastern North America, as in Europe, the dominant forest tree taxa have exhibited large-scale range changes in response to late-Quaternary environmental changes; Tom Webb provides a discussion of this type of migratory response. In contrast, the isolated archipelago of New Zealand offers limited scope for substantial shifts of geographical distribution and Matt McGlone argues that individual taxa have undergone only limited range adjustments; at some times they have been limited to spatially-restricted habitats (e.g. steep north- or south-facing slopes, gorges, etc.), whereas at others they have been able to occupy a major proportion of the landscape. Vera Markgraf and Ray Kenny document the speed with which the vegetation of southern South America responded to the rapid climate changes of the late-glacial; once again it seems that large-scale spatial shifts may have been less important than changes of dominance on the landscape. Despite the differences of scale of the documented responses, migration underlies all of these examples and, in the next paper, George King and Andrew Herstrom present a new approach to determining the migration rates of trees from the palynological record. The fifth of the palaeovegetation papers, by Pete Coxon and Stephen Waldren, discusses longer-term floristic changes that have accompanied the continuous changes of range seen amongst taxa in the flora of north-west Europe during the Quaternary period.

The next three papers, by Allan Ashworth, Alan Morgan and Philippe Ponel respectively, deal with the evidence of comparable range changes exhibited by beetles in different parts of North America and in Europe. These have been quite considerable, movements of thousands of kilometres not being uncommon; average rates probably are in the range of 500 m to 2 km per year. Beetles have exhibited little, if any, morphological evolution during the Quaternary. Species assemblages have remained relatively intact. A degree of individualistic response by some species, however, has resulted in somewhat different communities at various times; this being most pronounced at times of rapid climate change when generally short-lived, 'mixed' assemblages occur. Russ Graham provides an account of the large changes of range seen in North America amongst vertebrates, including especially those taxa with small body size. Finally, in this section, Richard Preece shows that even terrestrial molluscs have achieved surprisingly rapid and large-scale range changes in response to Quaternary climate changes.

A common feature of all of these studies, seen also in the palaeovegetation record, is the evidence that past assemblages of taxa frequently are unlike any modern assemblage. Taxa that co-occur in the fossil record often have non-overlapping biogeographic ranges at the present day. There also is a degree of consistency with respect to the rates of migration observed, especially of the large-scale range displacements of trees seen in Europe and eastern North America.

Mechanisms enabling a spatial response

The extent and speed with which plants and animals have migrated during the Quaternary having been documented, the eight papers in this section address the evidence that can be obtained from contemporary studies that may shed light upon the mechanisms whereby taxa attain such migrations.

In the first paper, Margaret Davis and Shinya Sugita re-examine the palynological record of tree migration in order to assess more critically the evidence relating to the rates and mechanisms of these past migrations. In contrast, Philip Grime considers principally community-level response and especially a research strategy designed to elucidate the general principles underlying such responses. Richard Mack presents a discussion of invasions by introduced plant species; he proposes that these provide an alternative model for the rate and character of future migratory responses and also argues that direct human impacts through introduction, cultivation, etc. may be a more important determinant of future migrations than will be the response of plants to anthropogenic climate changes. Rob Hengeveld and Frank Van den Bosch discuss an approach to modelling invasion, especially by mobile animals, and present an extension of their previous modelling approaches that now takes account of landscape heterogeneity.

In the next paper, attention is switched to birds as an animal group that undergoes regular annual migrations in response to seasonal climatic changes. Peter Evans discusses the various ways in which climate change may affect migratory birds by altering the overall length of their annual migrations and/or changing the quality and availability of feeding sites that may be of vital importance as 'staging posts' during migration.

The last three papers in this section consider different aspects of the population and structural changes in vegetation as a consequence of climate change. Dale Solomon and his co-authors consider the evidence that may be gleaned from repeated measurements of forest plots and how this may be used to reveal the nature and magnitude of the response of forests to current and ongoing climate change. Annika Hofgaard considers the historical ecological and palaeoecological evidence of dynamic structural changes at the forest–tundra ecotone in response to climate change. In the last paper in this section the use of vegetation dynamic models to investigate the structural responses of vegetation to climate change is discussed by Hank Shugart and his co-authors.

Evolutionary responses to past changes

Although, as the papers in the second section emphasise, many taxa have shown spatial responses to Quaternary climate change, a variety of taxa have undergone evolutionary changes during the Quaternary that may represent adaptive responses to climate change. The papers in this section consider different aspects of the morphological evolution seen especially amongst vertebrates during this time. In the first, Elisabeth Vrba concentrates upon the relationship between rapid environmental changes and periods of evolutionary change; she relates the evolutionary changes exhibited by various vertebrate groups to their long-term history of migration and also considers some of the implications for efforts to conserve taxa in the face of future climate change. Adrian Lister provides an account of various aspects of vertebrate evolution during the Quaternary, focusing upon the faunas of higher latitudes of the northern hemisphere and the extent to which the observed morphological evolution can be interpreted in terms of adaptive responses to a changing environment. Denis-Didier Rousseau considers the evidence for adaptive morphological evolution amongst terrestrial molluscs, and examines the extent to which the temporal changes observed at any given locality during the Quaternary mirror present-day spatial variation in morphology within the same taxon; his evidence suggests that the evolutionary responses seen within molluscs during the Quaternary involve principally differential selection among a constant range of genotypes.

Andrei Sher considers a somewhat different theme, addressing the causes of the late-Quaternary extinction of many large mammals in higher latitude areas of the northern hemisphere. These extinctions may be considered to be a consequence of failure to evolve the adaptations required to survive in the changed environment, whilst at the same time either failing to migrate so as to track suitable environmental conditions and/or suffering a catastrophic reduction or effective loss of areas of suitable environment. If, as Sher argues, large-scale habitat loss was the principal cause for these late-Quaternary extinctions, then this episode may provide a valuable analogue for the potential consequences of habitat losses brought about both directly by human land-use and indirectly as a consequence of anthropogenic climate change.

Mechanisms enabling an evolutionary response

Six papers in this section deal with various aspects of the mechanisms that might permit a taxon to adapt to a changing environment rather than migrating so as to maintain its position in environmental space. The first five papers deal with genetic mechanisms, whereas the sixth describes the extent to which apparent phenotypic plasticity may play a rôle in some cases.

In the first paper, Honor Prentice examines the way in which levels of genetic diversity and the spatial structuring of genetic variation may reflect both the consequence of chance events during range expansion or population fragmentation and the effects of selection under different environmental conditions. Csaba Mátyás then considers the evidence that populations from differing parts of the geographical range of a tree species show genetic variation that differentially adapts them to various climate régimes and thus may render some populations more, and others less, susceptible to climate change. Genetic variation within higher plants also is the subject of the third paper; Alan Gray, however, focuses upon aspects of plant reproductive biology that may be under genetic control and that exhibit apparently adaptive variation across species' geographical ranges.

In the fourth paper in the section, John Avise turns to animals and to the topic of intra-specific phylogeography. He shows how the recent phylogenetic and biogeographical history of taxa may be revealed by patterns of mitochondrial DNA variation in extant populations. This, in turn, indicates much about the response of species to past environmental changes. Paul Dolman also considers intra-specific genetic variations within vertebrate taxa; his paper deals with migratory birds and presents a simple model that enables the potential response of a species to changing environmental suitability to be assessed for various alternative genetic control mechanisms and/or mating and reproductive strategies.

The final paper in the section, by Jennifer Butterfield and John Coulson, considers invertebrate taxa that exhibit phenotypic plasticity with respect to their life-cycle attributes; the rôle of such plasticity in enabling these organisms to adapt rapidly to changing environmental conditions is discussed.

Predicted future environmental changes

The final section comprises five papers that consider both the nature and magnitude of forecast global environmental changes and the potential impacts of these changes on a series of different scales and levels of abstraction. Most of the results presented arise from the application of models of the environmental response of different components or functions of the biosphere. The first of these papers, by Wolfgang Cramer and Will Steffen, provides the general background by outlining the forecast global changes. Transformations of these changes into the likely consequent impacts upon ecosystems on various time scales are modelled, providing both an expression of the magnitude of the global changes that is more immediately comprehensible than the raw climatological changes, and at the same time representing one way to examine the biospheric response to the likely changes.

Martin Sykes provides an assessment of the potential impacts of climate change upon the geographic ranges of a series of higher plant species that are found today in northern Europe but that exhibit a variety of overall distribution patterns. Moving in scale to that of the stand or plant community, Harald Bugmann describes the results of modelling the impacts of climate change on the composition and dynamics of central European forests. Al Solomon addresses the dynamics of the migratory responses of trees that will determine their ability to attain the potential range changes exemplified earlier in the paper by Sykes; in particular he examines the consequences for global carbon balance of alternative scenarios of tree migration.

In the final paper, Alberte Fischer presents preliminary results from a comparative assessment of a series of global net primary productivity models. Such models represent a level of abstraction where only a global biospheric function is simulated. Such simulations, however, represent an important step beyond a modelling approach in which the biosphere primarily responds to changes in the global environment and towards an approach in which the feedbacks between the biosphere and the global environment are modelled mechanistically.

The final section of the volume addresses the topics discussed by the working groups and the plenary discussion sessions; these had as their general theme 'Predicting the response of terrestrial biota to future environmental changes'. Initially the discussion focuses upon a series of questions about the relative importance of the alternative mechanisms of response to environmental change, and especially upon the rates and magnitudes of environmental change to which each mechanism could provide a sufficient response. Plants and animals first are considered separately before being compared. Subsequently, the implications of environmental change are discussed with respect to:

i.   ecosystem function;
ii. the likelihood of extinctions; and
iii. the last deglaciation as a model for future rapid environmental change.

In relation to the last of these points, discussion focuses especially upon the conservation issues arising as a consequence of rapid, large-magnitude environmental change; an attempt is made to evaluate the likelihood of successful conservation measures being possible in the face of such rapid change.

Finally, the conclusions reached at the last plenary session of the workshop are presented. It was the clear and unanimous wish of the workshop participants that these conclusions be brought to the widest possible attention both amongst policy makers and the public. The editors hope that this volume may inform and stimulate the ongoing debate relating to the impacts of global environmental changes and the most appropriate policies to adopt if the causes of these changes are to be addressed.

Section 1

Past environmental changes – the late Quaternary

Past environmental changes:

Characteristic features of Quaternary climate variations

Patrick J. Bartlein

Department of Geography

University of Oregon

Eugene, OR 97403 -1251

U.S.A.

INTRODUCTION

Climate varies, both temporally and spatially, and the focus of the body of palaeoclimatic research has been to characterize and explain this variation. What has emerged is a view of climate as a multi-component *system* that varies across a range of temporal and spatial scales. Embedded in these variations are some characteristic features such as trends, abrupt changes, periodicities, and the formation of spatial mosaics of climatic changes; these features occur on many scales and are the subject of this paper.

Several approaches have been used to characterize the variability of climate at different temporal or spatial scales. These approaches include, for example:

1. the 'powers-of-ten' presentation (Webb, 1989) wherein climatic time series of different duration (e.g. the past 100,000, the past 10,000, the past 1000 years) are plotted and examined;

2. portrayal of the variance spectrum of climatic time series, which reveals the relative importance of variations at different timescales (Mitchell, 1976; Shackleton & Imbrie, 1990); or

3. 'scale diagrams' that indicate the characteristic temporal and spatial scales that the particular components of climate vary on (e.g. McDowell *et al.*; 1990; Webb, 1995).

NATO ASI Series, Vol. I 47
Past and Future Rapid Environmental Changes:
The Spatial and Evolutionary Responses of
Terrestrial Biota
Edited by Brian Huntley et al.
© Springer-Verlag Berlin Heidelberg 1997

Although these approaches indicate the general magnitude of climatic variations and the timescales that they occur on, the presentations tend to obscure the *particular* characteristics of the climatic changes that may have evoked spatial or evolutionary responses from terrestrial biota.

The plan of this paper is first to describe the hierarchy of controls and responses of components of the climate system, and then discuss some of the characteristic features of temporal and spatial variations of climate that recur on different scales. This discussion draws heavily on earlier papers by Barnosky (1987), Bartlein (1988), Bartlein and Prentice (1989), McDowell *et al.* (1990), Webb and Bartlein (1992), and Webb (1995).

THE HIERARCHY OF CLIMATE-SYSTEM CONTROLS AND RESPONSES

Climatic variations over time at a particular place are governed by a hierarchy of controls and responses (Table 1); these begin with the external controls of the climate system and end with the responses of a number of local climate components at a place. The responses at any one level of the hierarchy become the controls of the components at lower levels. For example, on timescales of 10^4 to 10^6 years, ice sheets are dependent variables in the climate system, governed by orbitally controlled variations in insolation. At shorter timescales (10^4 to 10^3 years), ice sheets act as independent variables that have an important influence on global temperatures and atmospheric circulation.

In general, the more slowly varying components of the climate system (e.g. ice volume, long-term trends in atmospheric composition) operate at higher levels of the hierarchy, while the more rapidly varying components (e.g. storm tracks, precipitation fields) operate at lower levels. The spatial and temporal scale of the potential impacts of the characteristic features can be gauged by considering which levels in the hierarchy control the climatic variations described below, and which are involved in the response.

The existence of this hierarchy also has implications for attempts to explain the variations at a particular place. For example, although the variations at a place are ultimately governed by global-scale controls, a specific palaeoclimatic record is generally not necessarily representative of the general state of the global system;

this situation arises because the intermediate controls and responses have the potential of reinforcing, cancelling, or even reversing the global trend. Gradual changes in large-scale controls may sometimes produce abrupt local changes when atmospheric circulation is reorganized. Conversely, abrupt changes in the large-scale circulation may produce warming in some regions and cooling or no change in others, as can be seen in the spatial anomaly patterns of year-to-year variations of climate. Consequently, it may be difficult or even impossible to ascribe a particular climate variation at a place to a specific set of higher-level controls.

Table 1. Hierarchy of climate system controls and responses

External Controls	astronomic solar (output, orbital variations) tectonic (continental distributions, mountain barriers, ocean gateways and barriers) atmospheric composition
Global-Scale Responses/Controls	latitudinal and seasonal distributions of insolation ice volume atmospheric composition zonal energy balance
Hemispheric-Scale Responses/Controls	atmospheric general circulation (e.g. westerlies) ocean general circulation ocean thermohaline circulation
Continental-Scale Responses/Controls	atmospheric circulation features (e.g. Aleutian low, E Pacific sub-tropical high) land-surface cover sea-surface temperature patterns
Regional-Scale Responses/Controls	storm track locations airmass distributions (heat and moisture fluxes) clouds
Mesoscale Responses/Controls	temperature, precipitation, and wind fields as modified by orography
Local-Scale Responses	precipitation, air temperature, incident shortwave radiation, net radiation, water vapour pressure, wind speed, soil moisture, atmospheric deposition, lightning

ILLUSTRATING THE CHARACTERISTIC FEATURES OF TEMPORAL AND
SPATIAL VARIATIONS OF PAST CLIMATE

The characteristic features of temporal variations of climate can be discussed using
a selection of representative time series, in the style of the powers-of-ten diagrams;
however, I will concentrate more on the kinds of variations that are evident than on
the amplitude or timescale of the variations. The characteristic features of the spatial
variations of climate are harder to describe, because, with the exception of the
present, the requisite dense networks of sites exist only in a few regions.
Consequently, these features will be described using the results of some
palaeoclimatic simulations, supplemented by some contemporary observations.

Representative Time Series

A selection of representative time series illustrates the kinds of features that occur in
palaeoclimatic variations (Figure 1). Series A in Figure 1 shows a composite record
of oxygen isotope variations over the Cenozoic (Miller *et al.* 1987). The variations
reflect both the average temperature of the oceans as well as the global ice volume.
Series B is a 5 Ma – long record of oxygen isotope variations from an ocean core
located off the coast of West Africa (Tiedemann *et al.* 1994). The short-term isotopic
variations displayed in the record reflect mainly the growth and decay of ice sheets,
while the trend in this series reflects the development and increase in size of
Northern Hemisphere ice sheets. Series C is the 'SPECMAP stacked-and-
smoothed' oxygen isotopic record (Imbrie *et al.* 1984). In this series, short-term and
local variations have been removed by the data analysis, and the record can be
viewed as an index of the general level of global ice volume over the past 800 kyr.
Series D displays the oxygen-isotopic variations contained in the Greenland Ice Core
Project (GRIP) Summit core (Dansgaard *et al.* 1993). This record provides a general
indication of climatic conditions over the North Atlantic and adjacent regions for the
last glacial-interglacial cycle, and comparison of this record with the SPECMAP
record (Series C) suggests that these variations generally reflect global-scale
variations. However, in detail, the isotopic variations in Series D probably also
represent local variations in air temperature and, in part, variations in the moisture
sources of precipitation over Greenland, both of which are controlled by atmospheric
circulation variations (Charles *et al.* 1994).

15

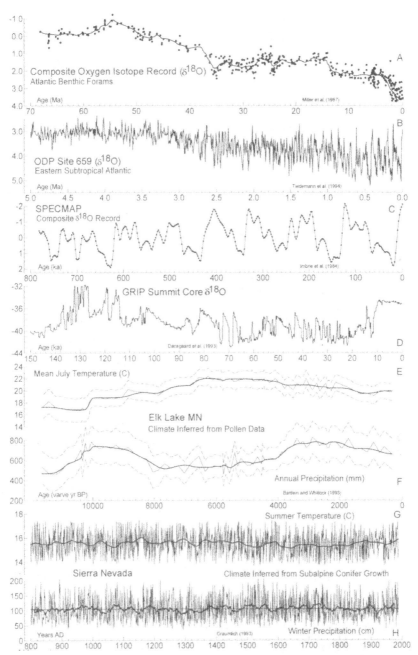

Figure 1: Representative time series that illustrate some characteristic features of palaeoclimatic variations (see text for discussion). Each series is plotted so that warm or wet conditions appear at the top of each graph, and cool or dry at the bottom.

The next sets of representative time series reflect regional- and local-scale variations more than they do global-scale ones. Series E and F display reconstructions of mean July temperature and annual precipitation based on pollen data in a varved core from Elk Lake, Minnesota (Bartlein & Whitlock 1993). The effective sampling interval is about 100 yrs in this core. The prediction error intervals (dashed lines), and smoothed values of the reconstructions are also shown on the figure. Series G and H are annual reconstructions of summer temperature and winter precipitation, based on the growth of subalpine conifers in the Sierra Nevada (Graumlich 1993). The thick solid lines superimposed on each series are smoothed values of the reconstructions, plotted to reveal the principal timescales of variability in each record, and the thin horizontal lines indicate the long-term mean of each series (Graumlich 1993).

Spatial Patterns of Past Climatic Variations

Only few regions, such as eastern North America (e.g. Webb *et al.* 1993), or western Europe (e.g. Huntley & Prentice 1988, 1993), have dense enough networks of palaeoclimatic observations to allow reconstruction of the spatial patterns of past climate at the regional or smaller scale. Consequently, I used the results of a set of palaeoclimatic simulations done with the National Center for Atmospheric Research (NCAR) Community Climate Model (CCM 1) to illustrate the continental-scale pattern of past climates for North America (Kutzbach *et al.* in prep). The simulations were completed for 21, 16, 14, 11 and 6 ka (calendar years) and present, and I interpolated the simulated anomalies onto a 25 km grid of modern climate values over North America (Bartlein *et al.* in prep.), adding the anomalies to the modern values. Maps of these data display the changing patterns of climate since the last glacial maximum (Figure 2). Although based on simulations that are likely inaccurate, such maps do illustrate the types of patterns of climatic change that arise over a continental-size region.

To further illustrate some of the features of the spatial variations of climate that occur, I determined the uniqueness of the simulated climate for each point and time (Figure 3), by finding the minimum dissimilarity (Euclidean distance) between the climate at that point, and the climates at the other grid points and times. These dissimilarities reveal those points without good analogues at other times, i.e. those with unique climates. Although the simulated climates are inaccurate, the resulting patterns of the uniqueness of the climate at each point and time nevertheless illustrate the kinds of patterns that likely occurred in the real climate.

Another perspective on spatial variations of climate can be gained by examining the pattern of precipitation seasonality using long-term average monthly precipitation (Mock 1996). A map of the season of occurrence of the precipitation maximum for climate stations in western North America appears as Figure 4, and this display reveals broad-scale patterns of precipitation seasonality as well as more local-scale patterns of heterogeneity (see also Whitlock & Bartlein 1993).

January Temperature (0 to 5°C)

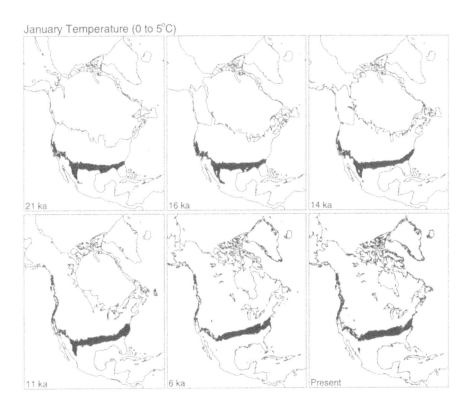

Figure 2: The changing location of the regions with mean January temperature between 0 and 5°C. In eastern North America, this region appears as a well-defined band that translates northward from 21 ka to present, while in the mountainous western North America, this region is discontinuous, with many outlying islands and inliers of different climates.

Climatic Uniqueness (Minimum Dissimilarities)

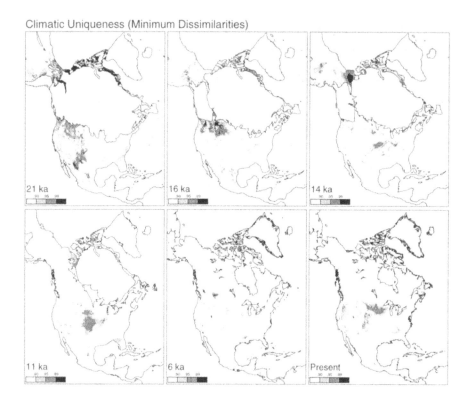

Figure 3: Uniqueness of simulated (21 – 6 ka) and observed present climate at individual 25 km grid points. For each grid point and time, the minimum dissimilarity between the climate of the point and the climates of all of the other points at the other times was determined. The shading indicates those points with climates that are relatively unusual (i.e. those with minimum dissimilarity values that exceed the 90th-percentile of the 1.6×10^5 minimum dissimilarities). Note the discontinuous patterns of unusual climates, or ones without analogous climates at other times.

CHARACTERISTIC FEATURES OF CLIMATIC VARIATIONS

Several distinctive features or patterns of climatic variations occur at a number of spatial and temporal scales. Depending on the temporal and spatial scales they occur at, these features can have pervasive influences on the terrestrial biota (as in the case of the larger-spatial, longer-temporal scale variations at higher levels in the hierarchy), or they may produce more localized responses (as in the case of the

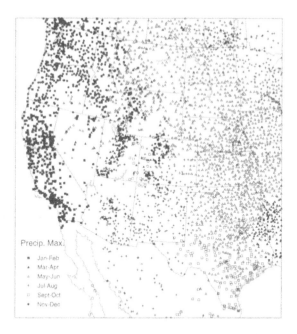

Figure 4: Seasonality of the precipitation maximum in western North America (Mock 1996). The broadscale pattern of the winter maximum of precipitation along the west coast and summer maximum in the interior is evident, but there is also considerable spatial heterogeneity in the precipitation regime in topographically diverse portions of Colorado, Utah, Wyoming, Montana and Idaho. This pattern of spatial heterogeneity has probably been stable during the Holocene, leading to heterogeneity in the response of vegetation to large-scale climatic changes (Whitlock & Bartlein, 1993; Whitlock *et al.* 1995).

smaller-spatial, shorter-temporal scale variations of lower-level components) (see Barnosky 1987; Bartlein & Prentice 1989; and Webb & Bartlein 1992 for further discussion). Table 2 lists and gives examples of the characteristic features, the levels in the climate system hierarchy that contain the controls of the feature, and those that are involved in the response. Table 2 also indicates the *predictability* and *reversibility* of the individual features. Features are

1. *predictable* (P), if the timing and amplitude of the variations can be specified;

2. *expectable* (E), if either the timing or amplitude can be specified, but not both; or

3. *unpredictable* (U), if neither the timing nor amplitude of the feature can be specified.

Features are *reversible* (R), if the climate returns to a previous general mean state following the occurrence of the feature, or *not reversible* (N), if the climate does not return. The table also shows for time series, the relative importance of trend, systematic (or periodic) and irregular components, and for spatial data, the relative importance of large-scale or small-scale patterns.

Table 2. Characteristic features of temporal and spatial variations of climate

Features	Characteristic examples	Levels in Hierarchy *		Predictability/ Reversibility †		Components of Temporal Variations ‡		
		Control	Response	Predict.	Reverse	Trend	System.	Irreg.
Trends								
era-long	Cenozoic global cooling and glacierization	E	G-L	E	N	H	L	L
glacial/ interglacial	last deglaciation	E,G	H-L	P	R	H	L	L
multi-millennial	late-Holocene summer cooling in northern midlatitudes	G-C	R-L	E	R	L	L	H
century-long	20th century warming	G-C	R-L	E	R	L	L	H
Steps								
inter-epoch	onset of N Hemisphere glaciation (ca. 2·65 Ma)	E	G-L	U	N	H	L	L
climate reversals	Younger Dryas climate reversal (ca. 13 - 11·5 ka)	G-C	R-L	E	R	H	L	L
Oscillations								
periodic (orbital)	ice sheet growth and decay, monsoon strengthen/weaken	E	G-L	P	R	L	H	L
changing-periodic (orbital)	strengthening of 100-ka cycle in past million years	E	G-L	U	N	L	H	L
quasi-periodic (sub-orbital)	Dansgaard-Oeschger 'cycles' /Heinrich events (75 - 10 ka)	E,G	H-L	E	R	L	M	H
interannual (10^0 - 10^1 yrs)	El Niño / Southern Oscillation variations, decadal-scale climate anomalies	H,C	R-L	E	R	L	M	M
Fluctuations								
Holocene (10^3 - 10^4 yrs)	Holocene temperature and precipitation variations	G,H	C-L	E	R	M	L	H
sub-millennial (10^3 - 10^4 yrs)	Medieval Warm Period/ Little Ice Age variations	G-C	R-L	E	R	L	L	H
interannual (10^0 - 10^1 yrs)	aperiodic interannual climate variations	H,C	R-L	E	R	L	L	H

Table 2. (continued)

Translations		Levels in Hierarchy * Control	Response	Predictability/ Reversibility † Predict.	Reverse	Components of Spatial Variations ‡ Large	Small
hemispheric (> 10^5 yrs)	high-latitude cooling, increase in N–S temperature gradients	E	G-L	U	N	H	L
continental (< 10^5 yrs)	latitudinal shifts of temperature zones, Last Glacial Maximum to present	E-H	C-L	P	R	H	L
local	elevational shifts of temperature zones, Last Glacial Maximum to present	E-R	M,L	P	R	L	H
Mosaics							
continental (> 10^6 yrs)	rainshadows; other continental-scale patterns	E	G-L	U	N	H	L
continental (< 10^6 yrs)	individualistic changes in climate variables since Last Glacial Maximum	E-H	C-L	E	R	H	L
regional (> 10^2 yrs)	regional contrasts (e.g. Pacific NW / SW-US contrasts since Last Glacial Maximum)	H,C	R-L	E	R	H	L
regional (10^1 - 10^2 yrs)	interannual variations (e.g. 1993 Mississippi River floods)	H,C	R-L	E	R	H	L
mesoscale (> 10^1 yrs)	spatial heterogeneity of precipitation regimes	E-R	M,L	P	R	L	H
mesoscale (10^0 - 10^1 yrs)	synoptic-scale climate anomalies	E-R	M,L	P	R	L	H

* Levels in climate system hierarchy (see Table 1: E = external, G = global, H = hemispheric, C = continental, R = regional, M = mesoscale, L = local) controlling (Control) characteristic features of climatic variation or responding (Response) to them.
† Predictability (Predict.) and reversibility (Reverse) of characteristic features of climatic variations: P = feature is predictable at the temporal or spatial scale indicated; E = feature is expectable, but not specifically predictable; U = feature is unpredictable.
‡ Relative amplitude (H = high, M = medium, L = low) of components of temporal (Trend, Systematic (Periodic), Irregular) or spatial (Large-scale, Small-scale) variations.

Trends, i.e. progressive increases or decreases in the levels of a particular climate variable, appear in palaeoclimatic variations on all timescales, but are particularly evident at certain scales. At the longest of timescales that are displayed in Figure 1 (Series A), cooling during the Cenozoic is evident in the general course of change in oxygen isotopic ratios toward heavier values (i.e. toward cooler oceans, more ice). This trend is likely driven by the external controls of the climate system, with all of the lower levels in the hierarchy responding. The general trend is broken, of course, by local increases in global temperature (as in the Eocene), and by locally more rapid decreases, but the overall impression one gets when viewing this series is of a progressive movement toward cooler conditions. The cooling trend is therefore expectable once underway (but its particular amplitude is not predictable), non-reversing overall, and dominated by the long-term changes (Table 2). Series B, oxygen-isotope data for the past 5 Ma, also shows a generalized trend, which is the same climatic change as the more rapid decrease in global temperatures that is evident in the last part of Series A. On a shorter timescale, the transition between the last glacial maximum (about 21,000 years ago) and present (Series C and D) can be viewed as a trend (predictable and reversible), although the higher temporal-resolution record from the GRIP Summit core (Series D) shows it broken by the Younger Dryas climate reversal (about 11·5 ka BP). During the latter half of the Holocene, many locations in the northern mid-latitudes experienced a cooling trend in summer; this is evident in the July temperature record at Elk Lake (Series E). Finally, during the last century, the global mean temperature, as well as that at individual stations has generally increased, and this trend is apparent in the summer temperature series from the Sierra Nevada (Series G). The climatic changes evident in Series E and G are expectable, but not predictable, and are dominated by the irregular component of variability.

Steps, i.e. abrupt (relative to the timescale of variations under consideration) transitions from one level to another, also appear on many different timescales. Notable examples of steps in the representative series include the unpredictable and non-reversing inter-epoch decreases in global temperature during the Cenozoic (as during the Eocene–Oligocene and Pliocene–Pleistocene transitions, Series A). Similarly, abrupt steps occur at the beginning and end of theYounger-Dryas climate reversal (around 11 ka BP) and at the terminations of the 'Dansgaard–Oeschger' cycles (see Bond *et al.* 1993; Bond & Lotti 1995) during the interval between 60 and 20 ka BP; these are evident in the Greenland ice core data (Series D). These steps are expectable, but not predictable, and have little irregular variation superimposed on the transition from one level to another.

Oscillations, i.e. either periodic or quasi-periodic variations about a stationary or slowly changing level, are one of the more prominent features of palaeoclimatic time series. Oscillations dominate the glacial–interglacial variations in global temperature, and the variations in the continent/ocean temperature contrast that govern the monsoons, each generated by the periodic variations of the earth's orbital elements

(Series B and C) (Imbrie *et al.* 1984; Prell & Kutzbach 1992). Consequently, these variations are highly predictable and systematic. One important characteristic of the oxygen-isotope variations apparent in the ODP Site 659 data (Series B) is the change in the relative importance of the different periodic components: prior to 1 Ma, the 41 ka cycle is relatively more important, whereas afterward, the 100 ka cycle becomes more prominent. These changes in periodicity seem inherently unpredictable and non-reversing. Quasi-periodic variations (i.e. variations that are less–regular than the strictly periodic ones of the SPECMAP data (Series C)) appear at 'sub-orbital' timescales, as in the aforementioned 'Dansgaard-Oeschger' cycles. Interannual climate variations, such as those constituting the El Niño/Southern Oscillation (ENSO) variations (Diaz & Markgraf 1992) also have been described as quasi-periodic. Although systematic variation is apparent in these series, they are only expectable, owing to the absence of strict periodicity.

Fluctuations, i.e. aperiodic variations of climate, appear at all timescales (unless they have been specifically suppressed, as in Series C), but tend to be more evident at shorter timescales. Variations in temperature and precipitation at Elk Lake (Series E and F), and in the Sierra Nevada (Series G and H), apart from the trends described above, can be described as fluctuations. Fluctuations are expectable, but not specifically predictable features in individual time series, and are distinguished from steps by their inherent reversibility.

Translations, i.e., lateral or elevational movements of climate zones, are apparent in sequences of palaeoclimatic simulations (e.g. Kutzbach *et al.* 1993; Webb *et al.* 1993) and may also be inferred from individual or global-scale palaeoclimatic records. For example, the movements toward cooler conditions apparent in Series A–D, like those observable in maps of the simulated climate of the Last Glacial Maximum, all involve greater cooling at high latitudes than at low, with a consequent equatorward displacement and latitudinal compression of individual isotherms, as in Figure 2 for eastern North America. In mountainous regions, these geographical shifts are expressed as vertical (downward in the case of cooling) movements of the elevations at which particular temperatures occur. The predictability and reversibility of translations depends on the timescale at which the changes occur.

Mosaics, i.e. heterogeneous patterns of climate or climatic changes, also develop on a variety of temporal and spatial scales. On the longer timescales, tectonism, and the changes in atmospheric circulation it induces, results in the development of rainshadows and belts of orographic precipitation (Ruddiman & Kutzbach 1989). These fundamental changes in the patterns of climate variables are inherently unpredictable and non-reversing. On the decadal or interannual timescales, variations in the configuration of the large-scale circulation of the atmosphere create alternating patterns of positive and negative anomalies of temperature and precipitation (Namias 1970). In regions of complex topography, even simple changes of a single climate variable will have a spatially heterogeneous expression (Figure 2, western North America). Moreover, in such regions, the influence of hemispheric- and continental-scale circulation changes may be registered in a systematic fashion on the regional or mesoscale patterns of individual climate variables (Figure 4), which may impart further spatial heterogeneity in the local response to changes in larger-scale controls (Whitlock & Bartlein 1993; Whitlock et al. 1995; Mock 1996; Mock & Bartlein, in press).

The variations of the simulated climate over North America since the Last Glacial Maximum (LGM) also illustrate how the simultaneous changes of several individual climate variables can produce changing regional-scale patterns of unique climates – climates that do not have analogues at other times and other locations. The patterns suggest that the relatively simple translations of individual climate variables at the continental scale can produce regional-scale patterns of unusual climates. For example, there is an area of unusual climate in the continental interior of North America at 11 ka BP (Figure 3), where the values of individual climate variables are not particularly unusual (relative either to present or to earlier times), but the combinations of the values of the variables are (in this case, the seasonality of temperature is relatively high). Figure 3 also indicates that our conception of the present as 'normal' or not unusual is mistaken. In overall terms, the climate of 6 ka BP is more representative of the climates at the other times during the past 21 ka BP than is the present climate (the area of unique climates is smaller at 6 ka than at other times).

The unusualness of the present climate can be further illustrated by considering the variations in the combinations over the past 800,000 years of the July insolation anomaly at 65°N and global ice volume, two indices of the general state of the climate system (Figure 5). One point stands out on this figure, the present. For

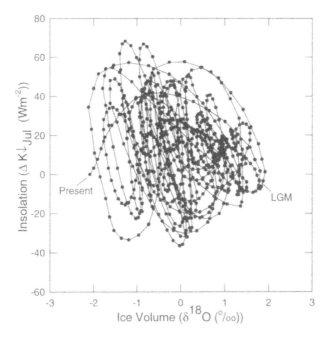

Figure 5: The trajectory of the climate system in a phase space defined by the July insolation anomaly (difference from present) at 65°N (Berger, 1978) and global ice volume as represented by the 'SPECMAP' oxygen-isotope record (Imbrie *et al.* 1984). Points are plotted at 1 ka intervals, and adjacent points are connected. The present and the last glacial maximum (LGM) are indicated.

nearly all of the time (allowing for some uncertainty in the oxygen-isotopic record), more ice has been present in the climate system than now, while about 70 percent of the time July insolation has been greater than present. The point in time that we know the most about is therefore quite unusual when viewed relative to the range of conditions that prevailed during the Quaternary. The Last Glacial Maximum, which is often regarded as having a climate 'opposite' to that of the present is also quite unusual; most of the time, the climate system has been in a state intermediate between the present and LGM. The tendency for the climate system to vary continuously, not dwelling in any particular state for long is also evident in the figure. Most of the time, the climate system is in the process of moving from one state to another, and only rarely do two of the points separated by 1 ka overlap one another. Therefore, our understanding of how the biosphere responds to climatic variations is predicated on observations of relatively unusual situations.

DISCUSSION AND CONCLUSIONS

The characteristic features of climatic variations described above represent potential sources of environmental change that may evoke in the terrestrial biota the kinds of spatial and evolutionary responses discussed in other chapters of this volume. At any given time or place, the prevailing climate is the product of the superimposition of all of these features and their operation across all of the different timescales. The efficacy of a particular feature in producing a response probably depends on the amplitude of the associated climatic variation and its duration, the number of features superimposed, the tendency for a particular variation to be predictable or reversible, and also on the existence of intrinsic thresholds, which when exceeded, produce a response.

When the predictability, reversibility, and relative importance of the components of temporal and spatial variations (Table 2) are considered in light of the kinds of spatial and evolutionary responses described in this volume, the following picture emerges:

- Those climatic variations that are unpredictable and non-reversing, as well as those that are dominated by long-term or large-scale changes, are those that are more likely to evoke an evolutionary response, because they confront species with new and unfamiliar environments.

- Those variations that are predictable or expectable, or are dominated by short-term or small-scale variability, instead promote mainly spatial adjustments, such as migration (Huntley & Webb, 1989), because they present to species environments that have been experienced before and have been adapted to.

How species will likely respond to the potentially great future changes of climate will therefore depend on how large, rapid, and unfamiliar those changes are. The persistence of species across many large (as in the glacial–interglacial oscillations), and sometimes rapid (as in the case of the step-like climate reversals and quasi-periodic, sub-orbital oscillations) past climate changes, argues for a primarily spatial response at first. As the future climate grows increasingly more unfamiliar, however, evolutionary (extinction and speciation) responses should begin to emerge.

ACKNOWLEDGEMENTS

I thank Nathalie de Noblet, Brian Huntley, Tom Webb, Cathy Whitlock and Sarah Shafer for their perceptive comments on the manuscript, and other conference participants for their stimulating ideas. Research was supported by National Science Foundation grants to the University of Oregon.

REFERENCES

Barnosky CW (1987) Response of vegetation to climatic changes of different duration in the late Neogene. Trends in Ecology and Evolution 2:247-250

Bartlein PJ (1988) Late-Tertiary and Quaternary palaeoenvironments. *in* Huntley B, Webb III T (eds) Handbook of Vegetation Science: Volume 7, Vegetation History, 113-152. Kluwer Academic Publishers Amsterdam

Bartlein PJ, Prentice IC (1989) Orbital variations, climate and palaeoecology. Trends in Ecology and Evolution 4:195-199

Bartlein PJ Whitlock C (1993) Palaeoclimatic interpretation of the Elk Lake pollen record. *in* Bradbury JP, Dean WE (eds) Elk Lake, Minnesota: Evidence for Rapid Climate Change in the North-Central United States, 275-293. Geological Society of America Boulder

Berger A (1978) Long-term variations of caloric insolation resulting from the Earth's orbital elements. Quaternary Research 9:139-167

Bond G, Broecker W, Johnsen S, McManus J, Labeyrie L, Jouzel J, Bonani G (1993) Correlations between climate records from North Atlantic sediments and Greenland ice. Nature 365:143-147

Bond G, Lotti R (1995) Iceberg discharges into the North Atlantic on millennial time scales during the Last Glaciation. Science 267:1005-1010

Charles CD, Rind D, Jouzel J, Koster RD, Fairbanks RG (1994) Glacial–interglacial changes in moisture sources for Greenland: influences on the ice core record of climate. Science 263:508-511

Dansgaard W, Johnson SJ, Clausen HB, Dahl-Jenson D, Gundestrup NS, Hammer CU, Hvidberg CS, Steffensen JP, Sveinbjörnsdottir AE, Jouzel J, Bond G (1993) Evidence for general instability of past climate from a 250-kyr ice-core record. Nature 364:218-220

Diaz HF, Markgraf V (1992) El Niño: historical and palaeoclimatic aspects of the southern oscillation. Cambridge: Cambridge Univ. Press

Graumlich LJ (1993) A 1000-year record of temperature and precipitation in the Sierra Nevada. Quaternary Research 39:249-255

Huntley B,Prentice IC (1988) July temperatures in Europe from pollen data, 6000 years before present. Science 241:687-690

Huntley B, Prentice IC (1993) Holocene vegetation and climates of Europe. *in* Wright Jr HE, Kutzbach JE, Webb III T, Ruddiman WF, Street-Perrott FA, Bartlein PJ (eds) Global Climates since the Last Glacial Maximum, 136-168. University of Minnesota Press Minneapolis

Huntley B, Webb III T (1989) Migration: species' response to climatic variations caused by changes in the earth's orbit. Journal of Biogeography 16:5-19

Imbrie J, Hays JD, Martinson DG, McIntyre A, Mix AC, Morley JJ, Pisias NG, Prell WL, Shackleton NJ (1984) The orbital theory of Pleistocene climate: support from a revised chronology of the marine $\delta^{18}O$ record. *in* Berger A, Imbrie J, Hays J, Kukla G, Saltzman B (eds) Milankovitch and Climate, 269-305. Reidel Dordrecht, Netherlands

Kutzbach JE, Guetter PJ, Behling PJ, Selin R (1993) Simulated climatic changes: results of the COHMAP climate-model experiments. *in* Wright Jr. HE, Kutzbach JE, Webb III T, Ruddiman WF, Street-Perrott FA, Bartlein PJ (eds) Global Climates since the Last Glacial Maximum, 24-93. University of Minnesota Press Minneapolis

Kutzbach J, Gallimore R, Harrison S, Behling P, Selin R, Laarif F (in prep) Climate and biome simulations for the past 20,000 years.

McDowell PF, Webb III T, Bartlein PJ (1990) Long-term environmental change. *in* Turner II BL, Clark WC, Kates RW, Richards JF, Mathews JT, Meyer WB (eds) The Earth as Transformed by Human Action: Global and Regional Changes in the Biosphere over the Past 300 Years, 143-162. Cambridge University Press Cambridge

Miller KG, Fairbanks RG, Mountain GS (1987) Tertiary oxygen isotope synthesis, sea level history, and continental margin erosion. Palaeoceanography 2:1-19

Mitchell JM, Jr. (1976) An overview of climatic variability and its causal mechanisms. Quaternary Research 6:481-493

Mock CJ (1996) Climatic controls and spatial variations of precipitation in the western United States. Journal of Climate 9:1111-1125

Mock CJ, Bartlein PJ (1995) Spatial variability of Late-Quaternary palaeoclimates in the western United States. Quaternary Research 44:425-433

Namias J (1970) Climate anomaly over the United States during the 1960's. Science 170:741-743

Prell WL, Kutzbach JE (1992) Sensitivity of the Indian monsoon to forcing parameters and implications for its evolution. Nature 360:647-652

Ruddiman WF,Kutzbach JE (1989) Forcing of Late Cenozoic northern hemisphere climate by plateau uplift in southern Asia and the American West. Journal of Geophysical Research 94:18409-18427

Shackleton NJ, Imbrie J (1990) The $\delta^{18}O$ spectrum of oceanic deep water over a five-decade band. Climatic Change 16:217-230

Tiedemann R, Sarnthein M, Shackleton NJ (1994) Astronomic timescale for the Pliocene Atlantic $\delta^{18}O$ and dust flux records of Ocean Drilling Program site 659. Palaeoceanography 4:619-639

Webb III T (1989) The spectrum of temporal climatic variability: current estimates and the need for global and regional time series. Bradley RS (ed) Global Changes of the Past. UCAR/Office for Interdisciplinary Earth Studies Boulder, CO

Webb III T (1995) Pollen records of Late Quaternary vegetation change: plant community rearrangements and evolutionary implications. *in* Effects of Past Global Change on Life, 221-232. Board on Earth Sciences and Resources, Commission on Geosciences, Environment, and Resources, National Research Council

Webb III T, Bartlein PJ (1992) Global changes during the last 3 Million years: climatic controls and biotic responses. Annual Reviews of Ecology and Systematics 23:141-173Webb T, III, Bartlein PJ, Harrison SP, Anderson KH (1993) Vegetation, lake levels, and climate in eastern North America for the past 18,000 years. *in* Wright Jr. HE, Kutzbach JE, Webb III T, Ruddiman WF, Street-Perrott FA, Bartlein PJ (eds) Global Climates since the Last Glacial Maximum, 415-467. University of Minnesota Press Minneapolis

Whitlock C, Bartlein PJ (1993) Spatial variations of Holocene climatic change in the Yellowstone region. Quaternary Research 39:231-238

Whitlock C, Bartlein PJ, Van Norman, KJ (1995) Stability of Holocene climate regimes in the Yellowstone Region. Quaternary Research 43:433-436

Modelling late-Quaternary palaeoclimates and palaeobiomes

Nathalie de Noblet
Laboratoire de Modélisation du Climat et de l'Environnement
DSM / Orme des Merisiers / Bat. 709
CE Saclay
91191 Gif-sur-Yvette cedex
France

INTRODUCTION

Climate models have been used to estimate the warming associated with future increases in greenhouse gases. Yet we have no evidence that these models, that have been calibrated against present-day data, are sensitive enough to represent a climate that is quite different from ours. One possible technique for evaluating their sensitivity is to simulate past climates. Indeed:

- palaeodata provide enough information to show a clear picture of the entire climatic system in the past and to document temporal evolution; and

- variations in insolation at the top of the atmosphere, that is the external forcing responsible for the glacial–interglacial cycles according to the Milankovitch theory (1941), can be calculated quite accurately (Berger & Loutre 1991) and used to force climate models.

An Earth System Model that encompasses all five components of the climate system (atmosphere, hydrosphere, biosphere, cryosphere and lithosphere) does not exist yet, and is far from being achieved due to both our limited understanding of all the interactions and the unrealistic computer requirements. Most efforts have then been devoted to models that account for only one component at a time, such as atmospheric general circulation models (AGCMs) similar to those used for weather forecasting. During the past few years there have been many attempts to couple ocean, and more recently terrestrial biosphere models, to AGCMs.

NATO ASI Series, Vol. I 47
Past and Future Rapid Environmental Changes:
The Spatial and Evolutionary Responses of
Terrestrial Biota
Edited by Brian Huntley et al.
© Springer-Verlag Berlin Heidelberg 1997

Within the scope of this paper we will focus on the biosphere/atmosphere interactions and will report on modelling studies only. The terms: *biosphere, vegetation, plants* and *biomes* will be extensively used in the following; they all refer to the terrestrial biosphere.

There is much evidence from the past that both vegetation and climate have changed through time (e.g. Wright *et al.* 1993), and many *in situ* experiments (crop fields, glasshouse, etc.) have shown how much a perturbation in weather conditions can influence both the seasonal behaviour of plants and their establishment, dispersal or replacement (e.g. Tamponnet *et al.* 1991; Körner & Arnone 1992; Sherwood & Kimball 1993).

Reciprocally a change in vegetation type can induce local, regional and eventually global changes in the atmospheric circulation, as observed over many deforested sites in the tropics (e.g. Dickinson 1987). Both components of the climate system therefore act significantly upon each other.

In the first section we will describe the modelling tools used to study these interactions. We will then discuss some of the numerical experiments that illustrate the sensitivity of biome distribution to climate change. In the next section we will show some numerical examples of vegetation feedbacks, and then we will introduce the reader to the latest developments in biosphere/atmosphere modelling and draw their attention to the problems that remain to be solved.

MODELLING TOOLS

Atmospheric General Circulation Models (AGCMs) are three-dimensional global models that solve the basic equations of atmospheric motion at the coarse spatial resolution of a few hundred kilometres, and with time-scales as short as an hour or less. Most of the processes occurring at a much smaller scale (referred to as sub-grid scale features), such as the formation of cloud systems or land-surface hydrology, are parameterized (i.e. they are described in the model using empirically-derived relationships). The largest differences between AGCMs reside in the number of sub-grid scale processes they include and in the formulae used to parameterize them.

Since these models consider solely the atmospheric component of the climate system, the state of the four others must be specified (hereafter referred to as *boundary conditions*). Indeed the distributions of land, ocean, terrestrial biosphere and ice sheets are prescribed, together with their associated characteristics that will interact with the overlying air. *Oceans* are defined by their surface temperatures and the seasonal evolution of sea-ice extent. The height and extent of continental *ice-sheets* are fixed. *Land* is described by means of topography, while albedo and roughness length are two of the main parameters driving the exchanges between the *biosphere* and the atmosphere. The chemical composition of the atmosphere (mainly CO_2 concentration) and insolation also need to be prescribed.

Starting from any *initial conditions* (i.e. the state of the atmosphere at a given instant) the numerical simulation then provides the atmospheric circulation in statistical equilibrium with the prescribed *boundary conditions*, as illustrated in Figure 1. Applied to past climates, such simulations are referred to as *snapshots* since they do not account for the time-evolution of insolation. Transient palaeoclimate simulations cannot be performed due to the very slow variation of insolation compared to the computer time such experiments would require.

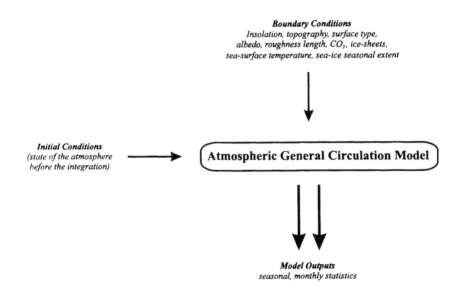

Figure 1. Schematic view of a numerical simulation using an atmospheric general circulation model (AGCM).

Most of the reported simulations of past climates using AGCMs concern the last glacial – interglacial cycle, with particular emphasis on the past 20,000 years (e.g. Kutzbach & Guetter 1986; Broccoli & Manabe 1987; COHMAP 1988; Kutzbach & Gallimore 1988; Lautenschlager & Herterich 1990; Liao *et al.* 1994) and on glacial inception (115 ka, e.g. BP Royer *et al.* 1983; Rind *et al.* 1989; Dong & Valdes 1995). Few others have addressed pre-Quaternary climatic conditions (e.g. Barron & Washington 1984).

Soil-Vegetation-Atmosphere Transfer Schemes (SVATs) are included within AGCMs and simulate all instantaneous fluxes exchanged between the biosphere and the atmosphere (i.e. evapotranspiration, sensible heat flux, etc.), as well as the surface hydrologic cycle. They all include a specific description of global vegetation distribution, together with a set of characteristics that is ascribed to each biome type: roughness length, albedo, leaf area index, canopy resistances and height, fraction of ground shaded by the foliage, etc. Some of them, referred to as *mosaic-SVATs*, allow the coexistence of several biomes within one AGCM grid-box (e.g. Koster & Suarez 1992; Ducoudré *et al.* 1993), while the others assume that only one type can occupy the space (e.g. Dickinson *et al.* 1986; Sellers *et al.* 1986). In the former case fluxes are computed for each vegetation type and then averaged before being transmitted to the atmosphere, while in the latter mean surface characteristics are used to compute those fluxes (Figure 2).

The SVATs generally differ in the number of vegetation types they include (from 8 to more than 20), the prescribed values of all characteristics, and the parameterizations used to represent phenomena such as evapotranspiration or the cycle of water in the soil.

Biome Models are used to translate climate change into biome change. They are of two kinds: *static* (time-independent) and *dynamic* (time-dependent). Static models predict global distributions of potential vegetation as a function of environmental variables (e.g. seasonal rainfall, temperature of the coldest and warmest months). Dynamic models predict how vegetation changes locally (seed dispersal, competition, mortality, nutrient cycling, etc.). Because we are dealing with snapshots and large-scale features in this paper we will restrict ourselves to static models.

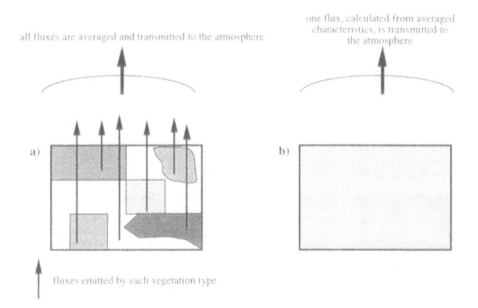

Figure 2. Schematic representation of vegetation within one AGCM grid-box. (a) a *mosaic SVAT:* the fluxes are computed for each vegetation type and then averaged before being transmitted to the atmosphere; (b) a *regular SVAT:* all land-surface characteristics are first averaged (i.e. as if we had only one *averaged biome*) and fluxes are computed from these mean parameters.

There are two broad categories of static models:

- *empirical classification schemes* that are based on transcontinental correspondence between geographic patterns of vegetation and climate (Thornthwaite 1948; Holdridge 1947; Köppen 1936; Troll & Paffen 1964; Friedlingstein *et al.* 1994). They are very popular owing to their ease of application;

- so-called *biome models,* that are based on more physiological considerations (Box 1981; Woodward 1987; Prentice *et al.* 1992). They do not translate climate directly into biomes but into plant functional types (PFTs). A dominance hierarchy is then applied and biomes emerge from the combination of equally dominant PFTs. Each PFT is assigned climatic tolerances (upper and/or lower bounds) for variables such as temperature and precipitation.

More detailed description of these models can be found in Shugart (1996).

VEGETATION: A DIAGNOSTIC TOOL TO VALIDATE CLIMATE MODELS

Validation of AGCMs can be performed in two ways:

- *the forward approach* translates the simulated climate into some proxy-data (e.g. pollen, lake levels) against which they will be compared. Biome modelling for example (hereafter referred to as *biomization*), using the tools described above, belongs to this approach;

- *the inverse approach* derives climatic variables (e.g. temperature or precipitation), from proxy-data, using for example transfer functions or analogue methods. Those variables are then compared to the ones calculated by the AGCM.

Jolly *et al.* (1996) used new compilations of African pollen data to compare with simulations for the last glacial maximum and early to mid-Holocene using the NCAR AGCM and BIOME1 (Prentice *et al.* 1992). Their results (Figures 3a-c) show that the model significantly underestimates the degree of monsoon expansion in West Africa 6,000 yr BP but, as noted by the authors, this is a rather robust feature shared by many AGCMs. In equatorial East Africa, however, the model reproduces well the northward migration of the intertropical convergence zone that translates into the development of tropical xerophytic bush/savanna, temperate xerophytic woods/shrub and warm grass/shrub. Such highlighted agreements/discrepancies between models and data, when carried out using several AGCMs and a single biome model, or one AGCM and several biome models, may help the modellers put error bars around the simulated estimates of climate and vegetation changes. Moreover, when all models agree in their failure to reproduce an observed shift (such as in West Africa), it may indicate either that the boundary conditions used to force the AGCMs are insufficient, or else that feedbacks from another component of the climate system (e.g. oceans, sea-ice) are crucially needed and not yet included.

Biomization can also be used to intercompare either results from different AGCMs, or different palaeoclimates simulated using the same AGCM. Guetter and Kutzbach (1990) for example used a modified Köppen classification and biomized their snapshot studies (carried out with the NCAR AGCM) of naturally-produced climatic extremes over the past 126,000 years. Their results show that about 70% of the land surfaces have experienced rather significant vegetation changes. This was the first

MODERN

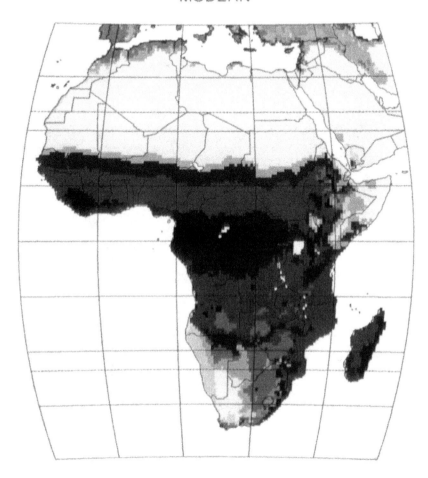

Figure 3a: Biome distributions simulated for present day using BIOME1 (Prentice *et al.* 1992) and a modern climatology. (after Jolly *et al.* 1996)

CCM1 6 ka biomes

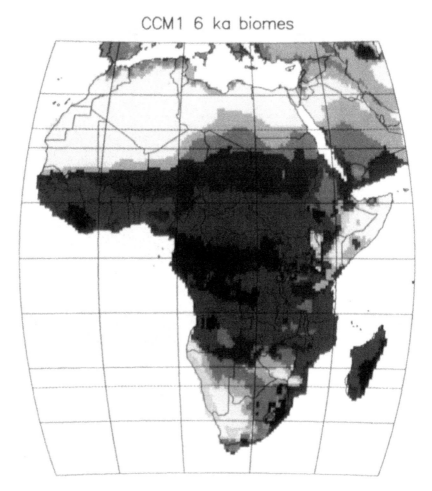

Figure 3b: Biome distributions simulated for 6,000 yr BP using BIOME1 and the NCAR AGCM. (after Jolly *et al.* 1996)

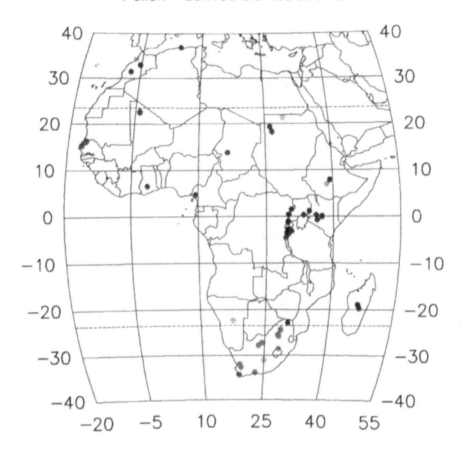

Figure 3c: Biome distributions derived from pollen data for 6,000yr BP. (after Jolly *et al.* 1996)

Figure 3d: Key to Biomes for Figures 3a, 3b and 3c.

attempt to provide quantitative estimates of the size of the area affected by biome change. No comparison with data was made at that point, but the authors raised two main problems that anyone should have in mind when undertaking such validation:

- discrepancies between simulated and observed vegetation types may be due to failures in both the climate and the biome models;

- past times vegetational units may have no analogue in the modern world.

More recently Harrison *et al.* (1995) studied the sensitivity of simulated climates, and inferred biome distributions (using BIOME1), to extreme insolation changes similar to the ones observed during the last interglaciation (marine oxygen isotope substage 5e). They demonstrated that the tundra/taiga limit in the northern hemisphere is very sensitive to changes in the seasonal contrast. Indeed this limit was displaced poleward when summers were warmer than today (as during the last interglacial, *ca.* 126 ka BP), and equatorward when they were cooler (as during glacial inception, *ca.* 115 ka BP). Such a qualitative contrast is expected to have occurred over a *ca.* 10,000 year interval and should be clearly recognizable in the pollen record. If it is, then their results strongly support the hypothesis that orbital forcing is the primary control over vegetation changes on a millennial time scale.

ON THE SENSITIVITY OF THE GLOBAL ATMOSPHERE TO VEGETATION CHANGES

It was only in the late 70's, when scientists started questioning the mechanisms of desertification in the Sahel and the possible climatic impacts (local, regional and global) of massive tropical deforestation, that experiments were performed to study the sensitivity of simulated climates to changes in the land-surface properties. Charney *et al.* (1975), for example, suggested and demonstrated, using a global climate model, that a desert feeds upon itself, implying that it cannot be naturally removed once it is installed. Up to now, as reviewed by Shukla (1992), no modeller has been able to contradict this mechanism.

There are two standard ways to perform such sensitivity studies:

- in the simplest method only one characteristic of the land-surface properties is changed, for example the albedo or roughness length of a specific region. These experiments are not meant to be realistic, but they help modellers quantify the relative importance of each land-surface parameter;

- in the more complex, but also more realistic, method a biome replaces another, all its characteristics are then changed simultaneously. In deforestation studies, for example grassland or bare soil replace tropical rainforest.

Changing only one characteristic at a time

Street-Perrott *et al.* (1990), using the UKMO AGCM, tried to simulate the northward penetration of summer monsoon rains into the Sahara, as evidenced from lake-level and palaeoecological data for 9,000 yr BP. The prescribed orbital forcing alone did increase rainfall in North Africa although the magnitude of the simulated changes was substantially underestimated when compared to observations. Given that pollen data indicate the presence of some vegetation at that time, and noting the mechanism of positive biogeophysical feedback in this area described by Charney *et al.* (1975), they carried out another sensitivity experiment prescribing a decreased surface albedo between 16°N and 24°N in Africa, Arabia and southern Asia, as if vegetation had grown. As a result of this change in land-surface characteristics, rainfall was significantly further increased by about 25% (Figures 4a-b).

Other simple studies have shown the importance of roughness length for the general circulation of the atmosphere. Sud *et al.* (1988) for example pointed out that decreasing roughness length over India (such as would result from deforestation or overgrazing) may considerably lower the intensity of the summer monsoon, having drastic economical consequences.

Changing a whole biome

Polcher and Laval (1994) compared results from three sensitivity studies regarding tropical deforestation. The changes simulated in Amazonia are summarized in Table 1. All three experiments display warmer surface temperatures, lower evapotranspiration rates and increased sensible heat flux when forests are removed, but they differ in their estimation of changes in rainfall intensity. Polcher and Laval (1994) show a considerable increase in precipitation and moisture convergence, whereas in the simulations carried out by Lean and Warrilow (1989) and Shukla *et al.*

(1990) those values are decreased, implying increased drought. Although changes in the fluxes emitted at the land/atmosphere interface are of the

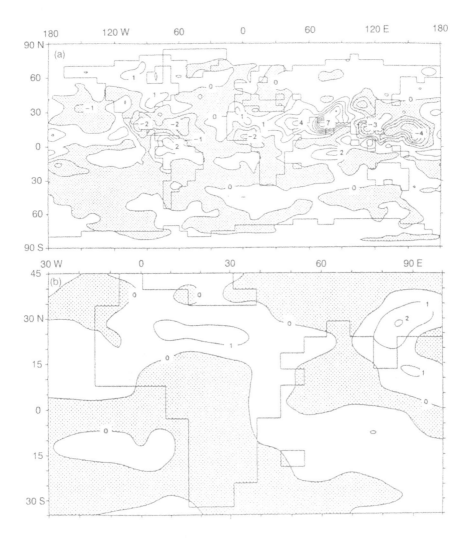

Figure 4: Differences in the simulated precipitation in June-July-August, with areas of decrease stippled. Redrawn from Street-Perrott *et al.* (1990). (a) 9,000 yr BP albedo-sensitivity experiment minus simulated present-day climate; (b) 9,000 yr BP albedo-sensitivity experiment minus 9,000 yr BP orbital alone experiment.

same order of magnitude in all three models, Polcher and Laval (1994) point out that atmospheric feedbacks can be quite different from one AGCM to another. We there-

Table 1. Mean changes simulated in Amazonia over the deforested sites.

Simulated climate variables	Polcher & Laval (1994)		Lean & Warrilow (1989)		Shukla et al. (1990)	
	Control	Deforested	Control	Deforested	Control	Deforested
Surface temperature (°C)	26·0	29·8	23·6	26·0	23·5	26·0
Evapotranspiration (mm day⁻¹)	5·3	2·6	3·12	2·27	4·53	3·18
Sensible heat flux (W m⁻²)	40	74	57	60	44	56
Total precipitation (mm day⁻¹)	7·17	8·25	6·60	5·26	6·75	5·0
Moisture convergence (mm day⁻¹) computed as the difference between precipitation and evapotranspiration	1·9	5·6	3·48	2·99	2·22	1·85

fore are not yet ready to predict confidently the climate response to tropical deforestation. Foley et al. (1994) simulated the mid-Holocene climate (ca. 6,000 yr BP) and concluded that orbital forcing alone does not produce sufficient warming in the high northern latitudes, especially during winter time, compared to the warming inferred from pollen data. They then prescribed a northward migration of the boreal forest/tundra limit and simulated an enhanced warming of the high northern latitudes, specially during spring time (Figure 5). Bonan et al. (1992, 1995) and Chalita and Le Treut (1994) also demonstrated the importance of positive vegetation feedbacks in the high northern latitudes, emphasizing the role played by albedo changes in winter and spring.

VEGETATION: A DYNAMIC COMPONENT OF GLOBAL CLIMATE MODELS

The studies reported above demonstrate that vegetation and atmosphere significantly influence each other. Coupled (interactive) models have then recently been developed to take better account of these interactions (Henderson-Sellers 1993; Claußen 1994; de Noblet et al. submitted). The general methodology of such coupling is illustrated in Figure 6. The objective is to iterate towards an equilibrium between simulated vegetation and climate, taking into account both the influence of

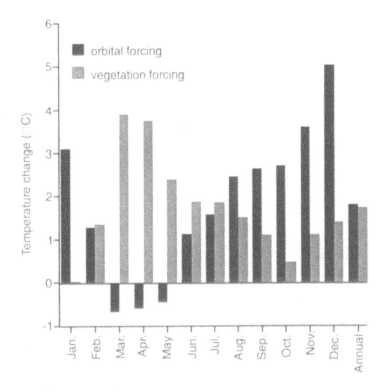

Figure 5: Changes in the simulated zonal-average (over land from 60° to 90°N) monthly surface temperatures. Redrawn from Foley *et al.* (1994).
orbital forcing = 6,000 yr BP (orbital forcing alone) minus present-day;
vegetation forcing = 6,000 yr BP (orbital + vegetation forcing) - 6,000 yr BP (orbital forcing alone).

climate on vegetation and the reciprocal influence (biogeophysical feedback) of vegetation on climate. Asynchronous coupling is always used because the biome distribution responds slowly to climatic forcing and is modelled, in the biome models, as a function of the average climate over a number of years, whereas the climate responds to vegetation changes 10^2–10^3 times faster.

When the coupling is applied to past climates, differences in its application appear that may influence the simulated climate/vegetation equilibrium:

- *frequency of iteration*: how many years of the simulated climate do we have to average before going through the biome model?

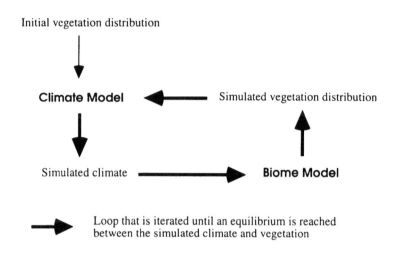

Initial vegetation distribution

Climate Model ◄———— Simulated vegetation distribution

Simulated climate ————► **Biome Model**

➤ Loop that is iterated until an equilibrium is reached between the simulated climate and vegetation

Figure 6. Schematic of the coupling between a Climate model and a Biome model.

- *sensitivity to the initial vegetation distribution*: does the equilibrium climate/vegetation system depend upon the initial distribution of vegetation? If the answer to this question is yes, then it may become crucial to have more data about past vegetation distribution if we are correctly to simulate past climates;

- *what climate is used to force the biome models:* most modellers (Guetter & Kutzbach 1990; Prentice *et al.* 1993; Harrison *et al.* 1995; Jolly *et al.* 1996; de Noblet *et al.* submitted) add their simulated climatic departures from present-day to a prescribed climatology of present climate. This procedure corresponds to a method routinely used in climate impact analyses to reduce the effects of possible biases in the simulated climate. Claußen (1994) on the other hand used his 'absolute' simulated palaeoclimate to force the biome model;

- depending upon what land-surface scheme is used in the climate model (mosaic vs. regular SVATs, see SVAT's above and Figure 2), we may either *reconstruct vegetation at the resolution of the climate model*, and end up with one biome per grid-point, or *reconstruct vegetation at a finer resolution*, and end up with several biomes that can share the same grid-point. In the latter case the simulated climate (or departures from present) have to be interpolated onto the finer grid before being used to force the biome model.

Claußen (1994) coupled version 3 of the ECHAM AGCM to BIOME1 (Prentice *et al.* 1992) and demonstrated that the simulated climate/vegetation equilibrium for present-day is very sensitive to:

- the initial vegetation distribution used to start the coupled experiment; and
- the prescribed frequency of the iterations (from one to several years).

Sensitivity to the initial vegetation distribution

Claußen (1994) started his integration of the coupled model with a biome map that drastically differs from today's; throughout the world hot desert was replaced by tropical rain forest, whereas tropical rain forest, tropical seasonal forest and savanna were replaced by hot desert. The simulated equilibrium present-day vegetation distribution, that he obtained after four iterations, was quite different from the actual one in many parts of the world. The western Sahara, for example, never reverted to desert, whereas most of India ended up covered by either desert or xerophytic woods/shrub.

This significant sensitivity of the coupled climate/biome model should therefore be kept in mind when applying this same strategy to past climates. For example the existence of some vegetated areas within the Sahara 6,000 yr BP (Jolly *et al.* 1996) may not be simulated if the initial biome distribution resembles that of today.

Sensitivity to the prescribed frequency of the iterations

Claußen (1994) also studied the sensitivity of the coupled climate/biome model to the prescribed frequency of the iterations (IF) using IF values of 1, 5 and 10 years. He initiated his simulations using the present actual vegetation distribution and found that, in all three cases, the simulated present-day climate/vegetation equilibrium was reached very soon after the first integration. However, the magnitude of the subsequent interannual/pentadal/decadal biome variability depended upon IF; variability was *ca.* 28% when IF = 1, *ca.* 13% when IF = 5 and *ca.* 12% when IF = 10. The biome maps computed from single and multi-year simulated climatologies differ significantly owing to the large interannual variability of moisture availability. Claußen (1994) therefore recommended that any biome distribution should be computed from climate variables averaged over at least 5 years.

Henderson-Sellers (1993) coupled the Holdridge scheme to the NCAR CCM and integrated it for 67 months of present-day climate, a new vegetation distribution being predicted and then incorporated at the end of each 12 month cycle (i.e. IF=1). As in the results of Claußen (1994) the largest vegetation change was obtained after the first simulated year and biome distribution was not significantly altered after that, interannual changes for each biome type being less than 3% (cf. the value of *ca.* 28% obtained by Claußen (1994)). She concluded that vegetation was a stable component of the global climate model. Continental surface climates, however, exhibited rather large seasonal changes, with, for example, temperature differences of ±10°C over Asia and Africa in July, and increased total runoff of about 8 mm day[-1] in India. The climate/vegetation equilibrium simulated was therefore rather different from the actual one because of the differing climatology. The most surprising point is that the climate change simulated in year 2 (resulting from the vegetation changes following year 1), although rather large, did not result in subsequent significant alterations of the biome distribution. This may indicate a drawback of the Holdridge scheme that accounts only for integrated annual climate variables to derive biomes, rather than making use of extremes and/or of information on the amplitude of the seasonal cycle as do models such as BIOME1 (Prentice *et al.* 1992).

CONCLUSION

This paper does not claim to be exhaustive. My intention was rather to give the reader an overview of the tools modellers use to simulate climate and biomes, and to show some examples of the considerable impacts one component has on the other. Coupled models are now under development in many groups and are the tools that will be commonly used for climate studies in the future.

Coupling atmosphere to ocean has also made considerable progress in the past few years. Indeed most AGCMs now include a 50 m mixed-layer ocean that allows for on-line calculations of sea-surface temperature. Dong and Valdes (1995), for example, showed that changes in the simulated sea-surface temperature due to orbital perturbation play a very important role in initiating the Laurentide and Fenno-Scandinavian ice sheets in their simulations.

Among the most recent progress in palaeoclimate modelling we may also report the development of new diagnostics for a better comparison with data:

- reconstructed biomes and inferred carbon storage (Friedlingstein *et al.* 1995);

- changes in the freshwater input to oceans, inferred from the simulated changes in continental drainage, as discussed in Coe (1995). These can be compared to salinity data;

- changes in the global chemical erosion inferred from simulated runoff and prescription of land-surface geology as described in Gibbs and Kump (1994). These can be compared to data on calcite compensation depth, Sr isotopic ratios and atmospheric pCO_2;

- changes in temperature extremes or growing degree-days, inferred from the simulated climates, can be compared to data derived from pollen (D. Texier & N. de Noblet, unpublished).

REFERENCES

Barron EJ, Washington WM (1984) The role of geographic variables in explaining paleoclimates: results from Cretaceous climate model sensitivity studies. Journal of Geophysical Research 89:1267-1279

Berger A, Loutre MF (1991) Insolation values for the climate of the last 10 million years. Quaternary Science Reviews 10:297-317

Bonan GB, Pollard D, Thompson SL (1992) Effects of boreal forest vegetation on global climate. Nature 359:716-718

Bonan GB, Chapin SF III, Thompson SL (1995) Boreal forest and tundra eco-systems as components of the climate system. Climatic Change 29:145-167

Box EO (1981) Macroclimate and plant forms: An introduction to predictive modeling in phytogeography. Junk Publishers, The Hague

Broccoli AJ, Manabe S (1987) The influence of continental ice, atmospheric CO_2 and land albedo on the climate of the last glacial maximum. Climate Dynamics 1:87-89

Chalita S, Le Treut H (1994) The albedo of temperate and boreal forest and the northern hemisphere climate: a sensitivity experiment using the LMD AGCM. Climate Dynamics 10:231-240

Charney J, Stone PH, Quirk WJ (1975) Drought in the Sahara: a biogeophysical feedback mechanism. Science 187:434-435

Claußen M (1994) On coupling global biome models with climate models. Climate Research 4:203-221

Coe MT (1995) The hydrologic cycle of major continental drainage and ocean basins: a simulation of the modern and mid-Holocene conditions and a comparison with observations. Journal of Climate 8:535-543

50

COHMAP members (1988) Climatic changes of the last 18,000 years: observations and model simulations. Science 241:1043-1052

Dickinson RE, Henderson-Sellers A, Kennedy PJ, Wilson MF (1986) Biosphere - Atmosphere Transfer Scheme (BATS) for the NCAR Community Climate Model. Technical note, NCAR / TN-275+STR, National Center for Atmospheric Research

Dickinson RE (ed.) (1987) The Geophysiology of Amazonia. Wiley, New York

Dong B, Valdes PJ (1995) The sensitivity studies of northern hemisphere glaciation using an atmospheric general circulation model. Journal of Climate 8:2471-2496

Ducoudré N, Laval K, Perrier A (1993) SECHIBA, a new set of parameterizations of the hydrologic exchanges at the land-atmosphere interface within the LMD atmospheric general circulation model. Journal of Climate 6:248-273

Foley JA, Kutzbach JE, Coe MT, Levis S (1994) Feedbacks between climate and boreal forests during the Holocene epoch. Nature 371:52-54

Friedlingstein P, Müller J-F, Brasseur G (1994) Sensitivity of the terrestrial biosphere to climate changes: impact on the carbon cycle. Environ Pollut, 83:143-146

Friedlingstein P, Prentice KC, Fung IY, John JG (1995) Carbon-biosphere-climate interactions in the last glacial maximum. Journal of Geophysical Research 100D4:7203-7221

Gibbs MR, Kump LR (1994) Global chemical erosion during the last glacial maximum and the present: sensitivty to changes in lithology and hydrology. Paleoceanography 9:529-543

Guetter PJ, Kutzbach JE (1990) A modified Köppen classification applied to model simulations of glacial and interglacial climates. Climatic Change 16:193-215

Harrison SP, Kutzbach JE, Prentice IC, Behling PJ, Sykes MT (1995) The response of northern hemisphere extratropical climate and vegetation to orbitally induced changes in insolation during the last interglaciation. Quaternary Research 43:174-184

Henderson-Sellers A (1993) Continental vegetation as a dynamic component of a global climate model: a preliminary assessment. Climatic Change 23:337-377

Holdridge LR (1947) Determination of plant formations from simple climatic data. Science 105:367-368

Jolly D, Harrison S, Damnati B, Bonnefille (1996) Simulated climate and biomes of Africa duing the Late Quaternary: Comparison with pollen and lake status data. Quaternary Science Reviews (in press)

Köppen W (1936) Das geographische System der Klimate. in Köppen W, Geiger G (eds) Handbuch der Klimatologie 1, Part C. Bornträger, Berlin

Körner C, Arnone JA III (1992) Responses to elevated carbon dioxide in artificial tropical ecosystems. Science 257:1672-1675

Koster RD, Suarez MJ (1992) Modeling the land surface boundary in climate models as a composite of independent vegetation stands. Journal of Geophysical Research 97D:2697-2715

Kutzbach JE, Gallimore RG (1988) Sensitivity of a coupled atmosphere/mixed layer ocean model to changes in orbital forcing 9,000 years BP. Journal of Geophysical Research 93:803-821

Kutzbach JE, Guetter PJ (1986) The influence of changing orbital parameters and surface boundary conditions on climate simulations for the past 18,000 years. Journal of Atmospheric Sciences 43:1726-1759

Lautenschlager M, Herterich K (1990) Atmospheric response to ice age conditions: climatology near the earth's surface. Journal of Geophysical research 95:22547-22557

Lean J, Warrilow DA (1989) Simulation of the regional climatic impact of Amazon deforestation. Nature 342:411-413

Liao X, Street-Perrott A, Mitchell J (1994) GCM experiments with different cloud parameterization: Comparisons with paleoclimatic reconstructions for 6,000 years B.P. Paleoclimates (Data and Modelling) 1:99-123

Milankovitch M (1941) Kanon der Erdbestrahlung. Royal Serbian Academy, Spec. Publ. 132, section of Mathematical and Natural Sciences, vol. 33

Noblet N de, Prentice C, Joussaume S, Texier D, Botta A, Haxeltine A (submitted) Possible role of Biosphere–Atmosphere interactions in triggering the last glaciation. Geophysical Research Letters

Polcher J, Laval K (1994) The impact of African and Amazonian deforestation on tropical climate. Journal of Hydrology 155:389-405

Prentice IC, Cramer W, Harrison SP, Leemans R, Monserud RA, Solomon AM (1992): A global biome model based on plant physiology and dominance, soil properties and climate. Journal of Biogeography 19:117-134

Prentice IC, Sykes MT, Cramer W (1993) A simulation model for the transient effects of climate change on forest landscapes. Ecological Modelling 65:51-70

Rind D, Peteet D, Kukla G (1989) Can Milankovitch orbital variations initiate the growth of ice sheets in a general circulation model?. Journal of Geophysical Research 94:12851-12871

Royer J-F, Deque M, Pestiaux P (1983) Orbital forcing for the inception of the Laurentide ice sheet. Nature 304:43-46

Sellers PJ, Mintz Y, Sud YC, Dalcher A (1986) A Simple Biosphere model (SiB) for use within general circulation models. Journal of Atmospheric Sciences 43:505-531

Sherwood BI, Kimball BA (1993) Tree growth in carbon dioxide enriched air and its implications for global carbon cycling and maximum levels of atmospheric CO_2. Global Biogeochemical Cycles 7:537-55

Shugart HH, Shao G, Emanuel WR, Smith TM (1996) Modelling the structural response of vegetation to climate change. in Huntley B, Cramer W, Morgan AV, Prentice HC, Allen JRM (eds) Past and future rapid environmental changes: The spatial and evolutionary responses of terrestrial biota, 265-271. Springer-Verlag, Berlin

Shukla J (1992) GCM response to changes in the boundary conditions at the earth's surface: a review. in Second International Conference on Modelling of Global Climate Change and Variability, Abstracts, 52. Max-Planck-Institut für Meteorologie, Hamburg

Shukla J, Nobre C, Sellers P (1990) Amazon deforestation and climate change. Science 247:1322-1325

Street-Perrott FA, Mitchell JFB, Marchand DS, Brunner JS (1990) Milankovitch and albedo forcing of the tropical monsoons: a comparison of geological evidence and numerical simulations for 9000 yBP. Transactions of the Royal Society of Edinburgh: Earth Sciences 81:407-427

Sud YC, Shukla J, Mintz Y (1988) Influence of land-surface roughness on atmospheric circulation and precipitation: a sensitivity study with a general circulation model. Journal of Applied Meteorology 27:1036-1054

Tamponnet C, Bino R, de Chambure D (1991) Higher plant growth in closed environment: preliminary experiments in life support facility at ESA-ESTEC. Proceedings of the 4th European Symposium on Space Environmental and Control Systems, 21-24 October, 1991. (ESA SP-324) Florence, Italy

Thornthwaite CW (1948) An approach towards a rational classification of climate. Geographical Review 38:55-94

Troll C, Paffen K (1964) Die Jahreszeitenklimate der Erde (Summary: the seasonal climates of the Earth). Erdkunde 18:1-28 plus map

Woodward FI (1987) Climate and plant distribution. Cambridge University Press. Cambridge

Wright HE, Kutzbach JE, Webb T III, Ruddiman WF, Street-Perrott FA, Bartlein PJ (eds) (1993) Global climates since the last glacial maximum. University of Minnesota Press, Minneapolis, London

Section 2

Spatial responses to past changes

Spatial response of plant taxa to climate change:
A palaeoecological perspective

Thompson Webb III
Department of Geological Sciences
Brown University
Providence, RI 02912-1846
U.S.A.

INTRODUCTION

Changes in climate over geological time have required that plant taxa move long distances. In responding to century- to millennial-scale climate changes, plant taxa have shifted their spatial distributions by extending and contracting their range boundaries and by altering the location of their abundance maxima. The spatial mosaic nature of climate patterns and climate change (see Table 2 in Bartlein 1996) has influenced the speed and direction of these motions but not prevented broad-scale movements. Because the different types of climate changes, whether they be trends, steps, oscillations, fluctuations, or translations (Bartlein 1996), extend back over millions of years, all modern (as well as ancient) taxa not only have evolved while their populations were moving but have also evolved mechanisms to deal with each type of change. Various dispersal mechanisms are therefore key to the existence of modern taxa and have allowed them to cope with the many large glacial–interglacial changes in climate of the last 2·5 million years as well as with short-term changes in climate. The large differences in seed size and generation time among taxa illustrate that many different dispersal and coping strategies have worked in the past. But how successful will the different taxa be in the future when a large global climate change may occur 3 to 10 times faster than past global changes? Can plant taxa move fast enough then and will modern ecosystems stay intact as the taxon populations relocate? Those are key questions facing ecologists, foresters and conservation biologists. Much research is now organised to answer

NATO ASI Series, Vol. I 47
Past and Future Rapid Environmental Changes:
The Spatial and Evolutionary Responses of
Terrestrial Biota
Edited by Brian Huntley et al.
© Springer-Verlag Berlin Heidelberg 1997

these questions (Peters & Lovejoy 1992; Houghton *et al.* 1990, 1992; Karieva *et al.* 1993).

Palaeoecological data are contributing to the answer by providing evidence for past vegetational responses to climate changes of many types. They document past rates of movement (King 1996), and past changes serve as a gauge for judging the magnitude and rate of predicted future change (Overpeck *et al.* 1991). The interpretation of past data also raises questions about the degree to which the individual taxa and the vegetation have been or might continue to be in equilibrium with climate. In light of other papers that discuss migration rates and modelling (King 1996, Davis 1996), I discuss the display and interpretation of spatial patterns in pollen data and then describe some of the long-term continental-scale changes in taxa that have accompanied the climate changes described by Bartlein (1996). These vegetational changes illustrate how large past climate changes have led to many independent spatial changes among taxa. The independence of these changes result both in biome and ecosystem rearrangements (Davis 1983; Huntley 1990a; Webb *et al.* 1983 a, b, 1987) and in the existence of past vegetational assemblages that are not analogous to any growing today (Overpeck *et al.* 1985, 1991; Baker *et al.* 1989; Huntley 1990b). The mapped record of these vegetation changes also shows how the taxa and vegetation responded to the times of fast and abrupt climate changes as well as to gradual changes. After describing the mapped fossil record of change, I discuss the issue of vegetation/climate equilibrium and the potential for disequilibrium conditions in the future.

INTERPRETATION AND DISPLAY OF POLLEN DATA

Interpretation of pollen data from sediments is similar to deciphering remotely–sensed data from satellites because pollen data are remotely–sensed data from plant populations and vegetation. They represent death assemblages of immature microgametophytes that differ manifestly from the vegetational assemblages of sporophytic plants that produced the pollen. Just as the current vegetation emits or reflects radiation that remote sensors on satellites intercept, so too does (and has) the current (and past) vegetation shed pollen that accumulates 'remotely' (i.e. away from the source) in lakes and bogs. Both types of 'remote' sensors (satellite

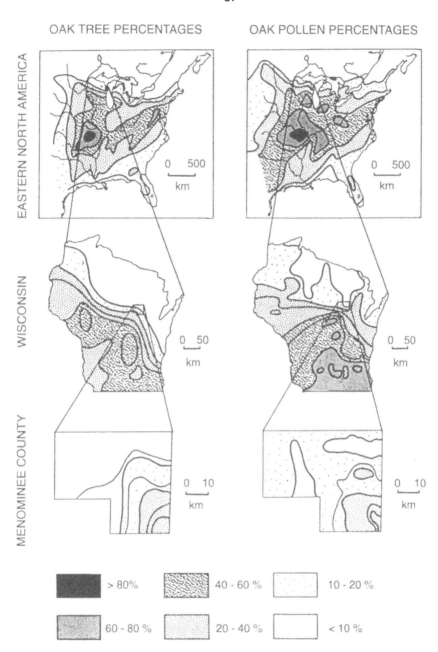

Figure 1: Relative abundance of oak trees (% basal area from forest inventory data) and oak pollen (% of tree pollen) at different spatial scales from that of 4 million km², 110,000 km², and 1000 km². Modified from Solomon and Webb (1985). The series of maps show how pollen data can provide a zoom lens view of vegetational patterns.

instruments and lake sediments) record data with certain sampling characteristics (e.g. spatial and temporal resolution), and their data need calibration and ground-truthing in terms of such vegetation attributes as composition (taxon abundances), structure (height and mixture of growth forms or plant-functional types), and pattern (mosaic and geographic gradients). Actualistic studies of modern data have provided this calibration and ground-truthing (Davis 1967; Webb 1974; Prentice 1988; Jackson 1994).

Maps of modern vegetation show patterns at a variety of spatial scales from the global distribution of biomes down to local communities along a lake shore (Kutzbach & Webb 1991; Webb 1993). These patterns result from a variety of controls and processes that vary from latitudinal gradients in climate at the global and continental scales (10^8–10^6 km^2) to soil and moderate topographic contrasts at broad regional scales (10^5–10^3 km^2) down to local differences in microhabitat and disturbance (10–10^{-1} km^2). Sensing the full spectrum of spatial vegetational patterns therefore requires data sets capable of recording several different spatial scales of variation. Pollen data have this capacity and can retrieve patterns at a variety of spatial scales from biomes and tree abundances across continents down to forest types and tree populations within 1000 km^2 (Figure 1), and even down to tree distributions within a woodland stand (Heide & Bradshaw 1982; Jackson 1994; Calcote 1995). Pollen data are therefore sensitive to a variety of spatial vegetation patterns and thus to the different processes that produce these patterns (Webb 1993). Vegetation also changes over a variety of time scales, and pollen data capture these different temporal variations. They vary from long-term species immigration after continental drift (Hoogheimstra 1984) and vegetation changes induced by orbital-scale climate variations of 10^4 years over the last 1·5 Ma. down to local successional changes over decades (Figure 2).

Time series of contoured maps and space-time plots can show taxon and vegetational variations in space and time simultaneously (Figure 3) and thus can ultimately allow construction of the full 4-dimensional aspects of the distribution changes in plant taxa and of the compositional and distributional changes in vegetation (Webb 1993). The main map displays of pollen abundance changes are for regional and continental sets of late-Quaternary pollen data in Europe and North America (Huntley & Birks, 1983; Webb et al. 1983a, b; Gaudreau 1988; Webb 1988;

Anderson *et al.* 1989; Anderson & Brubaker 1994). These illustrate changes in abundance, location, and association among plant taxa over the past 18,000 (radiocarbon) years when climates changed from full-glacial to interglacial conditions (Bartlein 1996). Separate studies at individual sites or clusters of sites have yielded either much longer time series or finer-scale resolution in selected areas (Huntley & Webb 1988).

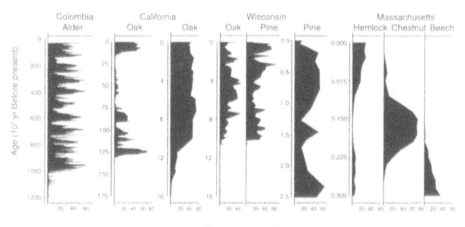

Percent of total pollen

Figure 2: Temporal variations in time series of pollen data from 1·2 million years down to 250 years to show a temporal zoom lens of past changes in the vegetation. The time series for 1·2 million to 16,000 years show climatically induced variations in alder and oak pollen percentages. Fire-, human-, and disease-induced changes appear in the time series for 2500 and 250 years. Modified from Webb (1993).

INDEPENDENT SPATIAL RESPONSES OF PLANT TAXA TO LARGE CLIMATE CHANGES

Maps in Webb (1988) and Huntley and Birks (1983) show the spatial nature of the taxon response to both gradual and abrupt climate change. These publications provide detailed descriptions of the changes. Here I focus on how the rates of geographic vegetational change vary with those in climate and then describe the general nature of the independent changes among taxa and the consequences for vegetational assemblages and communities over the past 18,000 years.

Maps of the 6 major pollen categories for eastern North America show a continuous sequence of spatial changes from 18,000 years ago to present (Figure 4), as the Laurentide ice sheet retreated and the seasonality of insolation increased and then decreased. They also show a time of more rapid change and rearrangement of abundance gradients between 12,000 and 10,000 years ago when the rate of climate change was highest. Webb *et al.* (1987) subdivided the sequence of changes into three periods: glacial (18,000 to 12,000 years ago), transitional (12,000 to 9000 years ago), and interglacial (9000 years ago to present). The north-south vegetational gradients for 18,000 to 12,000 years ago reflect the glacial mode when the climatic impact of the ice sheet and other glacial boundary conditions was dominant. Once solar radiation had become strong enough and the ice sheet had shrunk enough, the patterns among the taxa shifted to their interglacial patterns with a northeast-to-southwest gradient across Canada and a northwest-to-southeast gradient from Iowa to Florida. This transition occurred relatively quickly during a period of rapid climate change between 12 and 10 ka. After that the taxon distributions and abundance gradients continue changing but at a slower rate consistent with the rate of climate changes during the Holocene. These maps show that the taxon distributions are responsive to the changing climate gradients and that the distributions change fastest during the time of most rapid climate change.

Separate time-series studies of these data document the times and rates of change. Jacobson *et al.* (1987) and Overpeck *et al.* (1991, 1992) showed that large changes were occurring at most sites from 13 to 9·5 ka and the peak amount of change was about 10,000 years ago right at the time of most rapid climate change. Huntley (1990b) obtained similar results for Europe. These studies show the responsiveness of the continental-scale vegetation to climate change and that the vegetation has altered its rate of change as the rate of climate changes varied. Studies in selected areas and at individual sites (Huntley & Birks 1983; Peteet *et al.* 1994; Bernabo 1981; Gajewski 1988) show even higher frequency changes in vegetation that are responsive to climate changes, e.g. into and out of the Younger Dryas (which lasted 1500 calendar years) and the little ice age (which lasted 400 years) or other even shorter events (Gear & Huntley 1991; Levesque *et al.* 1993). The little ice

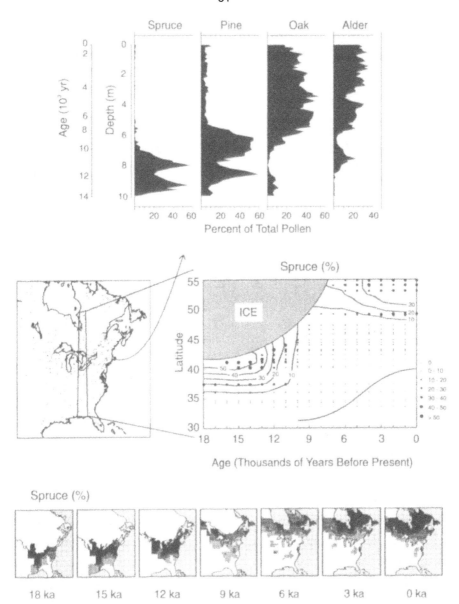

Figure 3: The translation of time series plots of pollen percentages (top diagram) into contoured plots of the spruce (*Picea*) pollen data on either a latitude-time diagram or on maps. The latitude-time diagram is a vertical cross-section from a 3 dimensional space–time box of the data (with time as the vertical axis), and the maps are horizontal cross-sections (modified from Webb, 1993). Black areas on the maps show where spruce (*Picea*) pollen percentages are > 20%, dark Grey 5-20%, and light Grey 1-5%. The retreating Laurentide ice sheet appears on the maps from 18 ka to 9 ka, and the shoreline is modified to show this time of lower sea levels.

age and shorter events are generally at a higher frequency than allowed for by the temporal resolution among sites in the continental data set (Webb, 1993), but their presence in the time series data indicates a finer scale of response in the vegetation to climate change over and above what Figure 4 shows.

The time series of maps (Figure 4) also show that the major tree and herb taxa changed independently of one another both in abundance and in location. These taxa migrated 100's of kilometres (Davis 1976, 1981) when responding to the glacial – interglacial changes in climate. Both the timing and the pattern of movement varied among taxa. For taxa with well marked climatic differences like *Picea*, *Pinus* and *Quercus*, such differences in timing and in migration routes are expected during times of major climate change. The mapped patterns of movement for these taxa seem linked to the changing combinations of climate variables that favour their growth. For other taxa like Fagaceous trees that all grow in the deciduous forest, differences in timing are also evident, and the dispersibility of the seeds was seen as a controlling factor (Davis 1976). Johnson and Webb (1989), however, showed that because blue jays *Cyanocitta cristata* disperse the soft-shelled nuts of each of these genera, all have a similar potential mobility, and thus that the migration differences are best related to the differing climate preferences among these taxa (Bartlein *et al.* 1986).

In eastern North America, the spatial patterns of the northward movement of taxa also illustrate the independent movement of taxa (Webb 1988). Some, like *Quercus,* first became abundant in the south and then moved north as abundant populations; others, like *Fagus,* became abundant in the south for a few thousand years and then shrank in abundance there before becoming abundant again to the north (Bennett 1985); and yet others, like *Alnus,* first became abundant at the glacial boundary, starting about 12,000 years ago, and then spread in relatively abundant populations from there. *Pinus* showed yet another pattern by being abundant in the south, then along the east coast, before its area of abundance spread westward and north at 10,000 years ago. None of these taxa moved northward in a zonal pattern or paralleled the movements of the other taxa with which they are associated today. Such different patterns of change show the highly independent nature of the responses of the different taxa to climate change.

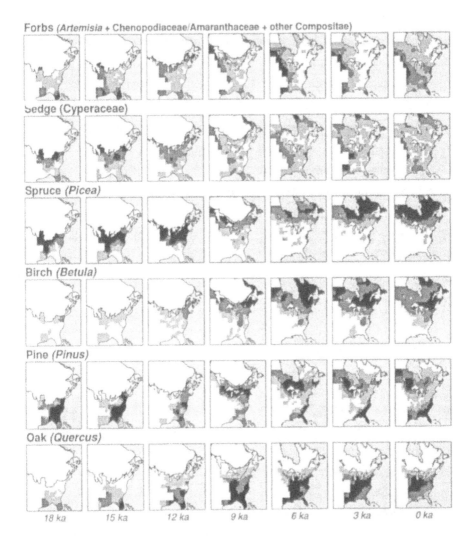

Forbs *(Artemisia* + Chenopodiaceae/Amaranthaceae + other Compositae)

Sedge (Cyperaceae)

Spruce *(Picea)*

Birch *(Betula)*

Pine *(Pinus)*

Oak *(Quercus)*

18 ka   *15 ka*   *12 ka*   *9 ka*   *6 ka*   *3 ka*   *0 ka*

Figure 4: Maps with contours of equal pollen percentages for forbs (*Artemisia* + other Asteraceae excluding *Ambrosia* + Chenopodiaceae + Amaranthaceae), sedge (Cyperaceae), spruce (*Picea*), birch (*Betula*), pine (*Pinus*), oak (*Quercus*) from 18,000 years ago until today (18 ka to 0 ka). Black areas are > 20%, dark Grey 5-20%, and light Grey 1-5%. The retreating Laurentide ice sheet appears on the maps from 18 ka to 9 ka, and the shoreline is modified to show this time of lower sea levels. (modified from Webb *et al.* 1993a).

The independent response of different taxa to climate change also manifests itself in the differences in timing of when each taxon is most abundant during the glacial–interglacial cycle. In eastern North America (Webb 1988), the maps show that some taxa (e.g. *Picea*, Cyperaceae and *Pinus*) seem well suited to full-glacial conditions (18,000 to 12,000 years ago), others (e.g. *Fraxinus, Ostrya-Carpinus, Populus, Ulmus* and *Betula papyrifera*) reach their maximum distribution and abundance during the transitional period between full-glacial and interglacial conditions (12,000 to 9000 years ago), and still others *(Quercus, Tsuga, Fagus, Betula, Castanea, Picea, Abies, Alnus* and *Pinus*) are most widespread and abundant during the interglacial period (9000 years ago to present). Even during peak interglacial conditions from 9 ka to present, individualistic differences in response are evident among tree taxa because the times of maximum abundance and distribution differ among taxa, with *Pinus* being most broadly distributed at 8000 yr BP, *Alnus* at 6000 yr BP, *Tsuga* at 5000 yr BP and *Castanea* at 2000 yr BP. The pollen maps indicate that different taxa are best suited to different parts of the glacial–interglacial cycle and that, as climates shift through their quasi-periodic cycles (see Bartlein 1996), different taxa flourish and/or are rare at different times.

VEGETATIONAL CONSEQUENCES

The pollen maps thus illustrate a continuously changing picture of vegetation dynamics in which vegetation change can be viewed as one set of variables (the different taxa) constantly chasing or responding to the changes in another set of variables (climate, soils, disturbance regimes, etc.). One direct consequence of the independent responses and resultant vegetational dynamism during the large climate changes of the last 18,000 years was that plant associations changed continuously in composition and ultimately became re-organised into new biomes and communities (Webb 1988; Huntley 1990a). In eastern North America, the modern deciduous forest first formed by 12,000 years ago in the south, the prairie by 10,000 years ago in the west, and the modern tundra by 8000 years ago in the north-east (Webb 1988). The mixed forest and the boreal forest did not begin to have their modern composition until 6000 years ago or later, and the forest dominated by southern pines only grew north of Florida by 8000 years ago. As the glacial biomes disappeared, so too did their ecotones, and modern ecotones developed along with the establishment of the modern vegetation regions (Webb *et al.* 1983a, b; Webb 1987). This interpretation of the data supports the view that plant assemblages are ephemeral and in continuous flux both geographically and compositionally. Such behaviour is a fundamental vegetational response to large climate changes involving changes in the seasonal interaction of different climate variables.

A second consequence of the independent responses is the appearance of vegetational assemblages unlike any that occur today (i.e. 'no-analogue' assemblages or vegetation). Beginning at 16,000 years ago in eastern North America, a widespread area with no-analogue vegetation developed (Overpeck *et al.* 1992). Abundant *Fraxinus* and *Ostrya-Carpinus* pollen, and later *Ulmus* pollen, combined with *Picea* and Cyperaceae pollen to create a large region whose pollen assemblages have no modern analogues (Jacobson *et al.* 1987). With the decrease in the abundance of *Picea* and Cyperaceae after 10,000 years ago and the spread of *Pinus* and *Quercus*, the area of no-analogue vegetation decreased rapidly in size and was mostly gone by 9 ka. Interestingly its time of demise was hastened by the period of rapid climate change 10,000 years ago, whereas its initiation matches the time when climates were most different from today. By 16,000 years ago, the seasonality of insolation was increasing while the impact of the ice sheets was still large, and climates developed that were highly unlike those of today. The no-analogue plant assemblages reflected the novel combinations of climate variables of that time.

All of the changes just described lead me to conclude that the extant plant taxa are well adapted to coping with the several types of past climate change, that their continental-scale abundance changes largely reflect climate changes, that their rates of change vary with the rates of climate change, that their independent responses lead to compositional changes and assemblage rearrangements, and that climates not analogous to current climates produce vegetational assemblages not analogous to current vegetation. The rapidity of past climate change therefore was not *per se* the cause of taxa moving independently, of past assemblages becoming disassociated, or of non-analogue vegetation developing. These are all normal types of taxon and vegetation responses to climate change whether the climate change is fast or slow.

The above set of conclusions, of course, depends on several assumptions about exactly what the past climates were like. Without a detailed independent picture of the changing climate patterns in eastern North America for the past 18,000 years, some of these assertions about the nature of past climates and past climate changes can be questioned. Climate model results are not yet either accurate or detailed enough to give full support to the above statements (Webb *et al.* 1993b). The climate model results for North America (Kutzbach 1987) described in Webb *et al.* (1987, 1993a), however, are not inconsistent with the above interpretations and conclusions. These climate model results show that the time of most rapid change was between 12,000 and 9000 years ago and that no-analogue climate conditions existed at 15,000 and 12,000 years ago.

IMPLICATIONS FOR THE FUTURE AND VEGETATION CLIMATE EQUILIBRIUM

The responsiveness of the past vegetation to large climate changes, as well as to rapid changes in climate, bodes well for the ability of the vegetation to respond to future climate changes. Concern, however, must derive from analysis of the eastern North American data that future climate changes may induce the vegetation to change at rates 3 to 10 times faster than those in the past (Overpeck *et al.* 1991). How might these predicted fast rates affect the vegetation? Will the vegetation be able to stay in dynamic equilibrium with the future climates, and will current ecosystems stay intact as they shift in location in response to the future climates?

Given the continuous and varied nature of climate change across time scales, Webb (1986) proposed applying the concept of dynamic equilibrium/disequilibrium to the response of vegetation to climate change. I then used theory applied to the response of thermometers to steps as well as periodic changes in temperature to show how the conditions for equilibrium depend on the ratio of the response time of a taxon (or the vegetation) to the period (or effective period) of forcing by the climate. Under a forcing–response model for vegetation and climate interactions, a taxon and the vegetation will always lag in their response to a given climate change. The lag arises from a variety of biotic factors and is best modelled as an exponential response (fast at first and slowing down as the taxon approaches appropriate conditions). The e-folding time (i.e. the time it takes to reach 1/e of the full response) is then used as a measure of the lag or response time.

Disequilibrium arises when the response time is longer than or approximately equal to the period of forcing, and disequilibrium results in a much damped response that is out of phase with the forcing. Equilibrium responses occur when the response times are much shorter than the period of forcing, e.g. 1/10 of the period or less. For such conditions, the amplitude of the response is full, and the taxa track the forcing with relatively short lags.

Individual taxa can vary in their response times from an order of years for herbs and shrubs to centuries for some trees. The vegetation as a mixture of taxa will have various response times depending upon what vegetation changes are required and what biotic processes (e.g. dispersal) are involved. For this reason, Webb (1986) distinguished between a Type A response in which changes occur among extant taxa after a climate change and a Type B response in which the full complement of taxa that should grow in the new climate respond. For the latter response, immigration of new taxa may be required, and the time for taxa to move must be added to the time for the taxa to grow and expand their populations at the new location. Immigration will therefore lengthen the response times for the Type B response in the vegetation.

My deduction from dynamic equilibrium theory (Webb 1986) is that the rapid climate changes of the future will favour the plant taxa with short climatic response times. If the future change involves an implied period of forcing of 200 to 2000 years, then taxa with 1000-year response times will not be favoured, but those with response times of 1 to 100 years should thrive. Rapid climate change, therefore, just shifts the selective forces to the fast end of the spectrum of response times. Herbs, shrubs, and fast-growing trees should thrive. To some extent, these taxa have already been favoured by logging and human land-use practices in many areas of the world. Where reforestation has occurred, these are the taxa that first appear. If we let go of the image of old growth or climax vegetation, then we can anticipate an earth covered by vegetation whose age structure is similar to much that now grows in many areas occupied by humans today. Evidence from recent plant invasions shows that many taxa faced with new areas of favourable habitat can spread across them quickly (Mack 1996). Selection for the fast-spreading taxa should also be favoured in the future.

If future changes favour taxa that are fast-spreading and have short generation times, then the vegetation, which is an emergent property of the assemblages of taxa, will change not just in location but also in character. Specific assemblages can only move across the landscape at rates set by the taxa with the slowest response times. Tree growth and dispersal rates that translate into response times of 50 to 400 yr will limit the spread of many extant forest types. The longer of the response times are similar to the 200 to 2000-year period of forcing for the predicted future climate changes and therefore imply disequilibrium for the forests dominated by these tree taxa even though certain individual taxa may be in dynamic equilibrium with the rapid climate changes. As a result, during the time of rapid change, new assemblages or ecosystems will emerge and certain current ones will disappear. Were the greenhouse-gas-warmed climates to stabilise, new vegetation regions might also result whose composition will reflect new mixtures of climate variables. Where the mixtures match current climates, then current vegetation assemblages will emerge, but where the climate conditions differ significantly from any extant today, then no analogue vegetation will emerge.

The prediction of the climate changes being more rapid than those of the past implies that dynamic models of the vegetation are needed to predict the changing vegetation and that during the time of rapid change simulations from equilibrium models will be inadequate (Prentice & Solomon 1991). No-analogue vegetation and independent motion among taxa will result from both the different climates and the rapid rate of climate change. By removing habitat and decimating populations of avial dispersal agents, human land-use will also influence the movement of plant taxa and affect the composition of the resultant vegetation. New plant assemblages and ecosystems should certainly develop if the climate changes as currently predicted during the next century or two (Houghton *et al.* 1992), and the rapid rate of the predicted climate changes should be a key factor in determining the composition, structure, and distribution of these new assemblages and ecosystems.

ACKNOWLEDGEMENTS

An NSF Climate Dynamics Grant supported this research. I thank PJ Bartlein, J Fuller, B Huntley, J McLachlan, and E Russell for critical comments and P Leduc for technical assistance.

REFERENCES

Anderson PM, Bartlein PJ, Brubaker LB, Gajewski K, Ritchie JC (1989) Modern analogues of late Quaternary pollen spectra from the western interior of North America. Journal of Biogeography 16:573-596

Anderson PM, Brubaker LB (1994) Vegetation history of north-central Alaska: a mapped summary of late Quaternary pollen. Quaternary Science Reviews 13:71-92

Baker RG, Van Nest J, Woodworth G (1989) Dissimilarity coefficients for fossil pollen spectra from Iowa and western Illinois during the last 30,000 years. Palynology 13:63-77

Bartlein PJ (1996) Past environmental changes: Characteristic features of Quaternary climate variations. in Huntley B, Cramer W, Morgan AV, Prentice HC, Allen JRM (eds) Past and future rapid environmental changes: The spatial and evolutionary responses of terrestrial biota, 11-29. Springer-Verlag, Berlin

Bartlein PJ, Webb T, III, Prentice IC (1986) Climatic response surfaces based on pollen from some eastern North American taxa. Journal of Biogeography 13:35-57

Bennett KD (1985) The spread of *Fagus grandifolia* across eastern North America during the last 18,000 years. Journal of Biogeography 12:147-164

Bernabo JC (1981) Quantitative estimates of temperature changes over the last 2700 years in Michigan based on pollen data. Quaternary Research 15:143-159

Calcote R (1995) Pollen source area and pollen productivity: evidence from forest hollows. Journal of Ecology 83:591-602

Davis MB (1967) Late-glacial climate in northern United States: a comparison of New England and the Great Lakes region. in Cushing EJ, Wright HE, Jr. (eds) Quaternary Palaeoecology, 11-43. Yale University Press New Haven, CT

Davis MB (1976) Pleistocene biogeography of temperate forests. Geoscience and Man 8:13-26

Davis MB (1981) Quaternary history and the stability of forest communities. in West DC, Shugart HH, Botkin DB (eds) Forest Succession, 132-177. Springer-Verlag New York

Davis MB (1983) Quaternary history of deciduous forests of eastern North America and Europe. Annals of the Missouri Botanical Garden 70:550-563

Davis MB, Sugita S (1996) Reinterpreting the fossil pollen record of Holocene tree migration. in Huntley B, Cramer W, Morgan AV, Prentice HC, Allen JRM (eds) Past and future rapid environmental changes: The spatial and evolutionary responses of terrestrial biota, 181-193. Springer-Verlag, Berlin

Gajewski K (1988) Late Holocene climate changes in eastern North America estimated from pollen data. Quaternary Research 29:255-262

Gaudreau DC (1988) Paleoecological interpretation of geographic patterns in pollen data: spruce and birch in northeastern North America. Bulletin of the Buffalo Society of Natural Sciences 33: 15-29.

Gear AJ, Huntley B (1991) Rapid changes in the range limits of Scotch Pine 4000 years ago. Science 251:544-547

Heide KM, Bradshaw RHW (1982) The pollen–tree relationship within forests of Wisconsin and upper Michigan, U.S.A. Review of Palaeobotany and Palynology 36:1-23

Hoogheimstra H (1984) Vegetational and Climatic History of the High Plain of Bogota, Colombia: A Continuous Record of the Last 3.5 Million Years. AR Gantner Verlag, Kommanditgesellschaft Vaduz, Germany

Houghton JT, Jenkins GJ, Ephraums JJ (eds) (1990) Climate Change: The IPCC Scientific Assessment. Cambridge University Press Cambridge

Houghton JT, Callender BA, Varney SK (eds) (1992) Climate Change 1992: The Supplementary Report to the IPCC Scientific Assessment. Cambridge University Press, Cambridge

Huntley B, Birks HJB (1983) An Atlas of Past and Preasent Pollen Maps for Europe: 0-13000 Years Ago. Cambridge University Press Cambridge

Huntley B, Webb T, III (1988) Vegetation History. Kluwer Academic Publishers Dordrecht

Huntley B (1990a) European post-glacial forests: compositional changes in response to climate change. Journal of Vegetation Science 1:507-518

Huntley B (1990b) Dissimilarity mapping between fossil and contemporary pollen spectra in Europe for the past 13,000 years. Quaternary Research 33:360-376

Jackson ST (1994) Pollen and spores in Quaternary lake sediments as sensors of vegetation composition: theoretical models and empirical evidence. in Traverse A (ed) Sedimentation of Organic Particles, 253-286. Cambridge University Press Cambridge

Jacobson GL, Jr, Grimm EC, Webb T III (1987) Patterns and rates of vegetation change during deglaciation of eastern North America. in Wright HE Jr, Ruddiman WF (eds) North America and Adjacent Oceans during the Last Deglaciation, 277-288. Geological Society of America Boulder, CO

Johnson WC, Webb T III (1989) The role of blue jays in the postglacial dispersal of Fagaceous trees in eastern North America. Journal of Biogeography 16:561-571

Kareiva P, Kingsolver J, Huey R (eds) (1993) Biotic Interactions and Global Change. Sinauer, Sunderland

King GA, Herstrom AA (1996) Holocene tree migration rates objectively determined from fossil pollen data. in Huntley B, Cramer W, Morgan AV, Prentice HC, Allen JRM (eds) Past and future rapid environmental changes: The spatial and evolutionary responses of terrestrial biota,91-101. Springer-Verlag, Berlin

Kutzbach JE (1987) Model simulations of the climatic patterns during the deglaciation of North America. in Wright HE Jr, Ruddiman WF (eds) North America and Adjacent Oceans during the Last Deglaciation, 425-446. Geological Society of America Boulder, CO

Kutzbach JE and Webb T, III (1991) Late Quaternary climatic and vegetational change in eastern North America: concepts, models, and dates. in Shane LCK, Cushing EJ (eds) Quaternary Landscapes, 175-217. University of Minnesota Press Minneapolis

Levesque AJ, Mayle FE, Walker IR, Cwynar LC (1993) The amphi-Atlantic oscillation: a proposed late-glacial climatic event. Quaternary Science Reviews 12:629-643

Mack RN (1996) Plant invasions: Early and continuing expressions of global change. in Huntley B, Cramer W, Morgan AV, Prentice HC, Allen JRM (eds) Past and future rapid environmental changes: The spatial and evolutionary responses of terrestrial biota, 205-216. Springer-Verlag, Berlin

Overpeck JT, Webb T III, Prentice IC (1985) Quantitative interpretation of fossil pollen spectra: dissimilarity coefficients and the method of modern analogs. Quaternary Research 23:87-108

Overpeck JT, Bartlein PJ, Webb T III (1991) Potential magnitude of future vegetation change in eastern North America: comparisons with the past. Science 252:692-695

Overpeck JT, Webb RS and Webb T, III (1992) Mapping eastern North American vegetation change over the past 18,000 years: no-analogs and the future. Geology 20: 1071-1074.

Peteet D, Daniels RA, Heusser LE, Vogel JS, Southon JR, Nelson DE (1994) Late-glacial pollen, macrofossils and fish remains in northeastern U.S.A. – the Younger Dryas oscillation. Quaternary Science Reviews 12:597-612

Peters RL, Lovejoy TE, (eds) (1992) Global Warming and Biological Diversity. Yale University Press New Haven, CT

Prentice IC (1988) Records of vegetation in time and space: the principles of pollen analysis. *in* Huntley B, Webb T III (eds) Vegetation History, 603-632. Kluwer Academic Publishers Dordrecht

Prentice IC, Solomon AM (1991) Vegetation models and global change. *in* Bradley RS (eds) Global changes of the past, 365-383. Office of Interdisciplinary Studies, University Corporation for Atmospheric Research Boulder, CO

Solomon AM, Webb T III (1985) Computer-aided reconstruction of late-Quaternary landscape dynamics. Annual Review of Ecology and Systematics 16:63-84

Webb T III (1974) Corresponding distributions of modern pollen and vegetation in lower Michigan. Ecology 55:17-28

Webb T III (1986) Is vegetation in equilibrium with climate? How to interpret late-Quaternary pollen data. Vegetatio 67:75-91

Webb T III (1987) The appearance and disappearance of major vegetational assemblages: long-term vegetational dynamics in eastern North America. Vegetatio 69:177-187

Webb T III (1988) Glacial and Holocene vegetation history: Eastern North America. *in* Huntley B, Webb T III, (eds) Vegetation History, 385-414. Kluwer Academic Publishers Dordrecht, The Netherlands

Webb T III (1993) Constructing the past from late-Quaternary pollen data: temporal resolution and a zoom lens space-time perspective. *in* Kidwell SM, Behrensmeyer AK (eds) Taphonomic Approaches to Time Resolution in Fossil Assemblages, 79-101. Paleontological Society Short Course in Paleontology, No. 6, University of Tennessee Knoxville, TN

Webb T III, Bartlein PJ and Kutzbach JE (1987) Climatic change in eastern North America during the past 18,000 years. *in* Wright HE Jr, Ruddiman WF (eds) North America and Adjacent Oceans during the Last Deglaciation, 447-462. Geological Society of America Boulder, CO

Webb T III, Cushing EJ, Wright HE Jr (1983a) Holocene changes in the vegetation of the Midwest. *in* Wright HE Jr (ed) Late-Quaternary environments of the United States: Volume 2, The Holocene, 142-165. University of Minnesota Press Minneapolis

Webb T III, Richard PJH, Mott RJ (1983b) A mapped history of Holocene vegetation in southern Quebec. Syllogeus 49:273-336

Webb T III, Bartlein PJ, Harrison S, Anderson KH (1993a) Vegetation, lake level, and climate change in eastern North America. *in* Wright HE Jr, Kutzbach JE, Webb T III, Ruddiman WF, Street-Perrott FA, Bartlein PJ (eds) Global Climate Change since the Last Glacial Maximum, 415-467. University of Minnesota Press Minneapolis

Webb T III, Ruddiman WF, Street-Perrott FA, Markgraf V, Kutzbach JE, Bartlein PJ, Wright HE Jr, Prell WL (1993b) Climatic changes during the past 18,000 years: regional syntheses, mechanisms, and causes. *in* Wright HE Jr, Kutzbach JE, Webb T III, Ruddiman WF, Street-Perrott FA, Bartlein PJ (eds) Global Climate Change since the Last Glacial Maximum, 415-467. University of Minnesota Press Minneapolis

The response of New Zealand forest diversity to Quaternary climates

Matt S. McGlone

Manaaki Whenua - Landcare Research

PO Box 69

Lincoln 8152

New Zealand

INTRODUCTION

The New Zealand archipelago lies at mid-latitudes in the south-west Pacific Ocean and has an oceanic, mainly moist, warm-temperate to cool-temperate climate. Evergreen forest forms the natural cover of *ca.* 85% of the landmass. The forests are not species rich; only some 420 species characteristically occur in them (Wardle 1991), *ca.* 210 of tree size. Approximately 70% of trees, 40% of shrubs and 30% of vines have bird-dispersed fruit, including all the podocarps; the remainder have wind- or gravity-dispersed fruits (Wardle 1991, Burrows 1994). Most terrestrial birds in New Zealand will eat fruit, but only 5 are regular consumers and effective dispersers, most important being the New Zealand pigeon *Hemiphaga novaeseelandiae* (Lee *et al.* 1991a). Seed of most forest species appears to be well dispersed over short distances of up to several kilometres. A recent study (Bray *et al.* 1994) found that gaps in lowland podocarp/angiosperm forest are colonised rapidly, most species arriving within 2 years. Although birds migrate to and from New Zealand, there are no migratory movements within the country. However, many fruit- and seed-eating birds move seasonally within a district, often between mountain and coast.

HISTORICAL BIOGEOGRAPHY

The New Zealand archipelago can be divided into 3 distinctive biogeographic landscapes (Figure 1). *Northern New Zealand* comprises the low-lying and tectonically stable northern third of the North Island, characterised by high levels of

NATO ASI Series, Vol. I 47
Past and Future Rapid Environmental Changes:
The Spatial and Evolutionary Responses of
Terrestrial Biota
Edited by Brian Huntley et al.
© Springer-Verlag Berlin Heidelberg 1997

forest endemics and diverse forests. *Central New Zealand* comprises the southern two-thirds of the North Island, and the northern third of the South Island. It has volcanically and tectonically active landscapes, and low numbers of forest endemics, but species-rich forests. *Southern New Zealand* comprises the southern half of the South Island, large areas of which have been affected by glacial activity and high rates of tectonic movement. It has low forest endemism and low forest species diversity.

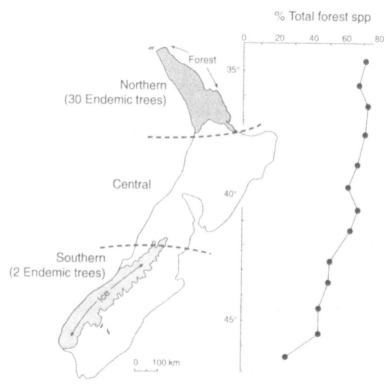

Figure 1: Map showing the biogeographical regions of New Zealand; full glacial shoreline, ice and forest cover. Graph of the percentage of total New Zealand forest flora occuring in 1° latitude bands.

Each biogeographic area has had a disinctive response to glacial-interglacial climate cycles, exemplified by the transition from the Last Glaciation to the Holocene. Northern New Zealand had continuous forest cover during the Last Glaciation with *Nothofagus* prominent in the interior and podocarp/angiosperm forests at the coast (Dodson *et al.* 1988; Newnham 1992). These continuous northern forests graded southwards into forest-scrub-grassland mosaics. Tall podocarp forest abruptly replaced these mosaics at around 14,500 yr BP (McGlone 1988; Newnham *et al.* 1989). Central New Zealand was in forest-shrubland-grassland mosaics during the peak of the Last Glaciation. The forest patches were largely of *Nothofagus* and rarer towards the south. Tall podocarp forest spread into Central New Zealand between 14,000 and 11,000 yr BP (McGlone & Neall 1995; Moar 1971). Southern New Zealand was in grassland, short shrubland, bare ground and glacial ice during the glacial maximum. Tall podocarp forest reoccupied lowland districts between 10,500 and 9500 yr BP, although low forest was present in some western districts earlier (Moar 1971; McGlone & Bathgate 1983). Opinion differs as to whether development of closed continuous forest tracts was the result of a colonising wave of forest species spreading south at the termination of the glaciation (Wardle 1988), or of forest taxa spreading from local patches (McGlone 1985, 1988).

MIGRATION VERSUS LOCAL PERSISTENCE

Arguments for the 'northern migration' model of forest colonisation at the end of the Last Glaciation are based on:

1. the absence of evidence for forest during the full glacial period in southern districts;

2. the supposition that colder annual temperatures (4·5°C to 5·0°C lower than present) during the Last Glaciation were sufficient to eliminate a large percentage of the current southern flora;

3. the presence of plant species in northern districts that can reproduce in the wild in the south as evidence for incomplete post-glacial migratory readjustment.

All of these arguments have some force, and therefore plant migration cannot be excluded as a significant element in the formation of the current distribution patterns. Nevertheless, the hypothesis that migration has played a minor role in the formation of regional floras appears more likely, and I here summarise the evidence for this view.

Pollen evidence from the Last Glacial Maximum attests to the presence of *Libocedrus* and *Nothofagus* forest stands in Central New Zealand, but good evidence for podocarp and temperate angiosperm trees is lacking. There is no clear evidence from southern New Zealand for forest presence during the maximum of the Last Glaciation; fossil wood is entirely lacking, and arboreal pollen levels are similar to those found at present on the subantarctic islands, 400 – 800 km from the nearest forested landmass (Moar 1980). However, three features of the pollen record argue for forest expansion from local forest patches in central and southern New Zealand.

1. During the late-glacial period in central and southern New Zealand (14,500 to 10,000 yr BP) traces of tree pollen are recorded long before the major expansion of forest. It is rare for any pollen taxon to make its first appearance late in a Holocene sequence; most are represented early, even if only as traces. If these traces resulted from long-distance pollen dispersal, it would be expected that they should have a similar composition. In fact, the late-glacial pollen traces are diverse, suggesting local sources rather than an undifferentiated New Zealand-wide record.

2. The sequence of dominance of pollen taxa in late glacial and Holocene sequences does not differ greatly from site to site, regardless of location. If migration from northerly sources controlled these patterns, an initial wave of well-dispersed trees would be expected, followed by more slowly moving taxa. This seems not to be the case. For example, among the tree podocarps the sequence is *Prumnopitys taxifolia* – *Prumnopitys ferruginea* – *Dacrydium cupressinum*. Among angiosperm trees, the sequence is nearly always *Nothofagus menziesii* before *Nothofagus* subgenus *Fuscospora*; and *Metrosideros umbellata* before *Weinmannia racemosa*. The tree fern dominance sequence is usually *Cyathea smithii* – *Dicksonia fibrosa* – *Dicksonia squarrosa*. Ecological and climatic preferences therefore appear to have played the major role. Taxa favouring raw soils spread first, those of leached soils later; taxa that resist heavy frosts and cool conditions first, less tolerant species later. Taxa that are advantaged by frequent disturbance and drought became common only late in the Holocene sequences.

3. Timing of arrival of tall forest appears to be controlled more by local site conditions than by latitude. For example: in the far south of the South Island, a pollen sequence records the presence of *Metrosideros* forest before 14,000 yr BP at a site immediately adjacent to the retreating western glacial ice margin (Pickrill *et al.*

1992); 1000 km north in lowland Taranaki, central North Island, forest of any sort did not begin to develop until after 13,000 yr BP (McGlone & Neall 1995)

Biogeographic evidence also supports forest survival in the south during the full glacial. Far southern New Zealand has disjunct areas of *Nothofagus fusca and N. truncata*, species that are not especially cold resistant. As *Nothofagus* has heavy, small-winged seeds and restrictive mycorrhizal requirements and therefore limited dispersal capabilities, it is most likely that it survived *in situ* in favoured locations, even in the interior of the southeastern South Island. If these *Nothofagus* species survived, it is highly likely that all but the most tender northern species were likewise capable of glacial survival.

I conclude that post-glacial forest spread was from local sources, and governed by the prevailing regional climate and local site conditions. The apparent simultaneous spread of forest within a region represents a filling in of a forest patch framework that persisted throughout the Last Glaciation. The repetitive post glacial sequences of forest taxa are therefore consistent with climate and soil mediated rise to dominance of individual species from a pool of taxa present from the beginning. This pattern corresponds to the Initial Floristic Model for succession, albeit on a longer timescale (Egler 1954; Bray 1989).

IMPLICATIONS FOR DIVERSITY AND DISPERSAL

Because Northern New Zealand retained an almost complete forest cover throughout the last glacial–interglacial cycle, endemic forest taxa are relatively common, and genetic diversity higher than in the southern region, where forest endemism is extremely low. In Central and Southern New Zealand, tectonically active landscapes and greatly fluctuating forest cover have ensured the flora is dominated by widespread species and genetically less diverse populations. The occasional transfer of genes via rare dispersal events or pollen drift is probably sufficient to prevent regional genetic differentiation of species. Severe population bottlenecks, such as those that occur when forest tracts are broken up into small isolated patches during glacial periods, may account for low genetic diversity (Haase 1992).

Fewer than 25% of the indigenous vascular flora are distributed as widely as their observed ecological tolerances would suggest (Wardle 1991). Many indigenous forest species will grow well and reproduce in cultivation hundreds of kilometres south of their current natural range, and there are now many documented instances of such species escaping and establishing in local forests (see e.g. Wardle 1988; Lee *et al.* 1991b). These invasions indicate that unassisted dispersal alone is insufficient to allow many species to make even moderate dispersal leaps to suitable habitats. The New Zealand archipelago is mountainous (over 40% of the terrain higher than 600 m), characterised by rivers with wide flood plains that flow more or less directly from the interior to the coast, and cut by two ocean straits. Suitable environments for many forest species are rarer and more disjunct southwards, because of increased landscape relief and youthful soils, lower average annual temperatures, and droughty conditions in the east. Chances of successful dispersal become correspondingly less. Not all barriers are topographical or climatic: the tall multilayered forests of lowland New Zealand are resistant to invasion. For example, although *Nothofagus* spp will invade lowland podocarp/angiosperm forest, their rate of penetration is slow and in some cases estimated at fractions of a metre per year (Haase 1990). In addition, despite 150 years of tree and vine introduction to New Zealand, only a handful of taxa have established in lowland forest (e.g. *Acer pseudoplatanus, Lonicera japonica, Clematis vitalba, Leycesteria formosa*) and it is rare for exotics to enter indigenous forests other than those disturbed by human activities (Webb *et al.* 1988). Biogeographic regions, although not having absolute barriers to dispersal, possess a combination of climatic, topographical and biotic factors that reduce the number and size of potential invasion points, and hence greatly reduce species interchange.

CONCLUSIONS

New Zealand regional forest floras have stable compositions despite major changes in climate and the extent of forested terrain in the course of a glacial-interglacial cycle. While forest invasion into non-forest vegetation is rapid with climatic warming, movement from one forested region to another is probably rare because of topographical, climatic and biotic barriers. Therefore, sustained warming of the climate should not of itself promote rapid large-scale changes in forest composition because of these natural barriers to dispersal. However, human movement of native plants into districts of New Zealand where they do not naturally occur may eventually result in major alteration of regional forests.

Perhaps a more serious threat is that of exotic species. There are nearly 1500 naturalised exotic vascular species in New Zealand at present, equivalent to the number of indigenous species (Webb *et al.* 1988), and an unknown but large number of species in cultivation. Should global warming increase annual temperatures beyond those experienced during interglacials, native trees and vines may lose their current competitive advantage. Exotic woody species from subtropical or tropical areas, currently held in check by low winter temperatures, may be able to rapidly invade and dominate indigenous forests. The problems presently experienced in Hawaii with aggressive tree weeds such as *Psidium cattleianum* and *Myrica faya* (Wagner *et al.* 1990) may be a model for what could happen with global warming in many temperate areas of the globe.

ACKNOWLEDGEMENTS

I thank Colin Webb and Neville Moar for their comments on the draft of this paper. This work was funded by the New Zealand Foundation for Research, Science and Technology, Contract No CO9405.

REFERENCES

Burrows CJ (1994) Fruit types and seed dispersal modes of woody plants in Ahuriri Summit Bush, Port Hills, western Banks Peninsula, Canterbury, New Zealand. New Zealand Journal of Botany 32:169-181

Bray JR (1989) The use of historical vegetation dynamics in interpreting prehistorical vegetation change. Journal of the Royal Society of New Zealand 19:151-160

Bray JR, Burke WD, Struik, GJ (1994) Regeneration dynamics of alluvial gaps in a warm temperate rain forest in New Zealand. Vegetatio 112:1-13

Dodson JR, Enright NJ, McLean RF (1988) A late Quaternary vegetation history for far northern New Zealand. Journal of Biogeography 15:647-656

Egler FE (1954) Vegetation science concepts. 1. Initial floristic composition - a factor in old-field vegetation development. Vegetatio 4:412-417

Haase P (1990) Environmental and floristic gradients in Westland, New Zealand, and the discontinuous distribution of *Nothofagus*. New Zealand Journal of Botany 28:25-40

Haase P (1992) Isozyme variability and biogeography of *Nothofagus truncata* (Fagaceae). New Zealand Journal of Botany 30:315-328

Lee WR, Clout MN, Robertson HA, Wilson JB (1991a) Avian dispersers and fleshy fruits in New Zealand. Acta XX, Congress Internationalis Ornithologici:1617-1623

Lee WR, Wilson, JB, Meurk CD, Kennedy PC (1991b) Invasion of the subantarctic Auckland Islands, New Zealand, by the asterad tree *Olearia lyallii* and its interaction with a resident myrtaceous tree *Metrosideros umbellata*. Journal of Biogeography 18:493-508

McGlone MS (1985) Plant biogeography and the late Cenozoic history of New Zealand. New Zealand Journal of Botany 23:723-749

McGlone MS (1988) Glacial and Holocene Vegetation History: New Zealand. *in* Huntley B, Webb T (eds) Handbook of Vegetation Science Vol. 7: Vegetation History, 558-599. Kluwer Academic Publishers

McGlone MS, Bathgate, JL (1983). Vegetation and climate history of the Longwood Range, South Island, New Zealand, 12 000 B.P. to the present. New Zealand Journal of Botany 21:292-315

McGlone MS, Neall VE (1995) The late Pleistocene and Holocene vegetation history of Taranaki, North Island, New Zealand. New Zealand Journal of Botany 32:251-269

Moar NT (1971) Contributions to the Quaternary history of the New Zealand flora, 6. Aranuian pollen diagrams from Canterbury, Nelson and north Westland, South Island. New Zealand Journal of Botany 9:80-145

Moar NT (1980) Late Otiran and early Aranuian grassland in central South Island. New Zealand Journal of Ecology 3:4-12

Newnham RM (1992) A 30,000 year pollen, vegetation and climate record from Otakairangi (Hikurangi), Northland, New Zealand. Journal of Biogeography 19:541-554

Newnham RM, Lowe DJ, Green JD (1989) Palynology, vegetation and climate of the Waikato lowlands, North Island, New Zealand, since *c.* 18,000 years ago. Journal of the Royal Society of New Zealand 19:127-150

Pickrill RA, Fenner JM, McGlone MS (1992) Late Quaternary evolution of a fjord environment, Preservation Inlet, New Zealand. Quaternary Research 38:331-346

Wagner WL, Herbst DR, Sohmer SH (1990) Manual of the Flowering Plants of Hawai`i. Bishop Museum Special Publication 83. University of Hawaii Press, Honolulu

Wardle P (1988) Effects of glacial climates on floristic distribution in New Zealand. 1. A review of the evidence. New Zealand Journal of Botany 26:541-55

Wardle P (1991) Vegetation of New Zealand. Cambridge University Press, Cambridge

Webb CJ, Sykes WR, Garnock-Jones PJ (eds) (1988) Flora of New Zealand. Vol IV. Botany Division, D.S.I.R., Christchurch

Character of rapid vegetation and climate change during the late-glacial in southernmost South America

Vera Markgraf and Ray Kenny[1]
Institute of Arctic and Alpine Research
University of Colorado
Boulder, Colorado 80309-0450
U.S.A.

INTRODUCTION AND BACKGROUND

A major question of future global change concerns the rate of environmental (e.g. vegetation) response to the predicted, rapid, shifts in future climate. Specific questions are: is the environment, specifically the vegetation, in equilibrium with climate, or does it lag behind the climate change? What is the effect of scale, spatial and temporal, on environmental response to climate change? Records of past vegetation change offer a possibility to address these questions of response time and magnitude (Prentice 1986; Huntley 1990, 1991). At large spatial (continental-scale) and temporal (century) scales dynamic equilibrium between vegetation and climate seems a reasonable assumption (Prentice 1986). At fine scales the existence of equilibrium or lags is disputed (Webb 1986). The most productive approach to this dispute is the study of high-resolution, multi-proxy records. Here we present a high-resolution palaeoclimate study based on combined pollen and stable isotope (δD) analyses for the late-glacial portion of a peat core from Tierra del Fuego, that allowed us to address these questions of character and timing of climate change and vegetation response.

Modern climate in southern South America is primarily related to the interaction between the thermal regimes of the tropical Pacific Ocean and Antarctica. In general

---

[1] Department of Geology, Highlands University, Las Vegas, New Mexico 87701, U.S.A.

NATO ASI Series, Vol. I 47
Past and Future Rapid Environmental Changes:
The Spatial and Evolutionary Responses of
Terrestrial Biota
Edited by Brian Huntley et al.
© Springer-Verlag Berlin Heidelberg 1997

terms this interaction determines the temperature gradient between pole and equator, which determines the intensity and latitudinal position of the subtropical high pressure cell and in turn the latitudinal position of the southern westerly stormtracks. This interaction is the cause of clear seasonal variability, with the stormtracks located farther poleward in summer and equatorward in winter. On longer time scales, this interaction also shows marked interannual and decadal climate variability, expressed most dramatically by the El Niño/Southern Oscillation events and their respective precipitation and temperature teleconnection modes.

The present distribution of vegetation in the temperate latitudes of South America reflects the seasonal precipitation patterns that result from the atmospheric circulation variations, and consequently occur along latitudinal as well as elevational gradients. The region west of the Andes between latitudes 45 and 55°S, where annual precipitation reaches between 2000 and over 5000 mm, is characterized by temperate rainforests. Under extreme windy conditions and poor drainage, Magellanic Moorland replaces rainforest vegetation. At the other end of the gradient, east of the Andes where annual precipitation is below 200 mm, steppe or steppe-scrub dominate, with *Empetrum* heath on exposed wind-swept ridges. Between these environmental extremes we find different types of deciduous and mixed evergreen forests, dominated by different species of *Nothofagus*.

All these major vegetation types can be distinguished in modern pollen assemblages. Changes in past pollen assemblages analysed from late Quaternary lake and peat bog sediments can be interpreted as reflecting changes along the environmental gradients described. The comparison of pollen records along and across the gradients yields information on the spatial and temporal vegetation response to changes in boundary conditions such as variations in insolation regimes, sea levels, circum-Antarctic sea-ice, etc., which govern the climate of the region.

First we discuss the major trends of vegetation change during the last 18,000 years for southern South America, with focus on latitudinal position of the westerly stormtracks, interpreted from presence of high precipitation and high winds. To address questions on amplitude and rate of vegetation response to climate change we present the results of combined pollen and hydrogen isotope analyses of a peat core in Tierra del Fuego (Harberton, eastern Beagle Channel, 54°53'S, 67°10'W), which had also yielded a decadal to multi-decadal resolution record of past atmospheric CO_2 content from carbon isotopes (White *et al.* 1994).

REGIONAL VEGETATION CHANGES

Although apparently synchronous within radiocarbon control, both magnitude and character of environmental change were different at different latitudes during the last 18,000 years (Markgraf 1991a, 1993a; Markgraf *et al.* 1992). Fullglacial environments were treeless throughout southern South America, except for the region between 41° and 45°S west of the Andes where *Nothofagus* woodland persisted (Villagran & Armesto 1993; Markgraf 1991a; Markgraf *et al.* 1995). Both north and south of this zone, steppe, steppe-scrub or *Empetrum* heath, respectively, characterized the environments on both sides of the Andes. This pattern implies that throughout southern South America, precipitation was greatly reduced. Even between latitudes 43° and 45°S, where woodland persisted and Magellanic Moorland is found at elevations lower than today, precipitation probably did not markedly increase. Instead, this vegetation combination was more likely an expression of greatly increased windiness under colder conditions (Markgraf *et al.* 1992). During the fullglacial, the westerly stormtrack, albeit bringing less moisture than today because of colder oceans, must have been focused year-round on the mid-latitudes, about 5 degrees latitude north of its modern mean summer position, and several degrees south of its mean winter position.

Emergence from fullglacial conditions occurred at 14,000 yr BP Andean glaciers had retreated into the mountains and ocean temperatures had reached modern levels even in high circum-Antarctic latitudes (Lowell *et al.* 1995; Clapperton *et al.* 1995). Sedimentation began in many lake and bog sites. The strongest vegetation response is documented at mid-latitudes, west of the Andes: Magellanic Moorland retreated to higher elevations and *Nothofagus* forest expanded (Villagran & Armesto 1993). All this suggests a coupled precipitation and temperature increase and a decrease in windiness. East of the Andes and at high latitudes (south of 45°S) on the other hand, steppe-scrub and *Empetrum* heathland, respectively, persisted, indicating that in these areas the temperature increase was not accompanied by a precipitation increase. Although decreased in intensity, westerly stormtracks continued to be located in the mid-latitudes.

The onset of Holocene-type environments occurred at mid-latitudes at 12,500 yr BP, with expansion of a species-rich forest vegetation at low elevations, and with Magellanic Moorland restricted to high elevations. Vegetation response to this change was also more widespread, affecting areas east of the Andes and south to latitude 50°S. This suggests a poleward extension of stormtrack activity, probably related to seasonality shifts. Latitudes south of 50°S, although continuing treeless, showed a shift to more mesic grassland environments. This moisture increase indicates that the effects of the stormtracks were felt at high latitudes as well, although not as strongly as at present.

Only after 9000 yr BP can Magellanic Moorland be traced in the high southern latitudes where it is found today and *Nothofagus* forests expanded throughout their modern distribution. Forests at mid-latitudes changed in composition to a higher proportion of summer-drought resistant taxa. This change implies that the centre of stormtracks had shifted and focused to higher latitudes.

Truly modern environments, characterized by large seasonal shifts in stormtracks (poleward in summer and equatorward in winter) and marked temporal and spatial variability, thought to relate to the development of the El Niño/Southern Oscillation pattern (McGlone *et al.* 1992), became established throughout southern South America only following a dry interval between 5000 and 4000 yr BP

A few palaeoenvironmental records from southern South America have been analysed with temporal resolution high enough to discuss rates of vegetation change (Markgraf 1991b 1993b; Moreno 1994). However, only the multiple indicator approach applied here for the first time in a record from southern South America can address the question as to the rate of vegetation response to climate change.

RATE OF ENVIRONMENTAL CHANGE

In order to more fully address the rate of vegetation response to climate change we present data from a high-resolution, multi-proxy study of the late-glacial portion of the Harberton peat core, analysed for pollen and hydrogen isotopes (each sample represents 10 years and intervals between samples range between 10 and 50 years). Palynological analyses show pronounced and repeated high amplitude palaeoenvironmental variability, especially between the base (at 14,000 yr BP) and 10,000 yr BP (Figure 1). The variability is expressed primarily by proportional

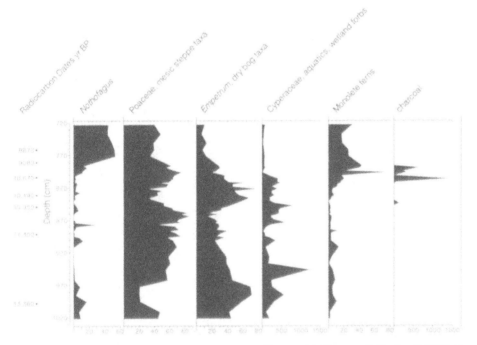

Figure 1. Pollen diagram from Harberton, Tierra del Fuego (latitude 54°53' S; longitude 67°10' W) representing the interval from 8000 to 14,000 yr BP, with major taxa (in per cent of total land pollen) and charcoal (in per cent out-of-pollen-sum).

fluctuations of non-arboreal taxa, Ericaceae (primarily *Empetrum*), Poaceae, herbaceous taxa, aquatic taxa, and traces of the only arboreal taxon *Nothofagus*. Peaks of *Empetrum* occur between 13,500 and 12,500 yr BP, 12,000 and 11,500, and 10,500 to 10,000 yr BP *Empetrum* maxima, especially the 10,500 to 10,000 yr BP peak, are accompanied by increases in herbaceous taxa characteristic of disturbance and of arid steppe environments (e.g. *Acaena*, *Plantago* and *Gunnera*) as well as in taxa characteristic of dry bogs (e.g. *Nanodea* and *Lycopodium*; all such taxa are included in the Ericaceae/dry bog taxa curve in Figures 1 and 2). Charcoal particles are also abundant in the sediment at these times, when dry vegetation dominates, especially between 10,500 and 10,000 yr BP Presence of charred moss fragments suggest that at times the surface of the bog itself must have burned. These intervals of high *Empetrum* heath pollen can be interpreted to represent repeated periods of extremely dry and variable conditions.

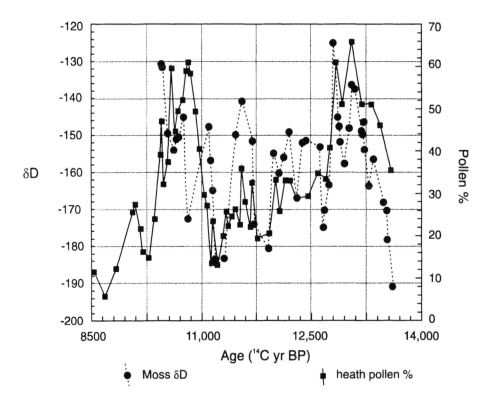

Figure 2. Heathland pollen (Ericaceae/dry bog taxa)(in per cent) and δD (in per mill) from Harberton. Every sample represents ten years.

During times of *Empetrum*-heathland peaks, such as between 13,500 and 12,700 yr BP and 10,500 and 10,000 yr BP, Poaceae, mesic herbaceous taxa and aquatics are present in low amounts. In contrast, between 12,500 and 10,500, and 10,000 and 9000 yr BP, when heathland taxa are present in low proportions, the mesic taxa (Poaceae and mesic forbs, including Apiaceae, Ranunculaceae, Caryophyllaceae and Asteraceae in Figures 1 and 3) are well represented and are interpreted to reflect intervals of increased effective moisture.

To define the actual climate change responsible for the vegetation response, and discuss rates of change, we analysed the deuterium/hydrogen (D/H) ratio in the mosses from the same samples analysed for pollen. Deuterium ratios have been widely used in reconstructing past temperature variations from ice cores (Ciais *et al.* 1992). By applying this technique to the peat components we expected to establish an independent measure of temperature change for the high southern latitudes.

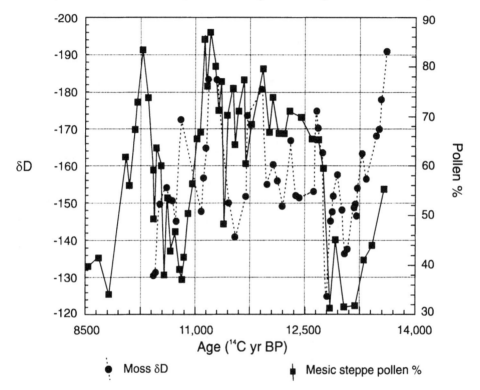

Figure 3. Mesic steppe pollen (Poaceae/steppe forbs)(in per cent) and δD (in per mill) from Harberton. Each sample represents ten years.

The stable hydrogen isotope composition of most vascular plant material reflects a complex combination of climatic information that is determined by the source water, leaf-water fractionation, and the fixation of hydrogen to organic material. Because mosses do not have large leaf-water fractionation effects (moss leaves lack stomata), the stable hydrogen isotopic composition faithfully records the isotopic composition of the source water. Additionally, D/H and $^{18}O/^{16}O$ ratios of bog source water plot near the meteoric water line, indicating only very little evaporative enrichment (Brenninkmeijer 1983). It has also been generally demonstrated that stable hydrogen isotopic composition of source water reflects the temperature at the site of precipitation (Dansgaard 1964; Taylor 1972), and that D/H values of non-exchangeable (carbon-bound) hydrogen in plant cellulose reflect primarily the D/H values of source water used by the plant during growth (Epstein *et al.* 1976). We have found that the D/H values of modern source water in southern South America are strongly controlled by temperature, yielding a δD/temperature gradient of about 4 ‰ °C^{-1}. The slope of the δD/°C gradient is less than the 5·6 ‰ °C^{-1} line for a maritime area close to a moisture source, but is in agreement with published gradients for the nearby Antarctic Peninsula (Aristarain *et al.* 1990). Based on the above considerations, we interpret the D/H moss record primarily as a temperature signal.

Fluctuations in δD show a close positive correlation with the abundance of heathland pollen taxa considered to represent dry conditions (Figure 2). This suggests that temperature fluctuations are primarily responsible for the high-amplitude environmental variability represented by proportional shifts between more arid and more mesic vegetation during the late-glacial. In terms of actual temperature change, the δD amplitude suggests that temperatures fluctuated repeatedly by 8 to 10 °C.

As discussed above, fluctuations of the combined Poaceae/steppe forb taxa are inferred to be responses primarily to the major changes in precipitation, related to stormtrack shifts. Thus we infer that during the intervals between 12,500 and 10,500 and after 10,000 yr BP, when mesic taxa are dominant (Figure 3), stormtracks had shifted to higher latitudes. In several instances, however, there are correlations between shifts in mesic taxa and those in δD values. When δD becomes less negative (i.e. higher temperatures), mesic taxa decrease (e.g. between 13,600 and 12,500 yr BP, and at 11,000 yr BP). When δD becomes more positive (i.e. temperatures decrease; e.g. between 12,500 and 11,500 yr BP) Poaceae/steppe forbs strongly increase (Figure 3). These variations in mesic taxa abundance may reflect the temperature-related change in effective moisture, with effective moisture increasing under cooler temperatures.

From this discussion it appears that both steppe and heathland vegetation were strongly responding to temperature fluctuations. However, the amplitude of vegetation response relates to the specific coupling between precipitation and temperature. Under relatively more mesic conditions, a temperature increase results in a smaller amplitude increase in heathland taxa than under more arid conditions. Higher amplitude responses occur amongst mesic taxa when a temperature decrease is coupled with a precipitation increase, or for arid taxa when a temperature increase is coupled with a precipitation decrease.

The time spans at which these fluctuations occurred were such that they were completed within periods on the order of two to five decades. Decreases in temperature occurred as fast as increases and vegetation response to shifts in either direction was apparently immediate, occurring without lags. The rate of temperature and vegetation change ranged between $0.4 - 0.5 \, °C \, yr^{-1}$ and $0.16 - 0.2 \, °C \, yr^{-1}$. Increased fire frequency may have enhanced the amplitude of heathland vegetation response to temperature increase, probably by virtue of changing the taxonomic composition toward a higher proportion of fast-responding annual plants.

The fact that non-arboreal vegetation responds immediately even to rapid climate change is perhaps not too surprising, given the length of the taxa's life cycles. In a forthcoming paper (Markgraf *et al.* in prep.) we will analyse the response of the arboreal component, *Nothofagus*, in the Harberton record during the Holocene using the same high-resolution, multi-proxy approach. Preliminary inspection of the data suggests that the mid-Holocene *Nothofagus* expansion occurred very fast, apparently responding as quickly as the non-arboreal taxa during the late-glacial climate fluctuations.

REFERENCES

Aristarain AJ, Jouzel J, and Poutchet M (1990) A 400 year isotope record of the Antarctic peninsula climate. Climate Change 8:69-89

Brenninkmeijer CAM (1983) Deuterium, oxygen-18 and carbon-13 in tree rings and peat deposits in relation to climate. PhD Thesis, University of Groningen, Netherlands

Ciais P, Petit JR, Jouzel J, Lorius C, Barkov NI, Lipenkov V, Nicolaiev V (1992) Evidence for an early Holocene climatic optimum in the Antarctic deep ice-core record. Climate Dynamics 6:169-177

Clapperton CM, Sugden DE, Kaufman DS, McCulloch RD (1995) The last glaciation in central Magellan Strait, southernmost Chile. Quaternary Research 44:133-148

Dansgaard W (1964) Stable isotopes in precipitation. Tellus 16:436-468

Epstein S, Yapp CJ, Hall JH (1976) The determination of the D/H ratio of non-exchangeable hydrogen in cellulose extracted from aquatic and land plants. Earth Planet Sc Let 30:241-251

Huntley B (1990) Studying global change: the contribution of Quaternary palynology. Palaeogeog Palaeoclimatol Palaeoecol 82:53-61

Huntley B (1991) How plants respond to climate change: migration rates, individualism and the consequences for plant communities. Ann Botany 67:15-22

Lowell TV, Heusser CJ, Andersen BG, Moreno PI, Hauser A, Heusser LE, Schlüchter C, Marchant DR, Denton GH (1995) Interhemispheric correlation of late Pleistocene glacial events. Science 269:1541-1549

Markgraf V (1991a) Late Pleistocene environmental and climatic evolution in southern South America. Bamberg Geograph Schriften 11:271-281

Markgraf V (1991b) Younger Dryas in southern South America. Boreas 20:63-69

Markgraf V (1993a) Paleoenvironments and paleoclimates in Tierra del Fuego and southernmost Patagonia, South America. Palaeogeog Palaeoclimatol Palaeoecol 102:53-68

Markgraf V (1993b) Younger Dryas in southermost South America - an update. Quat Sci Rev 12:351-355

Markgraf V, Dodson JR, Kershaw AP, McGlone MS, Nicholls N (1992) Evolution of late Pleistocene and Holocene climates in the circum-South Pacific land areas. Climate Dynamics 6:193-211

Markgraf V, McGlone M, Hope G (1995) Neogene paleoenvironmental and paleo-climate change in southern temperate ecosystems - a southern perspective. TREE 10:143-147

McGlone MS, Kershaw AP, Markgraf V (1992) El Niño/Southern Oscillation climatic variability in Australasian and South American palaeoenvironmental records *In* Diaz HF, Markgraf V (eds) El Niño: Historical and Paleoclimatic Aspects of the Southern Oscillation, 435-462. Cambridge University Press

Moreno PI (1994) Vegetation and climate near Lago Llanquihue in the Chilean Lake District. Ms Thesis, University of Maine, Orono

Prentice IC (1986) Vegetation responses to past climatic variation. Vegetatio 67:131-141

Taylor CB (1972) The vertical variations of the isotopic concentrations of tropo-spheric water vapor over continental Europe and their relationship to tropospheric structure. New Zeal Inst Nuclear Scie Res (DSIR), INS-12-107, 45 pp

Villagran C, Armesto JJ (1993) Full and late glacial paleoenvironmental scenarios for the west coast of southern South America. *in* Mooney HA, Fuentes ER, Kronberg BI (eds) Earth System Responses to Global Change. Contrasts between North and South America, 195-208. Academic Press

Webb T III (1986) Is vegetation in equilibrium with climate? How to interpret late-Quaternary pollen data. Vegetatio 67:75-91

White JCW, Ciais P, Figge RA, Kenny R, Markgraf V (1994) A high-resolution record of atmospheric CO_2 content from carbon isotopes in peat. Nature 367:153-156

Holocene tree migration rates objectively determined from fossil pollen data

George A. King and Andrew A. Herstrom

ManTech Environmental Research Services Corp.

National Health and Environmental Effects Research Laboratory

Western Ecology Division

200 S.W. 35th Street

Corvallis, OR 97330

U.S.A.

INTRODUCTION

The prospect of rapid, trace-gas induced climate change over the next 100 – 150 years has raised scientific and policy interest on the rates at which plant species, particularly tree species, can respond to these changes (Houghton *et al.* 1992; Davis & Zabinski 1992). Two particular responses are of interest over the next century: migration or movement of range limits and changes in abundance (i.e., tree density). This paper focuses on tree species migration, and particularly on what the fossil pollen record can tell us about potential rates of future migration.

The general patterns and rates of migration of major tree taxa in eastern North America during the Holocene have been reconstructed by Davis (1976; 1981a) and Delcourt and Delcourt (1987). Average migration rates for these taxa range between 200 and 400 m yr^{-1}. Elsewhere in North America, white spruce migrated at a rate of about 1000 m yr^{-1} into the Northwest Territories (Ritchie & MacDonald 1986). However, the projected northward movement of mean temperature isopleths (a simple metric of the geographic displacement of a suite of climate conditions) of 4 – 6 km yr^{-1} over the next 100 – 200 years is an order of magnitude greater than these average migration rates (Solomon *et al.* 1984; Huntley 1991; Davis & Zabinski 1992). Consequently, tree species will not be able to migrate at the same rate that climate could change, resulting in significant contraction of their actual range, with resulting effects on carbon fluxes and biodiversity (Davis & Zabinski 1992). Also of

NATO ASI Series, Vol. I 47
Past and Future Rapid Environmental Changes:
The Spatial and Evolutionary Responses of
Terrestrial Biota
Edited by Brian Huntley et al.
© Springer-Verlag Berlin Heidelberg 1997

ecological interest is that the observed rates of migration in the Holocene for tree species with wind-dispersed seeds, although slow compared to the rate of future climate change, are still faster than what one would predict from observational data collected on dispersal of seeds under average wind speeds (Davis 1976; Ritchie & MacDonald 1986; Godman & Lancaster 1990).

In this paper, the Holocene migration of two important eastern North America tree species, *Fagus grandifolia* and *Tsuga canadensis*, is investigated further by exploring a new spatial analysis approach to estimating migration rates. The objectives of the study were to estimate the variation in migration rates rather than just the overall average, and to estimate the frequency of maximum rates observed in the past, heretofore not attempted. The information on variability in migration rates will provide additional insight into the ability of tree species to respond to future climate change.

METHODS

Migration rates for *Fagus* and *Tsuga* were calculated by first estimating the arrival dates of these taxa at fossil pollen sites in eastern North America and then calculating rates of movement between grid cell centres to which the arrival dates were interpolated from the site data. The pollen data from the 464 sites used for these analyses were obtained from the North American Pollen Database (COHMAP 1988; Webb *et al.* 1993; NAPD 1994). The migration rate calculations were restricted to the current range limits of these two species in the region encompassing the Upper Midwest, the Upper Ohio Valley, the New England States and adjacent Canada, essentially the glaciated region of the eastern United States and southeastern Canada (Figure 1). Sites with arrival dates calculated later than 2,000 yr BP were excluded from the calculation of migration rates, since hemlock and beech had reached their modern range limits by that time (Davis 1976; 1981a). All spatial data analyses were conducted using ARC/INFO (ESRI 1992) and the Albers equal area projection.

The arrival date of a taxon at a site was defined by a pollen threshold value indicating the taxon was present near the site, and that threshold percentage being exceeded for three consecutive samples (Davis & Jacobson 1985). The consecutive sample requirement was added because some taxa (including *Fagus* and *Tsuga*) have low threshold values ($\leq 1\%$) indicating their presence at a site (Davis *et al.* 1986; 1991). These values can be exceeded relatively easily by chance alone since only one or two grains need be counted in a typical pollen count of 200 – 300 grains to meet that value. Furthermore, the consecutive sample requirement eliminates the possibility of selecting, as an arrival date, a single sample that exceeds the threshold percentage but is followed by a series of samples in which the threshold value is never exceeded. Again, the biological significance of this one occurrence above the threshold value is difficult to ascertain. The consecutive sample requirement encompasses in many instances the arrival criterion of a rapid increase in the deposition rate of a pollen type (Davis *et al.* 1991; MacDonald 1993). The threshold

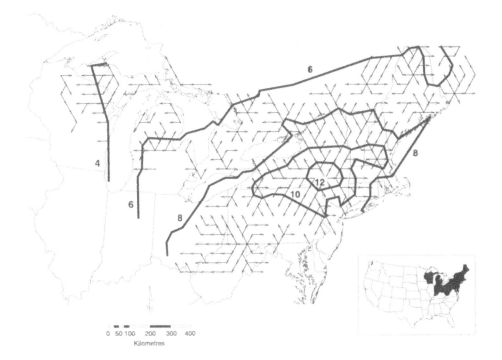

Figure 1. Migration map for *Tsuga canadensis* calculated using uncalibrated radiocarbon dates. Thick lines represent isochrones, and short arrows are migration vectors drawn to each hexagonal grid cell centre from the oldest adjacent grid cell.

values for *Fagus* and *Tsuga* were set at 0·5% and 1%, respectively, following the extensive analyses of modern surface samples and the current distribution of *Fagus* and *Tsuga* by Davis *et al.* (1986; 1991).

A grid-based approach was used to estimate migration rates for these two species. A hexagonal grid was used rather than a square grid since all the distances between grid centres are the same in a hexagonal grid system; this eliminated the potential bias of migration rates being estimated over a variety of distances on the spatial grid. The centres of the hexagonal grid cells were set at 50 km apart. The hexagonal grid data layers were created by first assigning the calculated arrival dates to the centre of those grid cells in which the fossil sites occurred. If more than one fossil site occurred in a cell the arrival dates were averaged together. Then, these arrival dates were interpolated to a 50 km rectilinear grid of points using an inverse distance squared algorithm. Values were not calculated for cells overlying the Great Lakes, nor for points outside the current range limits of *Tsuga* and *Fagus*. This set of points was used to interpolate values for the centres of the hexagonal grid cells using a bilinear interpolation procedure. Data from 149 and 127 fossil sites were used to construct the *Fagus* and *Tsuga* migration data layers, respectively.

To estimate migration rates, vectors were drawn to the centre of each hexagonal grid cell from the oldest adjacent grid cell. Migration rates were calculated for each vector by simply dividing the distance between cells (50 km) by the difference in arrival dates at each cell. All vectors with magnitudes < 4000 m yr^{-1} were averaged together to calculate the average migration for each species. (Two vectors, one each for *Fagus* and *Tsuga*, with magnitudes > 4000 m yr^{-1} were considered outliers.) Histograms were calculated to depict the distribution of the magnitudes of the vectors.

Isochrones depicting the range limits of *Tsuga* through time were also drawn for this analysis, using the spatial data layer in which the arrival dates from the fossil sites were aggregated to the hexagonal grid-cell centres. A triangulated irregular network (TIN) procedure (Burrough 1986) was used to generate the isochrones at 2,000 yr intervals.

RESULTS

The spatial interpolation of the arrival dates produced a reasonable map depicting the migration of *Tsuga*; the analysis indicated that *Tsuga* occupied much of the northeastern United States by 8,000 yr BP, reached its current northern limit in Canada by 6,000 yr BP, and reached its westward limit after 4,000 yr BP (Figure 1). The general direction of the migration vectors drawn on the hexagonal grid was to the north or west for both *Fagus* and *Tsuga*. The average migration rate for *Tsuga* and *Fagus* was 180 m yr^{-1} and 156 m yr^{-1} respectively, calculated by averaging the magnitudes of the migration vectors together (Table 1). The average migration rates

Table 1. Migration rates for *Fagus* and *Tsuga* in eastern North America

Time Interval (yr BP)	*Fagus*				*Tsuga*			
	Mean Migration Rate (m yr^{-1})	Stand. Dev.	Coeff. of Var.	Number of Vectors	Mean Migration Rate (m yr^{-1})	Stand. Dev.	Coeff. of Var.	Number of Vectors
2,000 – 2,999	35	15	42·9	3	117	82	70·1	8
3,000 – 3,999	83	73	88·0	18	169	240	142·0	16
4,000 – 4,999	327	472	144·3	77	110	91	82·7	12
5,000 – 5,999	109	88	80·7	62	245	269	109·8	55
6,000 – 6,999	116	95	81·9	86	120	101	84·2	54
7,000 – 7,999	220	200	90·9	186	145	115	79·3	79
8,000 – 8,999	127	157	123·6	82	103	70	68·0	48
9,000 – 9,999	79	52	65·8	20	185	194	104·9	95
10,000 – 10,999	111	42	37·8	4	158	264	167·1	37
11,000 – 11,999	NA	NA	NA	NA	34	14	41·2	9
12,000 – 12,999	NA	NA	NA	NA	35	4	11·4	3
13,000 – 13,999	NA	NA	NA	NA	57	6	10·5	4
Overall Average	180	240	133·3	538	156	181	116·0	420

Migration rates (m yr^{-1}) were calculated using uncalibrated ^{14}C dates. (Stand. Dev. = Standard Deviation; Coeff. of Var. = Coefficient of Variability, NA = Not Applicable)

calculated for 1000 yr intervals during the Holocene ranged from 35 m yr^{-1} (2,000 – 2,999 yr BP) to 327 m yr^{-1} (4,000 – 4,999 yr BP) for *Fagus*, and from 34 m yr^{-1} (11,000 – 11,999 yr BP) to 245 m yr^{-1} (5,000 – 5,999 yr BP) for *Tsuga*. The migration rate for *Tsuga* declined between 5,000 and 4,000 yr BP, coincident with its observed decline throughout its range (Davis 1981b). The distribution of calculated migration rates was highly skewed; about 45% of the migration vectors for both *Fagus* and *Tsuga* had magnitudes less than 100 m yr^{-1} whereas about 5% of the vectors had magnitudes > 499 m yr^{-1} and < 4,000 m yr^{-1} (Figure 2). The median migration rate was 109 m yr^{-1} for *Fagus* and 101 m yr^{-1} for *Tsuga*.

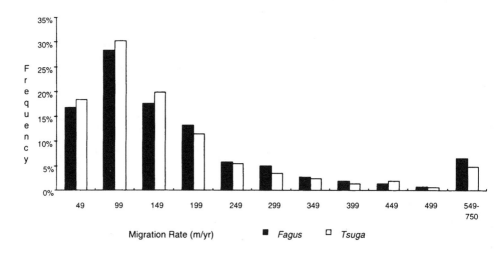

Figure 2. Histogram of estimated migration rates calculated for *Fagus* and *Tsuga*. The rates were grouped in 50 m yr^{-1} classes (e.g. 0 – 49 m yr^{-1}). The last class contains all vectors with rates over 499 m yr^{-1}. (See text for a description of how the migration rates were calculated.)

DISCUSSION

The grid-based approach used here differs from previous more subjective methods for estimating tree migration rates, in which rates were calculated from isochrone maps depicting the displacement of a taxon's range limit through time (Davis 1981a; Delcourt & Delcourt 1987). In these earlier analyses, average migration rates were calculated by measuring the distance between two isochrones and dividing by the difference in time between them. Delcourt and Delcourt (1987) calculated migration rates along five separate migration tracks drawn on migration maps for tree species in eastern North America. Davis (1981a) estimated the average migration rates for *Fagus* and *Tsuga* as 200 m yr^{-1} and 200 – 250 m yr^{-1}, respectively, whereas Delcourt and Delcourt's (1987) estimates are 169 m yr^{-1} and 202 m yr^{-1}, respectively. The estimates of average migration rates calculated in this study (Table 1) using the more quantitative grid-based approach are similar to these earlier estimates.

As a further comparison to earlier work, the migration map for *Tsuga* generated in this study (Figure 1) is similar to that published by Davis (1981a), although the 8000 yr BP and 6000 yr BP isochrones are southeast of those in Davis' (1981a) map. Migration rates calculated between the *Tsuga* isochrones (Figure 1) are similar to both Davis' (1981a) estimates using the same approach and to the grid-based estimates made using the migration vectors.

The key finding of this new analysis is that the average migration rate is a misleading representation of how fast these two tree species responded to climatic change during the Holocene. The median values of the migration vectors for these species were 35 – 40% less than the average rates, meaning that most of the time these species were moving at rates less than the overall average. Conversely, about 10% of the vectors had rates more than twice the average, indicating that these species can respond rapidly to climatic shifts if ecological conditions are highly favourable. What those conditions are is of considerable interest in terms of understanding the mechanisms of range expansions. The faster rates could be due to rare long-distance dispersal events, rare conditions for establishment of mature trees far beyond the species range limit, or a combination of these two (see Davis *et al.* 1986 for a discussion of long distance dispersal and establishment by *Fagus* and *Tsuga*). To fully explore the relative importance of dispersal vs. establishment would require a structured modeling environment in which the trade-offs between the frequency of a long distance dispersing agent (high wind events or long distance dispersal by a vertebrate) and the distance of dispersal can be evaluated together with constraints placed on establishment. However, establishment conditions may have been more limiting for the migration of these species than long-distance dispersal (Davis 1981a; Delcourt & Delcourt 1987), given that blue jays *Cyanocitta cristata*, a common consumer of *Fagus* nuts, can cache nuts at distances greater than 1 km (Darley-Hill & Johnson 1981; Johnson & Adkisson 1985), and winds greater than 25 m sec[-1] are not that uncommon during the life span of a tree in the eastern United States (unpublished analyses of wind data from Michigan; see also Canham & Loucks 1984).

The implications of this analysis for our understanding of future global change are important. Applying an average rate of migration to estimates of the future range limits of tree species (Davis & Zabinski 1992) is somewhat misleading and will overestimate the range expansion in most areas, as the median rate of migration can be considerably less than the average rate. Episodic events apparently increased the rate of migration over the past 14,000 yr, but over the next 100 yr these would likely have minimal effect on the range expansion of tree species into new areas where the climate becomes favourable for their growth. The fact that the landscape in temperate North America is now heavily impacted by human land use also suggests that rates of migration will be less than those observed in the past (Davis 1989; Schwartz 1992).

Another interesting result of this and previous analyses is that tree species with very different modes of seed dispersal have similar migration rates (Davis 1981a; Webb 1986; Delcourt & Delcourt 1987). *Fagus* nuts are animal dispersed while *Tsuga* seeds are wind dispersed. Moreover the distribution of their observed migration rates was similar (Figure 2), suggesting that the different means of dispersal did not affect the pattern of range expansions.

The important assumptions made in this analysis were that,

1. the radiocarbon dates provided unbiased estimates of the age of stratigraphic horizons,
2. the establishment criteria provided consistent and accurate estimates of the arrival time of the species at a site,
3. the spatial scale of the analysis was appropriate,
4. the number of fossil sites was sufficient to conduct the analysis, and
5. climate was not limiting the observed rate of migration during the Holocene (in terms of estimating the potential rates of migration in the future).

The first assumption is known to be incorrect, as the relationship between radiocarbon years and calendar years is not constant (e.g., Bard *et al.* 1990; Stuiver & Reimer 1993). Future analyses using this grid-based approach will be made using calibrated radiocarbon dates and chronologies to calculate the effect of the calibration on migration rates. Individual radiocarbon dates can also be biased depending upon a wide range of depositional factors such as the contamination of a sample with modern ^{14}C or old ^{14}C relative to its true age (Olsson 1986; MacDonald *et al.* 1991). At a regional scale these biases may not affect the overall average migration rate, but could lead to the calculation of a few relatively extreme migration rates between grid cells and thus affect the distribution of observed migration rates.

There are many uncertainties in delimiting the arrival of a taxon at a site (e.g., Davis *et al.* 1991; MacDonald 1993). For instance, the threshold percentage can be expected to vary with the pollen productivity of the surrounding vegetation. Also, the lag between when a taxon arrives near a site and when it produces enough pollen to exceed the threshold percentage in the stratigraphic record may vary regionally depending on a wide range of climatological and ecological factors (Davis 1976; Birks 1989; MacDonald 1993). The effect on estimated migration rates of varying, firstly the threshold percentage for these species, and secondly the number of consecutive samples required to consider the species established at a site, warrants additional study.

The spatial scale of analysis also warrants additional evaluation to determine the effect of grid cell size on the results. However, a grid cell size of less than 20 km would be inappropriate based on the analysis of the relationship between the modern range limits of *Fagus* and *Tsuga* and their representation in the modern pollen record (Davis *et al.* 1991). Likewise, a grid cell size > 100 km would probably smooth the data to the extent that real variation in migration rates would be difficult to detect. Whether there are sufficient fossil sites to conduct this analysis could be evaluated by a jack-knife type simulation in which one to five fossil sites are randomly removed from the data set, the analysis repeated, and the resulting effects on estimated migration rates evaluated.

In conclusion, the grid-based approach produced results consistent with earlier analyses, but with the additional capability of estimating the variability of the rate of range expansion. The distribution of migration rates was highly skewed, with the median rate about 40% less than the average rate. This approach should be tested on other species in eastern North America and elsewhere, and an evaluation should be made of the sensitivity of the results to changes in the criteria used to estimate when a species became established at a site and to changes in the size of the grid cells.

ACKNOWLEDGEMENTS

We thank Al Solomon for suggesting that we investigate the distribution of observed migration rates and providing useful comments on the data analysis and interpretation, Henry Lee for suggesting appropriate statistical techniques to analyze the data, and Mark Schwartz, Shinya Sugita and two anonymous reviewers for useful reviews of an earlier draft of the manuscript. The U.S. Environmental Protection Agency through its Office of Research and Development funded the research described here under Contract 68-C4-0019 to ManTech Environmental Research Services Corp. It has been subjected to Agency review and approved for publication.

REFERENCES

Bard E, Hamelin B, Fairbanks RG (1990) U/Th ages obtained by mass spectrometry on corals from Barbados: Sea level during the past 130,000 years. Nature 346:456-458

Birks, HJB (1989) Holocene isochrone maps and patterns of tree-spreading in the British Isles. J Biogeogr 16:502-540

Burrough, PA (1986) Principles of Geographical Information Systems for Land Resources Assessment, Monographs on soil and resources survey No 12. Clarendon Press, Oxford

Canham CD, Loucks OL (1984) Catastrophic windthrow in the presettlement forests of Wisconsin. Ecology 65:803-809

COHMAP (Cooperative Holocene Mapping Project) (1988) Climatic changes of the last 18,000 years: Observations and model simulations. Science 241:1043-1052

Darley-Hill S, Johnson WC (1981) Acorn dispersal by the blue jay (*Cyanocitta cristata*). Oecologia 50:231-232

Davis MB (1976) Pleistocene biogeography of temperate deciduous forests. Geoscience and Man 13:13-26

Davis MB (1981a) Quaternary history and the stability of forest communities. *in* West DC, Shugart HH, Botkin DB (eds) Forest Succession: Concepts and Application. Springer-Verlag, New York

Davis MB (1981b) Outbreaks of forest pathogens in Quaternary history. Proc 4th Int Palynol Conf Lucknow, India 3:216-227

Davis MB (1989) Lags in vegetation response to greenhouse warming. Clim. Change 15:75-82

Davis MB, Schwartz MW, Woods K (1991) Detecting a species limit from pollen in sediments. J Biogeogr 18:653-668

Davis MB, Woods KD, Webb SL, Futyma RP (1986) Dispersal versus climate: Expansion of *Fagus* and *Tsuga* into the Great Lakes. Vegetatio 67:93-103

Davis MB, Zabinski C (1992) Changes in geographical range resulting from greenhouse warming: effects on biodiversity in forests. *in* Peters RL, Lovejoy TE (eds) Global Warming and Biological Diversity, 297-308. Yale University Press

Davis RB, Jacobson GL Jr (1985) Late Glacial and Early Holocene landscapes in Northern New England and adjacent areas of Canada. Quat Res 23:341-368

Delcourt PA, Delcourt HR (1987) Long-term forest dynamics of the temperate zone. Springer, New York

ESRI (Environmental Systems Research Institute, Inc) (1992) Understanding GIS, The Arc/Info Method. Revision 6. Redlands, CA

Godman RM, Lancaster K (1990) *Tsuga canadensis* (L.) Carr. Eastern Hemlock. *in* Burns RM, Honkala BH (eds) Silvics of North America. Volume 1, Conifers, 604-612. Agriculture Handbook 654. U.S. Forest Service, U.S. Department of Agriculture. Washington, DC

Houghton JT, Callander BA, Varney SK (eds) (1992) Climate change 1992. The supplementary report to the IPCC scientific assessment. Cambridge University Press, Cambridge

Huntley B (1991) How plants respond to climate change: Migration rates, individualism and the consequences for plant communities. Annals of Botany 67:15-22

Johnson, WC, Adkisson CS (1985) Dispersal of beech nuts by blue jays in fragmented landscapes. Am Mid Nat 113: 319-324

MacDonald GM (1993) Fossil pollen analysis and the reconstruction of plant invasions. Adv in Ecol Res 24:67-110

MacDonald GM, Beukens RP, Keiser WE. (1991) Radiocarbon dating of limnic sediments: a comparative analysis and discussion. Ecology 72:1150-1155

NAPD (North American Pollen Database) (1994) National Oceanic and Atmospheric Administration, Paleoclimatology Program

Olsson IU (1986) Radiometric dating. *in* Berglund BE (ed) Handbook of Holocene Palaeoecology and Palaeohydrology, 273-312. Wiley, Chichester

Ritchie JC, MacDonald GM (1986) The patterns of post-glacial spread of white spruce. J of Biogeogr 13:527-540

Schwartz MW (1992) Modelling effects of habitat fragmentation on the ability of trees to respond to climatic warming. Biodiv and Conserv 2:51-61.

Solomon AM, Tharp ML, West DC, Taylor GE, Webb JM, Trimble JC (1984) Response of unmanaged forests to CO_2-induced climate change: Available information, initial tests and data requirements. US Dept of Energy, Washington, DC

Stuiver M, Reimer PJ (1993) Extended ^{14}C data base and revised Calib 3.0 ^{14}C age calibration program. Radiocarbon 35:215-230

Webb SL (1986) Potential role of passenger pigeons and other vertebrates in the rapid Holocene migration of nut trees. Quat Res 26:367-375

Webb T III, Bartlein PJ, Harrison SP, Anderson KH (1993) Vegetation, lake levels and climate in eastern North America for the past 18,000 years. *in* Wright HE Jr, Kutzbach JE, Webb T III, Ruddiman WF, Street-Perrott A, Bartlein PJ, (eds) Global Climates since the Last Glacial Maximum, 415-467. Univ of Minn Press, Minneapolis

Flora and vegetation of the Quaternary temperate stages of NW Europe: Evidence for large-scale range changes

Peter Coxon and Stephen Waldren[1]
Department of Geography
Museum Building
Trinity College
Dublin 2
Ireland

INTRODUCTION

Oxygen isotope records from deep marine sediments provide ample evidence that large scale fluctuations in global climate have occurred since about 2·6 Ma (the Gauss–Matuyama boundary, e.g. Shackleton 1987; Ruddiman & Raymo 1988). These climatic fluctuations have resulted in alternating periods of cold, cool and warm conditions in NW Europe which can be clearly identified, albeit in fragmentary fashion, from terrestrial sedimentary sequences. The degree of climate change during the Pleistocene has been variable in magnitude and frequency with earlier cyclicity (over the periods 1·4 Ma – 2·0 Ma and 0·9 Ma – 1·4 Ma) having a shorter wavelength and amplitude than later cyclicity. During the period between 0·9 Ma and 1·4 Ma isotopic fluctuations suggest a predominant 41,000 year cycle (Ruddiman *et al.* 1986) whilst from 0.9Ma to the present the cold stages occur over 100,000 year intervals (Shackleton & Opdyke 1976). Figure 1 shows an oxygen isotope record and the inherent complexity of climate change since 2·6 Ma.

In NW Europe the temperate stages, or interglacials, during the Pleistocene can be recognised and defined as periods when climatic conditions were similar to those of the Holocene at the same latitude for a substantial time interval (West 1984; Watts 1988). Interglacial periods are complex events which show distinctive vegetational

---

[1] Trinity College Botanic Gardens, Palmerston Park, Dartry, Dublin 6, Ireland

NATO ASI Series, Vol. I 47
Past and Future Rapid Environmental Changes:
The Spatial and Evolutionary Responses of
Terrestrial Biota
Edited by Brian Huntley et al.
© Springer-Verlag Berlin Heidelberg 1997

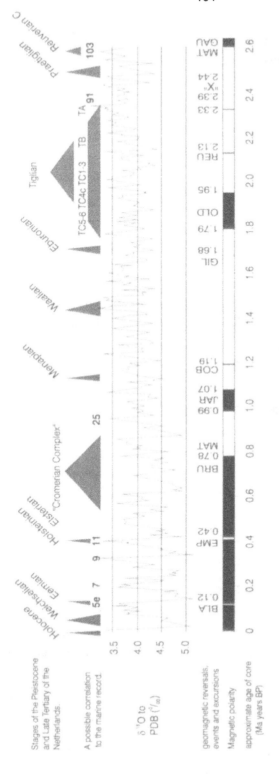

Figure 1. An oxygen isotope record from core ODP 677 (after Shackleton *et al.* 1991) with possible correlation to the Pleistocene sequence of the Netherlands (after Funnell 1995). The correlations are tentative only and given to provide a timeframe and to show the complexity of the Pleistocene climate record from the ocean cores and the incomplete nature of the terrestrial record. (Geomagnetic abbreviations from left: BLA – Blake event; EMP – Emperer event; BRU – Brunhes chron (base); MAT – Matuyama chron (top); JAR – Jaramillo sub-chron; COB – Cobb Mountain event; GIL – Gilsa event; OLD – Olduvai subchron; REU – Reunion subchron; "X" – 'X' event; MAT – Matuyama chron (base); GAU – Gauss chron (top).

development related to climate change. The changing periodicity of climate change (clear from Figure 1) can be recognised within terrestrial sequences, e.g. the complexity and duration of the Tiglian temperate stage and the short cold episodes of the Early Pleistocene relative to the shorter Middle and Late Pleistocene interglacials with their long cold stages (Figure 1 and Watts 1988).

Recent research has allowed correlation of terrestrial interglacial sequences to be made with the GRIP ice core studies (Field *et al* 1994) allowing a comparison of the climatic information held within the two records.

THE EUROPEAN INTERGLACIAL RECORD

From the onset of the Pleistocene taxa have been reoccupying NW Europe after cold stages with varying degrees of success and at different rates. Some taxa have successively recolonised the region during interglacials producing long continuous Pleistocene records (Figure 2; e.g. *Quercus, Corylus, Ulmus, Fraxinus, Alnus....*) whilst others, discussed below, disappeared at the Pliocene–Pleistocene boundary.

The records of climate change available to us (e.g. Figure 1) show that during the last 900,000 years the climate of NW Europe has been predominantly cold with 'glacial' phases lasting 100,000 years and interglacial phases lasting *ca.* 15,000 years. This has meant that the flora recolonising NW Europe has had to do so after very substantial episodes of cold climate. Thermophilous taxa are absent from NW Europe during the long cold stages but are found in low frequencies in southern Europe (e.g. Suc 1984; Leroy *et al.* 1994) and it is these areas that must act as refugia from which thermophilous taxa recolonise northern Europe during the temperate stages. Huntley (1993) provides evidence that the southern European lowlands were subject to summer aridity, not conducive to the formation of temperate forest. He argues that the likely refugia for northern temperate forests during glacial periods were at moderate altitude on southern European mountains, where orographic effects relieved summer drought.

The range chart (Figure 2) compiled from European (specifically, that from the Netherlands) biostratigraphy shows the disappearances quite clearly, but does not take into account spatial variation throughout NW Europe, nor does it deal with more detailed information, such as less biostratigraphically important taxa. The record

Range of Genera in Pleistocene interglacials of NW Europe (Netherlands)

Timeline stages (left to right): Reuverian · Praetiglian · Tiglian · Eburonian · Waalian · Menapian · Cromerian · Elsterian · Holsteinian · Saalian · Eemian · Weichselian · Holocene

Epochs: Late Tertiary — Pleistocene — Holocene

No.	Genus	Modern Distribution	Spp. per genus	Disjunct genus	Habit	Modern European spp.
1	Aesculus	N. America; SE. Europe (?); E. Asia	15	Y	tree	1
2	Diospyros	N. & S. America; Africa; Asia	475	?	tree/shrub	0
3	Elaeagnus	N. America; Asia	45	Y	shrub	0
4	Epipremnum	SE. Asia; W. Pacific	8	N	climber	0
5	Halesia	China; Eastern N. America	4 or 5	Y	shrub	0
6	Karwinskia	SW. USA; Caribbean; S. America	14	N	tree/shrub	0
7	Liquidambar	N. America; E. & W. Asia	4	Y	tree	0
8	Meliosma	Trop. America; Trop. & Temp. Asia	20-25	Y	tree	0
9	Pistacia	Mediterranean; W., C. & E. Asia (dis); C. America	9	Y	tree/shrub	3
10	Pseudolarix	E. China	9	N	tree	0
11	Stewartia	Eastern N. America; E. Asia	9	Y	tree/shrub	0
12	Styrax	Americas; Asia; SE. Europe	100	Y	tree/shrub	1
13	Zelkova	Crete; W. & E. Asia; Japan; Taiwan	5	(?)	tree/shrub	0
14	Nyssa	Eastern N. America; E. Asia	5	Y	tree/shrub	0
15	Actinidia	E. Asia	40	N	climber	0
16	Euryale	N. India; China; Japan; Taiwan	1	Y	aquatic	0
17	Liriodendron	N. America; Indo-China	2	Y	tree	0
18	Magnolia	N. America; C. & E. Asia	~125	Y	tree/shrub	0
19	Proserpinaca	N. & C. America	5	N	aquatic herb	0
20	Taxodium	N. America	3	N	tree	0
21	Sciadopitys	Japan	1	N	tree	0
22	Sequoia	N. America	1	N	tree	0
23	Phellodendron	E. Asia	10	N	tree	0
24	Castanea	N. Temperate (except W. USA)	12	Y	tree	1
25	Ostrya	America; Europe; Asia	9	Y	tree	1
26	Juglans	N. & S. America; SE. Europe; Asia	15	Y	tree	0
27	Tsuga	W. & E. N. America; Himalaya; China; Japan	9 or 10	Y	tree	1
28	Carya	E. USA; C. & SE. China	25	Y	tree	0
29	Parthenocissus	Tropics; S. Europe; N. America	10	Y	climber	0
30	Celtis	C. China	70	?	tree	4
31	Eucommia	Caucasus to E. & SE. Asia	1	?	tree	0
32	Pterocarya	N. Temperate	10	N	tree	0
33	Staphylea	N. Temperate	11	N	shrub	1
34	Fagus	Cosmopolitan	10	N	tree	1
35	Ilex	N. Hemisphere	400	N	tree/shrub	2
36	Vitis	Eurasia; Tropical & S. Africa; C. America	65	N	shrub/climber	1
37	Buxus	N. Temperate; C. Malesia	70	N (?)	shrub/tree	2
38	Taxus	N. Hemisphere; Andes	2 or 3	N	tree	1
39	Abies	N. Hemisphere	50	N	tree	5
40	Alnus	Europe; N. & C. America; E. Asia	35	N	tree/shrub	5
41	Betula	N. Temperate	60	N	tree/shrub	4
42	Carpinus	N. Hemisphere	35	Y?	tree	2
43	Corylus	N. Hemisphere	15	N	tree/shrub	3
44	Picea	Widespread	35	N	tree	2
45	Pinus	Cosmopolitan except Australia	110	N	tree	11
46	Quercus	N. America; Europe; Asia	600	N	tree/shrub	24
47	Salix	N. Temperate	300	N?	tree/shrub	65
48	Tilia	N. America; Europe; Asia	45	N?	tree	5
49	Ulmus	N. Temperate	45	N	tree	6?

Table 1. Irish Late Tertiary taxa no longer native

Family	Name	Irish modern	British modern	Biogeography	Ecology
Pinaceae	*Tsuga* sp.	A	A	America, C. & E. Asia	Montane forest
Sciadopityaceae	*Sciadopitys* sp. (type)	A	A	E. Asian	
Taxodiaceae	*Sequoia* sp. (type)	A	A	American	
Taxodiaceae	*Taxodium* sp.	A	A	American	
Taxodiaceae	*Taxodiaceae* undiff.	A	A	American	
Juglandaceae	*Pterocarya* cf. P. *fraxinifolia* (Poiret) Spach	A	A	W. Asia	Moist Woodland
Juglandaceae	*Carya* sp.	A	A	America, 1 in E. Asia	
Vitaceae	*Vitis* sp.	A	A		

KEY: G – Gortian; A – Absent; H – Holocene record; R – Rare; (V)L – (Very)Local; W – Widespread; Int – Introduced

from the Netherlands was chosen as it provides the most complete terrestrial Pleistocene record from NW Europe. The sheer volume of fossil records from the Pleistocene of NW Europe precludes a complete coverage in a short review paper.

Table 1 shows the progressive loss of NW European tree taxa from the Late Pliocene to the Holocene. This appears not to be due to a replacement of genera, for example a replacement of a Late Pliocene/Early Pleistocene element by a Middle Pleistocene/Holocene element. Instead, genera present in the Holocene were also present since the Late Pliocene, so the Quaternary period shows a progressive diminution in the number of genera present. The largest number of genera were lost at the Pliocene/Pleistocene boundary, indicating catastrophic effects on the vegetation (Watts 1988); subsequent glaciations throughout the Pleistocene resulted in fewer extinctions from NW Europe.

Figure 2. (*facing page*) The Pleistocene range and modern distribution of selected taxa from the Pleistocene of the Netherlands (after van der Hammen *et al.* 1971; Tallis 1990)

Fourteen genera are known only from the Late Tertiary, being wholly absent from the Quaternary. One of the most remarkable features of this group is their disjunct modern distribution between Asia and North America. Three of the 14 (*Halesia, Stewartia, Nyssa*) are now restricted to eastern North America and eastern Asia, with a further two (*Eleagnus, Liquidambar*) rather more widespread but still disjunct between Asia and North America. *Liquidambar* is also disjunct between western and eastern Asia, and it has been suggested that the modern west Asian species (*L. orientalis*) may have occurred as a Pliocene fossil in NW Europe (Watts 1988). A further two genera (*Aesculus, Pistacia*) also show this American/Asiatic disjunction, but have a limited modern representation in southern Europe. *Aesculus* has a single south-eastern European species, and is better represented in North America than Asia. *Pistacia* has three southern European species, but is not confined to the temperate regions; in North America there are species in Mexico, and in Asia the genus extends to Malesia. This extension of range to tropical zones is also seen in *Stryrax*, which has one species in south east Europe but is well represented in tropical Asia and the Americas. *Diospyros, Epipremnum, Karwinskia* and *Meliosma* are also mainly tropical in modern distribution. *Diospyros* has one widespread species in temperate eastern North America (*D. virginiaca*), but it is not known whether the European Pliocene fossils refer to this or some other, possibly extinct, taxon. The genus is most diverse in south-east Asia.

Of the remaining Pliocene genera, *Psuedolarix amabilis* is known from China, while *Zelkova* has three species in eastern Asia, one from the Caucasus and one from Crete. Ten of the genera known in NW Europe only from Pliocene fossils have disjunct modern distributions, indicating considerable reduction in former range.

Contraction of distributions to North America and eastern Asia also occur in genera which survived until the Eburonian glaciation. *Liriodendron* and *Magnolia* are now both disjunct between N America and central and eastern Asia. Some genera, such as *Sequoia* and *Taxodium,* are now entirely American (and there highly restricted), while others such as *Actinidia, Phellodendron* and *Sciadopitys* are restricted to eastern Asia. The aquatics *Euryale* and *Proserpinaca* show different retreats, the latter is now found only in central and northern America, and to some extent shows a similar distribution to *Brasenia* (which is mainly restricted to central and north America, but has several widely disjunct locations throughout the tropics). If

Proserpinaca has retreated west, it might be expected to occur in Early Pleistocene deposits from Ireland, though it has not yet been detected. Of the nine genera mentioned, three are now monospecific.

A further eight genera survived in NW Europe until the Middle Pleistocene (i.e. with Waalian and Cromerian records), and three of these (*Tsuga, Carya* and *Parthenocissus*) show the modern American/Asiatic disjunction. *Parthenocissus* is better represented in Asia than North America, while the opposite is true for *Carya*. A further three genera (*Castanea, Juglans* and *Ostrya*) show similar modern disjunctions but have single species present in Europe. *Celtis* is also disjunct; it has three local species in south-east Europe and one widely distributed in southern Europe; there are also several temperate American species, but the genus is essentially tropical. The single extant species of *Eucommia* is restricted to central China.

Pterocarya and *Staphylea* both have fossil records which extend to the Holsteinian interglacial in the middle Pleistocene. *Pterocarya* has most species restricted to temperate east Asia (China, Japan), but one species occurs in the Caucasus and northern Iraq (*P. fraxinifolia*). *Staphylea* has 11 species widely distributed through the northern temperate zone, including one species (*S. pinnata*) from southern Europe and south west Asia; Pleistocene fossils from NW Europe have been assigned to this taxon.

Thus of 31 terrestrial genera which survived until the Middle Pleistocene, nine now have highly disjunct distributions between North America and east-central Asia, while a further ten are restricted to one or other of these regions. Of the remainder, about six show similar disjunctions, but have a limited (mainly south eastern) representation in Europe, and at least five are mainly tropical genera with poor representation in temperate zones. Apart from the tropical genera, the majority of the others are small genera each with less than 15 species. NW European Quaternary fossil taxa which now show disjunct generic distributions between North America and eastern Asia are generally thought to most closely resemble, or be identical with, the Asiatic species rather than the North American (Watts 1988).

Of the genera present more or less continuously into the Holocene, none show wide disjunctions. Most are widespread throughout the northern temperate zones, none is entirely restricted to Eurasia. Most of these genera are medium sized, with some (e.g. *Quercus*) moderately large; quite distinct from the small genera with disjunct modern distributions which show Quaternary extinction from NW Europe.

Accurate identification of fossil plant material to species is difficult for several reasons. Few characters are usually available in fossil fruit, seed, pollen and leaves for detailed comparison with extant taxa, where a larger number of such characters can readily be observed in a single specimen. A further problem is presented by morphological differences between fossil material and extant taxa. European Pliocene and Early Pleistocene fossils of *Eucommia* have a larger fruit size than the extant *E. ulmoides* (see Watts 1988). This may represent an extinct taxon, which has been named as *E. europaea*, but may equally be part of a more varied *E. ulmoides* than exists at present. Two fossil *Pterocarya* species have been recognised, particularly from Dutch Reuverian deposits (Zagwijn 1963). One is referable to *P. fraxinifolia* from Asia Minor, the other has been named *P. limburgensis* (Reid & Reid 1915), but closely resembles modern Chinese *P. hupehensis*. Small variations in morphological features between fossil and extant material may simply be the result of part of the ancient variation of taxa being reduced by catastrophic local extinctions during the Pleistocene.

Despite such taxonomic difficulties, the data show clearly a progressive restriction of certain components of a formerly more widespread northern temperate flora. Many of the continuously present genera have a wide northern temperate distribution, although they are represented by several species, often with sympatric distributions. Those genera which have been eliminated by successive glaciations have modern distributions in North America, central East Asia, or are disjunct between both of these areas. Apart from a few mainly tropical genera, most of which were eliminated from NW Europe at the onset of the Pleistocene, the flora has consistently been northern-temperate. There are, therefore, clear biogeographic trends in the Pleistocene record.

Reasons for the decline in the number of genera present are less easy to determine, but it seems clear that the more ancient fossil tree genera have not been replaced by more recent colonists, and this probably also holds true at the species level. It may be that continuously present genera have undergone speciation which usurped the Pliocene-Early Pleistocene element, but this is unlikely, many of the former are represented in Europe by few species (e.g. *Abies, Carpinus, Taxus, Corylus, Tilia*).

Huntley (1993) has argued that the evolution of vicariant species (i.e. sympatric, closely related taxa) in southern Europe has occurred, following glacial retreat of ancestral taxa to different refugia in southern Europe, and this may explain the higher number of species to genera in Europe compared to temperate western or eastern North America. However, none of the vicariant taxa mentioned by Huntley (in *Quercus*, *Abies* and *Pinus*) extend as far as NW Europe where only two species of *Quercus* and one of *Pinus* occur. It seems more likely that the temperate forest vegetation of NW Europe, and possibly Europe as a whole, was considerably more diverse in species in the Late Pliocene/ Early Pleistocene than today. The small size of genera from the Pliocene/Early Pleistocene element suggests either:

A. that these genera have genetic limits on speciation, which may affect their overall variation and hence ability to cope with adverse conditions; or

B. that these genera were formerly more species-rich and that widespread extinction in the Pleistocene has restricted the size of the genera.

The latter seems much less likely, as the fossil taxa are similar if not identical with extant taxa with more restricted modern distributions.

The detailed nature of the changing ranges of taxa throughout the Pleistocene is not easy to assess as data are not available from every interglacial (or interstadial) from all parts of NW Europe (the correlation of interglacial sequences is also the subject of considerable argument). Further, the biostratigraphic zones of the interglacials have not been dated and their time transgressive nature prevents significant comparisons of taxa distributions within individual interglacial episodes. However, it is possible to recognise trends within European interglacials and important information can be gleaned by analysing the more detailed records.

THE IRISH INTERGLACIAL RECORD

The Irish record of Pleistocene interglacials is a poor one (Coxon 1993). However, the disappearance of certain taxa that failed to recolonise after the onset of the Pleistocene mirrors that of the Netherlands (Coxon & Flegg 1987; Coxon & Waldren 1995; Table 1). From this list, only *Pterocarya* has been recorded from the Irish Middle Pleistocene (Coxon *et al.* 1994) and this tree appears to have similarly survived into the Middle Pleistocene throughout Europe (Oldfield & Huckerby 1979; Table 1)

The wealth of palaeobotanical detail available from the Middle Pleistocene Gortian Interglacial stage of Ireland (Coxon 1993; Coxon & Waldren 1995) allows the ranges of some biogeographically interesting species to be analysed. Two hundred and fifty one taxa have been recorded from the Gortian, although it must be realised that some identifications are tentative, many are to genus level and a number are to family only (Coxon & Waldren 1995).

The Gortian is not firmly dated but can probably be placed in the Middle Pleistocene with an estimated age lying between 428 ka BP and 198 ka BP. If the Gortian is the equivalent of the continental Holsteinian then it probably represents oxygen isotope stage 9 (302 – 338 ka BP) or 11 (352 – 428 ka BP) (Bowen *et al.* 1986). It is also possible that the Gortian represents oxygen isotope Stage 7 and hence lies between 252 ka BP and 198 ka BP, but this is considered less likely as that stage appears poorly developed (Coxon 1993). As such the Gortian presents a glimpse of plant distributions in the Middle Pleistocene which is summarised below:

The taxa in Table 2 have significantly changed their ranges since the Middle Pleistocene and certain elements can be identified from the taxa listed: Montane/Arctic–Alpine, recent extinctions (i.e. those that survived until the Holocene, but are now absent as native species), American species and Eurasian species. The restriction in range of the first two elements is due, in part, to ecological requirements as well as the fact that the interglacial records cover the whole interglacial 'cycle' whilst modern records catalogue present, temperate and disturbed distributions. The American element, *Azolla filiculoides*, *Nymphoides cordata* and *Brasenia schreberi*, can be thought of as species which once had an amphi-Atlantic distribution, but are now more restricted. *Brasenia* may have been very widespread and now has undergone a large fragmentation of range.

The twenty Eurasian species in Table 2 that were present in the Irish Middle Pleistocene have retracted their ranges to the south and east. A good example is that of *Rhododendron ponticum*. The range of *R. ponticum* is now disjunct and from fossil evidence this is a taxon that has been receding throughout the Pleistocene but has maintained a disjunct distribution in southern and eastern Europe. Other taxa on the list have reached southern England and northern France

Table 2. Irish Middle Pleistocene (Gortian) taxa no longer native

Family	Name	Irish modern	British modern	Biogeography	Ecology
Lycopodiaceae	*Lycopodium annotinum* L.	A	L	Circumpolar	Montane
Lycopodiaceae	*Diphasiastrum complanatum* (?) (L.) Holub	A	VL	N. Europe	Montane
Woodsiaceae	*Athyrium distentifolium* Tausch ex Opiz	A	L	Amphi-Atlantic	Arctic-Alpine
Azollaceae	*Azolla filiculoides* Lam.	A (int)	A (int)	America	Aquatic
Pinaceae	*Abies* cf. *A. alba* Miller	A	A	C. & S. Europe	Montane forest
Pinaceae	*Picea abies* (L.) Karsten	A	A	N. Europe, NW Asia	Boreal forest
Pinaceae	*Picea* sp.	A	A		
Pinaceae	*Pinus sylvestris* L.	A/H (+ int)	L	Europe N.Asia	Woodland
Nymphaeaceae	*Nuphar* cf. *N. pumila* (Timm) DC.	A	L	Amphi-Atlantic?	Aquatic
Nymphaeaceae	*Brasenia* cf. *B. schreberi* J.F. Gmel.	A	A	Widespread disjunct	Aquatic
Juglandaceae	*Pterocarya* cf. *P. fraxinifolia* (Poiret) Spach	A	A	W. Asia	Moist Woodland
Fagaceae	*Fagus sylvatica* L.	A (int)	L (int)	Europe	Woodland
Betulaceae	*Betula nana* L.	A/H	L	Circumpolar	Arctic–Alpine
Betulaceae	*Carpinus betulus* L.	A (int)	L (int)	Europe, W. Asia	Woodland
Caryophyllaceae	*Dianthus* sp.	A			
Polygonaceae	*Persicaria bistorta* (L.) Samp.	A (int)	L	N. & C. Europe, W. & C. Asia	Grassland
Tiliaceae	*Tilia* sp.	A (int)	L		Woodland
Brassicaceae	*Thlaspi caerulescens* J.S. & C. Presl.	A	VL	Circumpolar	Arctic–Alpine
Ericaceae	*Rhododendron ponticum* L.	A (int)	A (int)	C. Europe, Portugal	Woodlands
Ericaceae	*Bruckenthalia spiculifolia* (Salisb.) Reichenb.	A	A	E. Europe, W. Asia	Montane
Primulaceae	*Lysimachia punctata* L.	A (int)	A	C. Europe, W. Asia	Wetland
Elaeagnaceae	*Hippophae rhamnoides* L.	A/H (int)	L (int)	Europe, Temperate Asia	Open scrub
Loranthaceae	*Viscum album* L.	A	L	Europe, Temperate Asia	Parasite
Buxaceae	*Buxus sempervirens* L.	A	VL	Europe, N. Africa, W. Asia	Scrub/ Woodland
Apiaceae	*Astrantia* cf. *A. minor* L.	A	A	Europe, W. Asia	Montane
Menyanthaceae	*Nymphoides* cf. *N. cordata* (Ell.) Fern.	A	A	America	Aquatic
Menyanthaceae	*Nymphoides* cf. *N. peltata* Kuntze	A (int)	L (int?)	Europe, Asia	Aquatic
Plantaginaceae	*Plantago media* L.	A (int?)	W	Europe ?	Grassland
Cyperaceae	*Eleocharis ovata* (Roth) Roemer & Schultes	A	A	C. Europe (W. Asia?)	Wetland
Cyperaceae	*Eleocharis* cf. *E. carniolica* Koch	A	A	E. Europe, W. Asia	Wetland

KEY: A – Absent;   H – Holocene   record;   R – Rare;   (V)L – (Very)Local;   W – Widespread;

Int – Introduced

but have failed to recolonise Ireland, yet others have restricted Holocene ranges not necessarily due to marine barriers (see Coxon and Waldren (1995) for a full discussion).

Of equal interest are the taxa that have maintained Atlantic and amphi-Atlantic distributions from the Middle Pleistocene to the present day and have not disappeared from Ireland nor in some cases from the Atlantic seaboard of France (Table 3). Of the taxa listed in Table 3, only 2 are now absent from Ireland. This long continuity of part of the flora along the western seaboard of Europe is of considerable importance biogeographically.

CONCLUSIONS

As further analyses are made of the climate during European interglacial sequences (e.g. Field *et al.* 1994), and as detailed compilations are made of plant distributions, then our understanding of the pattern of range contraction throughout the Pleistocene and its underlying meaning should become clearer. The range changes themselves are a function of the ability of the taxa to migrate and re-occupy areas that they are driven from during cold stages. The barriers to such migration are not always understood (Coxon & Waldren, 1995) and the reasons for the failure of species to recolonise areas that they previously occupied are not straightforward and not necessarily linked to climate; biotic interactions may also be important.

The late Pliocene flora of NW Europe indicates the presence of a temperate, species-rich forest (Zagwijn 1960; Watts 1988) very different in composition from any modern European forest vegetation. Major extinctions occurred at the Pliocene/Pleistocene boundary; although the extinct genera are tolerant of some frost, Watts (1988) pointed out that none are likely to tolerate prolonged, severe winter cold. The change from the warm, relatively stable climate of the late Pliocene to the first Pleistocene cold period therefore had a major effect on the European flora. The later Pleistocene oscillations between temperate interglacial or interstadial stages and cold glacial events had a much smaller effect on the flora, probably because the most cold-sensitive taxa were eliminated at the onset of the Pleistocene. By the Middle Pleistocene (e.g. the Holsteinian) the composition of the flora was essentially similar to that of modern NW Europe, further emphasising the

Table 3. Irish interglacial Atlantic and amphi-Atlantic taxa (34 taxa), including interglacial presence in the Pays Basques.

Family	Name		Irish modern	British modern	Biogeography	Ecology
Lycopodiaceae	*Diphasiastrum alpinum* (L.) Holub		L	L	Amphi-Atlantic	Arctic–Alpine
Selaginellaceae	*Selaginella selaginoides* (L.) P. Beauv.	>M<	W	L	Amphi-Atlantic	Arctic–Alpine
Isoetaceae	*Isoetes lacustris* L.		W	L	Amphi-Atlantic	Aquatic
Isoetaceae	*Isoetes echinospora* Durieu		VL	L	Amphi-Atlantic	Aquatic
Osmundaceae	*Osmunda regalis* L.	M	W	L	Amphi-Atlantic/ Widespread	Wetland
Marsileaceae	*Pilularia globulifera* L.	M	L	L	Sub Atlantic	Lake margin
Hymenophyllaceae	*Trichomanes speciosum* Willd.		VL	VL	Atlantic	Humid microclimate
Hymenophyllaceae	*Hymenophyllum tunbrigense* (L.) Smith		L	L	Atlantic (amphi-Atlantic?)	Humid microclimate
Hymenophyllaceae	*Hymenophyllum wilsonii* Hook.		L	L	Atlantic	Humid microclimate
Woodsiaceae	*Athyrium distentifolium* Tausch ex Opiz		A	L	Amphi-Atlantic	Arctic–Alpine
Nymphaeaceae	*Nuphar* cf. *N. pumila* (Timm) DC.		A	L	Amphi-Atlantic?	Aquatic
Ranunculaceae	*Ranunculus flammula* L.		W	W	European(amphi-Atlantic?)	Wetland
Ranunculaceae	*Ranunculus hederaceus* L.		W	W	Amphi-Atlantic	Aquatic
Myricaceae	*Myrica gale* L.	M	W	W	Amphi-Atlantic	Bogs
Caryophyllaceae	*Silene vulgaris* (Moench) Garcke		W	W	Amphi-Atlantic?	Grassland/ Woodland
Ericaceae	*Daboecia cantabrica* (Hudson) K. Koch	M	L	A	Atlantic	Acid open
Ericaceae	*Erica ciliaris* L.	M	R	VL	Atlantic	Bogs
Ericaceae	*Erica mackaiana* Bab.	M	L	A	Atlantic	Bogs
Ericaceae	*Erica cinerea* L.		W	W	Sub-Atlantic	Bogs
Saxifragaceae	*Saxifraga stellaris* L.		L	W	Amphi-Atlantic	Arctic–Alpine
Saxifragaceae	*Saxifraga granulata* L.		R (int)	L	Sub-Atlantic	Meadow
Saxifragaceae	*Saxifraga* cf. sect Gymnopera		L	A	Atlantic	Montane/ rocks
Fabaceae	*Ulex* sp.				Sub-Atlantic	Scrub
Haloragaceae	*Myriophyllum alterniflorum* DC.	M	W	W	Amphi-Atlantic	Aquatic
Lamiaceae	*Clinopodium vulgare* L.		Int?	W	Amphi-Atlantic	Scrub/ Grassland
Caprifoliaceae	*Viburnum opulus* L.	M	W	W	Amphi-Atlantic?	Woodland/ scrub
Potamogetonaceae	*Potamogeton filiformis* Pers.		L	L	Amphi-Atlantic	Aquatic
Najadaceae	*Najas flexilis* (Willd.) Rostkov & W. Schmidt	M	VL	VL	Amphi-Atlantic	Aquatic
Eriocaulaceae	*Eriocaulon* cf. *E. aquaticum* L.		L	VR	Amphi-Atlantic	Aquatic
Juncaceae	*Juncus bulbosus sens. lat.* L.		W	W	Amphi-Atlantic	Wetland
Juncaceae	*Juncus effusus* L.	M	W	W	Amphi-Atlantic	Wetland
Juncaceae	*Juncus conglomeratus* L.		W	W	Amphi-Atlantic	Wetland
Juncaceae	*Luzula campestris/L. multiflora*	>M	W	W	Amphi-Atlantic?	Grassland
Cyperaceae	*Cladium mariscus* (L.) Pohl	M	W	L	Amphi-Atlantic?	Wetland

KEY: A – Absent; (V)L – (Very)Local; R – Rare; W – Widespread; Int – Introduced; M – Marbellan record;

>M – Pre Marbellan; M< – Post Marbellan

greater effects of Pliocene/Pleistocene and Early Pleistocene climate changes on the European flora.

Global warming would, in theory, allow thermophilic and currently southern species to move northwards. However, human disturbance of natural plant populations prevents a realistic model of what might happen under such conditions, as natural populations have been severely fragmented. Under such a scenario of future climate change, it is possible that many colonists may result from adventive spread of naturalised populations and human introductions, rather than expansion of extant native populations.

Despite current indications of global warming the pattern of climate change revealed by the oxygen isotope record suggests that future change may involve a rapid drop in temperatures in NW Europe. Climate deterioration would cause a repeat of the disappearance of thermophilous taxa from more northerly locations, and their replacement by more cold-tolerant taxa currently occurring in arctic and alpine habitats. Future recolonisation in subsequent temperate interstadial and interglacial periods may not follow that of temperate Pleistocene stages because of widespread anthropogenic disturbance of southern European refugia, as discussed above.

A crucially important objective of future research should be to compile records of interglacial plant distributions from sites where macrofossil work has been carried out and begin to compare these data on a spatial and temporal basis.

REFERENCES

Bowen DQ, Richmond GM, Fullerton DS, Sibrava V, Fulton RJ, Velichko AA (1986) Correlation of Quaternary Glaciations in the Northern Hemisphere. Quaternary Science Reviews, 5:509-510

Coxon P (1993) Irish Pleistocene biostratigraphy. Irish Journal of Earth Sciences 12:83-105

Coxon P, Flegg AM†(1987) A Late Pliocene/Early Pleistocene deposit at Pollnahallia, near Headford, Co. Galway. Proceedings of the Royal Irish Academy 87B:15-42

Coxon P, Waldren S (1995) The floristic record of Ireland's Pleistocene temperate stages. in Preece RC (ed) Island Britain; a Quaternary Perspective, 243-268. Geological Society Special Publication No. 96

Coxon P, Hannon G. Foss P. (1994) Climatic deterioration and the end of the Gortian Interglacial in sediments from Derrynadivva and Burren Townland, near Castlebar, County Mayo, Ireland. Journal of Quaternary Science 9(1):33-46

Field MH, Huntley B, Müller H (1994) Eemian climate fluctuations observed in a European pollen record. Nature 371:779-783

Funnell BM (1995) Global sea-level and the (pen-)insularity of late Cenozoic Britain. *In* Preece RC (ed) Island Britain; a Quaternary Perspective,3-13. Geological Society Special Publication No. 96

Hammen T, van der, Wijmstra TA, Zagwijn WH (1971) The floral record of the Late Cenozoic of Europe. *In* Turekian KK (ed) Late Cenozoic Glacial Ages,391-424. Yale University Press, New Haven, Connecticut

Huntley B (1993) Species-richness in north-temperate zone forests. Journal of Biogeography 20:163-180

Leroy S, Ambert P, Suc J-P (1994) Pollen record of the Saint-Macaire maar (Hérault, southern France): a lower Pleistocene glacial phase in the Languedoc coastal plain. Review of Palaeobotany and Palynology 80:149-157

Oldfield F, Huckerby E (1979) The Quaternary palaeobotany of the French Pays Basque: A summary. Pollen et Spores 21(3):337-360

Reid C, Reid EM (1915) The Pliocene floras of the Dutch-Prussian border. Mededelingen van de Rijksopsporing van Delfstoffen, 6

Ruddiman WF, Raymo ME (1988) Northern Hemisphere climate regimes during the last 3Ma: possible tectonic connections. Proceedings of the Royal Society of London B318:411-430

Ruddiman WF, Raymo ME Mcintyre (1986) Matuyama 41,000-year cycles: North Atlantic Ocean and northern hemisphere ice sheets. Earth and Planetary Science Letters 80:117-129

Shackleton NJ (1987) Oxygen isotopes, ice volume and sea level. Quaternary Science Reviews 6:183-190

Shackleton NJ Opdyke ND (1976) Oxygen isotope and palaeomagnetic stratigraphy of Equatorial Pacific core V28-239, Late Pliocene to Latest Pleistocene. Geol Soc Amer Mem 145:449-464

Shackleton NJ, Berger A, Peltier WR (1991) An alternative astronomical calibration of the Lower Pleistocene timescale based on ODP Site 677. Transactions of the Royal Society of Edinburgh 81:252-261

Suc J-P (1984) Origin and evolution of the Mediterranean vegetation and climate in Europe. Nature 307:429-432

Tallis JH (1990) Plant Community History. Long-term changes in plant distribution and diversity. Chapman and Hall.

Watts WA (1988) Europe. *In* Huntley B, Webb T III (eds) Vegetation History, 155-192. Kluwer Academic Publishers

West RG (1984) Interglacial, interstadial and oxygen isotope stages. Dissertationes Botanicea 72:345-357

Zagwijn WH (1960) Aspects of the Pliocene and Early Pleistocene vegetation in the Netherlands. Mededelingen van de Geologische Stichting CIII (1):5, 1-78

Zagwijn WH (1963) Pollen-analytic investigations in the Tiglian of the Netherlands. Mededelingen van de GeologischeStichting, N.S.16:49-71

The response of beetles to Quaternary climate changes

Allan C. Ashworth
Department of Geosciences
North Dakota State University
Fargo, ND 58105
U.S.A.

INTRODUCTION

Desender and Turin (1989) estimated that 142 (34%) of the 419 ground beetle species in the Netherlands and adjoining countries were endangered to varying degrees. Changes in geographic distribution which have occurred during the last century are attributed to either climate change (Hengeveld 1985) or habitat fragmentation caused by increased urbanization and intensification of agriculture (Turin & den Boer 1988). Desender *et al.* (1994) concluded that it was the loss of specific habitat types associated with the intensification of agriculture, and not climate change, that had caused rare species to become rarer and common species to become more abundant.

Beetles are the most diverse of all organisms and changes in their geographic distribution and abundance are bound to effect all other organisms. As cold-blooded animals, they are especially dependent on environmental temperatures so concerns about the effects of global warming are especially relevant. Modern data sets similar to the one from the Low Countries (the best that exists) are exceptionally rare, and knowledge about the response of beetles to climate change is mostly based on their Quaternary fossil record. Three questions relevant to global warming are: (1) Will climate change lead to adaptation? (2) Will climate change lead to extinction? (3) Will climate change lead to changes in geographic distribution?

1. Will climate change lead to speciation?

The climatic instability of the Earth during the Quaternary Period has resulted in numerous glacial/interglacial cycles. Intuitively, the repeated fragmentation of

NATO ASI Series, Vol. I 47
Past and Future Rapid Environmental Changes:
The Spatial and Evolutionary Responses of
Terrestrial Biota
Edited by Brian Huntley et al.
© Springer-Verlag Berlin Heidelberg 1997

landscapes and isolation of populations should have favoured allopatric speciation. Fossil evidence, however, does not support that supposition. The *Checklist of beetles of Canada and Alaska* (Bousquet 1991) lists 640 species of ground beetles of which 142 (22%) have been described from North American fossil assemblages that predate the Holocene. Several of these species, or close relatives, have existed since the Late Tertiary (Figure 1). From the Late Miocene to the Mid-Pleistocene the representation of extant species, or their morphologically close relatives, increased from 82 to 100% (Figure 1).

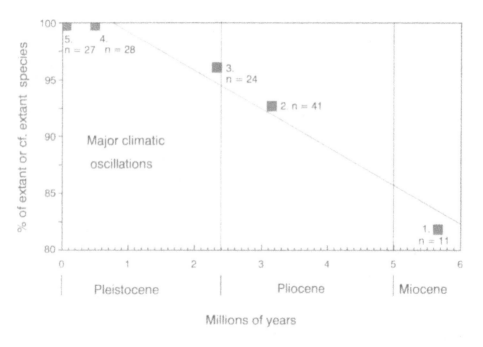

Figure 1. The percentage of extant or cf. extant ground beetle species in fossil assemblages from North America and Greenland (1. Lava Camp, Matthews 1979; 2. Meighen Island, Matthews 1979; 3. Kap København, Böcher 1989; 4. Cape Deceit, Matthews 1974; 5. Titusville, Cong, Ashworth, Schwert & Totten, in prep.).

The percentage of existing ground beetle species with known Tertiary ancestors is about 8%. The number is small but so is the number of sites that have been examined. Another example of stasis involves geographic races or subspecies. Subspecies are interpreted as intermediate steps in the speciation process and those from glaciated regions are assumed to be of recent origin. Angus (1973) reported that two races of the European hydrophilid water beetle, *Helophorus aquaticus,* have persisted for at least 100,000 years. He commented that whatever selection pressures maintain races within an intergrading cline, they are not only stable for tens of thousands of years but also are sufficiently robust for the geographic distribution of the races to have changed without the loss of clinal structure.

2. Will climate change lead to extinction?

The spectacular extinctions of Pleistocene mammals have deeply influenced our view of the response of organisms to climate change. Our expectation is that extinction rates are bound to increase at times of climatic change. For beetles, however, that expectation is false. A few Pleistocene species described as extinct have been determined later to be extant. For example, the hydrophilid *Helophorus wandereri*, described as an extinct species from Pleistocene deposits in Germany and England, was later found in a collection from a remote part of Siberia under the name of *H. obscurellus* (Angus 1970). Similarly, the staphylinid *Micropeplus hoogendorni*, described as an extinct species in a Late Miocene assemblage in Alaska, was later discovered as a Mid-Pleistocene fossil in England, and is possibly the extant Siberian species *M. dokuchaevi* (Coope 1987). Matthews has described a few extinct species from Late Tertiary assemblages. Initially, he believed that most Miocene and Pliocene fossils were extinct even though they had close modern relatives. He now believes that most would fall within the range of variation for them to be considered extant species (Matthews, pers. comm.). The only known late Quaternary extinctions are from the 40,000 yr BP asphalt deposits at Rancho La Brea in southern California. Miller (1983) noted that W. D. Pierce, who described a large, mostly extinct beetle fauna from La Brea, had erred in his taxonomy. Miller reassigned most of Pierce's extinct species to existing Californian species. The two extinct species, *Copris pristinus* and *Onthophagus everestae*, are both dung beetles.

3. Does climatic change result in change in geographic ranges?

There can be no better example of a shift in geographic distribution of a species resulting from climatic change than that of the dung beetle *Aphodius holdereri*. Presently, this species is restricted to the high plateaus of Tibet and western China. From about 40,000 to 25,000 years ago, this species inhabited the British Isles

(Coope 1975). While individual examples of range disjunctions are spectacular they neither convey the scale of disruption of the fauna nor the rapidity of the faunal response. To illustrate these points the faunal response at the end of the last glaciation on three continents is briefly described (Figure 2).

The most complex of these responses was in the British Isles (Figure 2a). The highlights of this response are rapid faunal turnover at about 13,000 and 10,000 yr BP, respectively (Coope 1987). At both times, cold-adapted species were extirpated over large areas and warmer-adapted species took their place on the landscape. The amount of temperature change was in the range 4 – 8°C. Transition times were at most a few centuries in length and possibly just decades. What is striking about this response in the British Isles is that within a 3000-year period the fauna underwent massive reorganizations on at least two, and possibly three occasions.

Following each of these events, species associations were reestablished. Non-synchroneity between beetle and plant responses to climate change during this time has been attributed to factors such as population density and soil development (Pennington 1986). Beetles with their superior dispersal mechanisms colonized climatically suitable landscapes several centuries before birch trees.

In the continental interior of North America, climate change at the end of the last glaciation was simpler than that in the British Isles (Figure 2b). At 21,500 yr BP, arctic and subarctic species replaced boreal forest species in Iowa and Illinois. The arctic species persisted until 14,500 yr BP when they were replaced along the southern margin of the ice sheet by boreal species. Both the beginning and the end of the last ice age are characterized by faunas for which there are no existing analogues. The faunas consist of mixtures of specialized arctic species and boreal species of open habitats. Many of these species are associated with cold meltwater streams. The most noteworthy event for beetles was the regional extinction of arctic and subarctic species that occurred about 14,500 yr BP. Large numbers of cold-adapted species became regionally extinct as the climate warmed. Northward dispersal, which would have ensured survival, was blocked by the ice sheet. Perhaps caused by climatic warming, major lobes on the southern margin of the ice sheet continued to advance until about 14,000 yr BP. Schwert and Ashworth (1988)

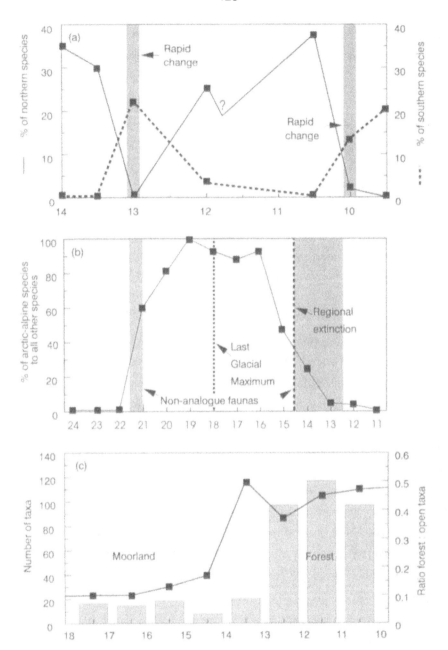

Figure 2. The response of beetles to climate change at the end of the last glaciation. 2a. The British Isles (Ashworth 1973; Coope & Brophy 1972, Osborne 1980); 2b. The Midwest of North America (Schwert & Ashworth 1988, Schwert 1992); 2c The Chilean Lake Region (Ashworth & Hoganson 1993). The x-axis represents ^{14}C yr BP $\times$ 10^3.

proposed that the existing arctic fauna was derived in postglacial time from Beringia and not from species which had survived south of the ice sheet (see Postscript).

In the temperate latitudes of South America, the Chilean beetle fauna was also perturbed in a major way by climatic warming (Figure 2c). Within 1500 years a fauna consisting of relatively few moorland species was replaced entirely by a species-rich forest fauna. Unlike the British Isles and North America, the fauna of the glacial interval never received any immigrant species. Northward dispersal was blocked by marine and mountain barriers emphasizing the importance of landscape in shaping faunas. As the climate cooled, species became regionally extinct leaving a depauperate fauna of climatic generalists and species that had dispersed down slope from montane refugia. The amount of climatic warming which resulted in the rapid replacement of the glacial fauna is estimated to have been about 4 – 5°C.

Postscript: The consequences of Quaternary climate change

Events which occur in geological time and their effects on the biota are difficult for biologists to assess and consequently are often assigned a low priority in their interpretations. Events, such as the climate changes described in this paper left a genetic stamp on the fauna that is decipherable using molecular population genetics techniques. During the last (Wisconsinan) glaciation in North America, the arctic biota was split by the growth of the Laurentide ice sheet into two regions. For about 75,000 years the ice sheet was a barrier to gene flow between populations in Beringia (Alaska, Yukon Territory and northern British Columbia) and those south of the ice sheet in the continental interior. With climatic warming, most populations south of the ice sheet became regionally extinct. Some survived by dispersing to higher elevations in the Rocky Mountains in the west and the northern Appalachians in the east. By about 10,000 years ago montane populations within regions were isolated from one another on their alpine islands. Schwert and Ashworth (1988) proposed that: (1) populations in Beringia have been separated from populations in the Rocky and northern Appalachian mountains for at least 75,000 years.; (2) populations in the Rocky and the northern Appalachian mountains have been isolated from each other for at least 14,500 years; and (3) Hudson Bay and eastern Arctic populations were derived during postglacial time from populations in Beringia.

The molecular population genetics of the ground beetle *Amara alpina* are being studied from arctic and alpine habitats throughout its North American range. (Reiss, Ashworth, & Schwert, in prep.). There is no significant difference in morphology between individuals from geographically widely separated populations. Using Restriction Fragment Length Polymorphism analysis of mitochondrial DNA (mtDNA RFLP) 15 haplotypes have been defined. A summary of the results is:

1. The highest amount of genetic diversity (8 haplotypes) occurs in Beringian populations supporting the inference that the tundra habitats of Alaska and Yukon Territory have been inhabited for longer than anywhere else in the North American Arctic.

2. Hudson Bay populations show the least amount of genetic diversity supporting the inference that this region was colonized most recently. The area became ice-free only about 6000 years ago.

3. Hudson Bay populations share the same haplotype as populations in northern British Columbia indicating a common ancestor that probably survived in Beringia.

4. Rocky Mountain and Appalachian populations have different haplotypes but are more closely related to one another than to all other haplotypes.

5. Age divergence estimates from the genetic data generally agree with those from the palaeontological evidence. The exception is that the genetic data suggest a much older divergence time for the Rocky and Appalachian mountain populations than was inferred from the palaeontological data.

CONCLUSION

1. Quaternary entomologists are united in their conviction that the repeated climate changes of the Pleistocene were not a major forcing factor in speciation. Coope (1978) argued that the environmental instability of the Pleistocene is paradoxically the cause for stasis. Stasis in beetles may be the result of interactions between a a cold-blooded physiology, excellent dispersal capabilities, and climate change. Global warming would not be expected to lead to enhanced rates of speciation.

2. Species extinction was exceptionally rare during the Quaternary. Regional extinction on different scales, however, was probably common. The question of

whether global warming will lead to enhanced rates of species extinction is a difficult one. The modern world with its fragmented landscape is very different from the Quaternary world. My intuition is that species extinction will be higher, especially for species with poor dispersal capabilities (den Boer 1977).

3. Large population size and active dispersal have ensured that beetle species survive climatic change. Dispersal is probably stochastic although behavior might give direction to dispersal in some species. As some members of populations were extirpated, others survived by dispersing into climatically suitable habitats. Shifts in geographic range were the end result. Species associations in beetles are amazingly robust. They may break down at times of climatic change but re-establish themselves rapidly. There is no doubt that global warming would result in changes in the geographic range of species. Populations of alpine beetles would be especially endangered.

REFERENCES

Angus RB (1970) A revision of the beetles of the genus *Helophorus* F. (Col. Hydrophilidae), subgenera *Orphelophorus* d'Orchymont, *Gephalophorus* Sharp and *Meghelophorus* Kuwert. Acta Zool Fenn 129:1-62

Angus RB (1973) Pleistocene *Helophorus* (Coleoptera, Hydrophilidae) from Borislav and Starunia in the western Ukraine, with a reinterpretation of M. Lomnicki's species, description of new Siberian species, and comparison with British Weichselian faunas. Phil Trans Roy Soc B 265:299-326

Ashworth AC (1973) A Late-Glacial insect fauna from Red Moss, Lancashire, England. Ent Scand 3:211-224

Ashworth AC and Hoganson JW (1993) The magnitude and rapidity of the climate change marking the end of the Pleistocene in the midlatitudes of South-America. Palaeogeography, Palaeoclimatology, Palaeoecology 101:263-270

Böcher J (1989) Boreal insects in northernmost Greenland: palaeoentomological evidence from the Kap København Formation (Plio-Pleistocene), Peary Land. Fauna Norv Ser B 36:37-43

Bousquet Y (ed) (1991) Checklist of beetles of Canada and Alaska. Agriculture Canada, Ottawa.

Coope GR (1975) Mid-Weichselian climatic changes in western Europe, re-interpreted from coleopteran assemblages. *in* RP Suggate & MM Creswell (eds) Quaternary Studies, 101-108. Roy Soc New Zeal

Coope GR (1978) Constancy of insect species versus inconstancy of Quaternary environments. *in* LA Mound & N Waloff (eds) Diversity of Insect Faunas, 176-187. Roy Entomol Soc Lond Symposium 9.

Coope GR (1987) The response of Late Quaternary insect communities to sudden climatic changes. *in* JHR Gee & PS Giller (eds) Organization Of Communities Past And Present, 421-438. The 27th Symposium of the British Ecological Society.

Coope GR and Brophy JA (1972) Late Glacial environmental changes indicated by a coleopteran succession from North Wales. Quaternary insects and their environments. Boreas 1:97-142

den Boer PJ (1977) Dispersal power and survival. Carabids in a cultivated country-side (with a mathematical appendix by J Reddenguis). Misc Pap Lanbouwhogeschool Wageningen,14, Wageningen

Desender K and Turin H (1989) Loss of habitats and changes in composition of the ground and tiger beetle fauna in four west European countries since 1950 (Coleoptera:Carabidae, Cicindelidae). Biol Conserv 48:277-294

Desender K Dufrene M and Maelfait J-P (1994) Long term dynamics of carabid beetles in Belgium: a preliminary analysis on the influence of changing climate and land use by means of a database covering more than a century. *in* K Desender M Dufrene M Loreau M L Luff & J-P Maelfait (eds), Carabid Beetles Ecology and Evolution, 247-252. Kluwer Academic Publishers, Dordrecht

Hengeveld R (1985) Dynamics of Dutch beetle species during the twentieth century (Coleoptera, Carabidae). J Biogeog 12: 389-411

Matthews JV (1974) Quaternary environments at Cape Deceit (Seward Peninsula, Alaska): evolution of a tundra ecosystem. Geol Soc Amer Bull 85:1353-1384

Matthews JV (1979) Late Tertiary carabid fossils from Alaska and the Canadian Archipelago. *in* TL Erwin GE Ball DR Whitehead & AL Halpern (eds), Carabid Beetles: Their Evolution, Natural History, and Classification, 425-446. Dr. W. Junk bv, The Hague

Miller SE (1983) Late Quaternary insects of Rancho La Brea and McKittrick, California. Quat Res 20:90-104

Osborne PJ (1980) The Late Devensian-Flandrian transition depicted by serial insect faunas from West Bromwich, Staffordshire, England. Boreas 9:139-147

Pennington W (1986) Lags in adjustment of vegetation to climate caused by the pace of soil development:evidence from Britain. Vegetatio 67: 105-118

Schwert DP (1992) Faunal transitions in response to an ice age: the Late Wisconsinan record of Coleoptera in the North-Central United States. Coleopts Bull 46:68-94

Schwert DP and Ashworth AC (1988) Late Quaternary history of the northern beetle fauna of North America: a synthesis of fossil and disributional evidence. Mem Entomol Soc Can 144:93-107

Turin H and den Boer JP (1988) Changes in the distribution of carabid beetles in the Netherlands since 1880. II. Isolation of habitats and long-term time trends in the occurrence of carabid species with different powers of dispersal (Coleoptera, Carabidae). Biol Conserv 44:179-200

Fossil Coleoptera assemblages in the Great Lakes region of North America: Past changes and future prospects

Alan V. Morgan
Quaternary Sciences Institute
Department of Earth Sciences
University of Waterloo
Waterloo
Ontario
Canada N2L 3G1

INTRODUCTION

This paper is intended as a summary which specifically addresses certain fossil Coleoptera assemblages from an area centred on the Great Lakes region of central and eastern North America – a region which has a relatively large number of sites with well-preserved fossil assemblages. It is an area which was crossed several times by ice advances during the last glacial stage, and relict periglacial features are present in the region. Most importantly, it has seen climatic fluctuations which represent temperature extremes in both warm and cold episodes of the late Quaternary. However, the questions to be addressed by this NATO ARW, namely the spatial and evolutionary responses of terrestrial biota in response to past and future climate change, necessitate a somewhat broader discussion.

Well over 250 sites have now been examined for fossil Coleoptera in North America and adjacent Greenland (Figure 1). The bulk belong to the Quaternary Period (both Pleistocene and Holocene) but there are a few sites which are believed to date from the Miocene and Pliocene epochs of the late Tertiary. From this record spanning more than 5 million years some general comments can be made about the nature of the responses of Coleoptera to past climate changes.

NATO ASI Series, Vol. I 47
Past and Future Rapid Environmental Changes:
The Spatial and Evolutionary Responses of
Terrestrial Biota
Edited by Brian Huntley et al.
© Springer-Verlag Berlin Heidelberg 1997

EVOLUTION, MOVEMENTS AND ECOLOGIES

There has been remarkably little morphological evolution of Coleoptera during late Tertiary and early Quaternary time, and there has been no obvious morphological evolution during the Quaternary. We recognise that the lack of morphological evolution does not necessarily preclude physiological evolution; however, insect species in community assemblages remain quite similar throughout these time frames and for this reason it has been assumed that there has not been major physiological evolution. For example, the most recently described Coleoptera assemblages dating back to early Quaternary time are cited in Morgan *et al.* (1993). The faunas from the Isortoq River and Flitaway Lake sites (Figure 1) are most likely of early Quaternary, or possibly Late Tertiary age. In this paper the authors point out that the two north-central Baffin Island sites contain "...ecologically coherent insect assemblages...more typical of modern western, rather than eastern, Arctic assemblages." One of the beetles recovered, *Pterostichus punctatissimus* Randall, is a species regularly found inside the Boreal forest limit. Preservation of this specimen was remarkable, with only slight abrasion on the elytron, and the microsculpture allowed identification as a female. Similar morphological details preserved on beetles which have gone through at least one million generations between the fossil and modern forms can be cited from specimens in Europe, Asia and elsewhere in North America and Greenland. Comments on the longevity of Coleoptera as species are given in papers by Coope (1978), Angus (1973), Matthews (1974, 1979) and Böcher (1989).

Although there has been little or no apparent evolution of Coleoptera there have been considerable shifts in the geographic ranges of the species found as fossils both in Europe and in North America. This suggests that Coleoptera are sensitive to changing ecologies brought about by the temperature fluctuations which also resulted in the growth and demise of Quaternary glaciers. There have been a number of publications which outline the vast geographic shifts in Quaternary insect faunas, and some of the more striking examples are cited by Coope (1973), Morgan and Morgan (1980) and Morgan *et al.* (1983a, b).

131

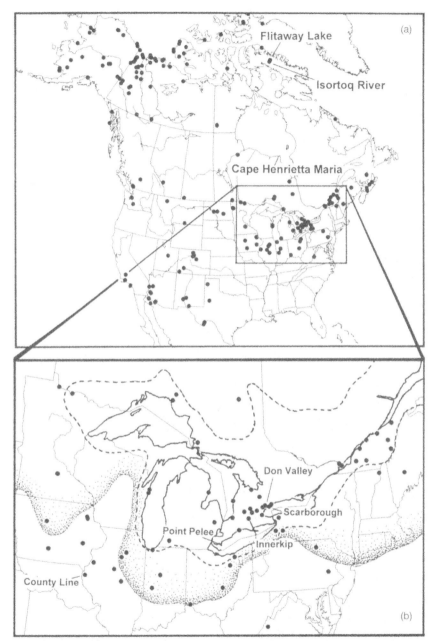

Figure 1. (a) Fossil Coleoptera sites in North America (b) Fossil Coleoptera in the Great Lakes Basin. The watershed is shown (dashed line) as is the approximate maximum Wisconsinan Laurentide Ice Sheet position (increasingly stippled area to the ice margin). Sites and locations mentioned in the text are marked.

Fossil insect assemblages generally reflect local conditions and are derived from the immediate vicinity of the site. In the case of small lakes or kettle ponds, these are species which lived at the margins of the water bodies, both in the water and in the littoral zone. Individuals and species groups can provide valuable information about the size of the water body, and whether the water was still, affected by wave action or was fast-flowing, as in the case of streams. Individuals are also incorporated which lived in drier areas away from the pond margins, but which were likely brought into the site by storm runoff, or occasionally by misadventures while flying. It seems most probable that, in the case of lacustrine sequences, practically all of the species trapped in the sediments would be living within the catchment basin, and since most sites represent small lakes (ca.10,000 m^2), most beetles are probably derived from an area of a few kilometres, or less, around the lake. In fluvial systems, where insects are occasionally found in detrital organic lenses, the Coleoptera could come from anywhere along the system upstream from the site. Unpublished laboratory experiments conducted at Waterloo show that beetles can travel considerable distances with very little signs of breakup or abrasion. However, again it is likely that the Coleoptera concerned probably originated within the stream divide of the basin.

Coleoptera typically found in a site are represented by carnivores and herbivores as well as omnivores. Beetles will colonise landscapes which appear to be barren and open, for example in periglacial regimes immediately following ice retreat. At the microscopic level there are small plants and animals living on and within these substrates. These provide adequate food sources for other insects (especially Collembola and Diptera) as well as arachnids and predatory Coleoptera. Later in time, as the landscape becomes colonised by plants, phytophages invade the region. Trophic habits can be fairly specific, with certain species feeding on certain plants, and these may vary from herbs to specific portions of mature trees. Still others take advantage of vertebrates occupying an area by living in burrows created by small mammals, in the dung of larger mammals, such as bear, deer, or large herbivores, or on various carcasses. Many beetle species provide indirect evidence of the types of substrates in the area, since they often have preferences for sand, silt, clay, or organic-rich surfaces.

It is of interest to Quaternary geologists and modellers that many modern beetles usually have restricted climatic tolerances, and this allows accurate reconstruction of former climates within restricted geographic areas (Atkinson *et al.* 1986, 1987). In long, well-dated sediment sequences, recovered from sections which can be adequately sampled, it is often possible to indicate temperature fluctuations through time. Coleoptera have provided some surprises about the speed and magnitude of climatic changes, especially in certain areas of Europe (Coope *et al.* 1971; Coope & Brophy 1972; Ashworth 1973; Coope 1975).

THE GREAT LAKES REGION OF NORTH AMERICA

Forty fossil insect assemblages have been described in the Great Lakes catchment basin shown in Figure 1. A further 17 sites are situated in the upper headwaters of the Ohio-Mississippi system. The two in the Arctic catchment basin and 5 more near the head of the Atlantic drainage are not considered in this summary.

One of the problems in resolving the ages of different Coleoptera faunas is that there is no easy way of accurately dating sites in excess of about 35,000 yr BP. In a few situations materials can be bracketed by palaeomagnetic reversals, by volcanic ash horizons, or by relative dating using fossil vertebrates or inter-digitated fossiliferous marine sediments. In most cases uranium series dates, thermoluminescence, or amino-acid racemisation ratios have been unsuccessful in resolving fossiliferous sequences. There is a degree of circularity used in attributing ages to many sites. If the biotic indicators are as warm or warmer than present in the same region the site is assumed to be interglacial; if not, it is interstadial. The stratigraphic sequence dictates whether it is regarded as last interglacial (Sangamonian) or older. Undetected breaks in stratigraphy, easily missed in mapping slumping river banks or borehole sequences, may lead to a site being mis-assigned in age. Furthermore, the lack of knowledge of which part of an interglacial is preserved may also lead to misinterpretation. These matters have been discussed at length elsewhere (West 1968; Morgan 1990a), but they do serve as warnings about the correct ages and locations of particular faunas through time.

The oldest site in the region is the County Line assemblage in Illinois (Miller *et al.* 1994). The sediments have a reversed palaeomagnetism and a fossil vertebrate assemblage which place them around 800 ka BP. Although the site contains two extinct microtine mammals, all of the identified beetles are extant, living today at the southern margins of the Boreal forest in eastern North America.

Supposed Sangamonian (*ca.* 125 – 75 ka BP) sites are known from 16 different locations in North America with a further 6 sites which are more tentatively assigned to Sangamonian or Early Wisconsinan time (*ca.* 80 – 70 ka BP). Four of the Sangamonian sites are in the region of the mid-continent considered in this paper. Two of these contain relatively small specifically-identified faunas, but the Innerkip and Don Valley sites in southern Ontario have fairly large and well-preserved faunas. The two assemblages represent different local conditions. Innerkip was a small pond setting whereas the Don Valley site was located at the mouth of a relatively large river. Both sites have a number of species in common, and the majority of the species are not recorded again until early Holocene time. Innerkip (Pilny & Morgan 1987) contains at least 120 genera from 19 different Coleoptera families which could live in the region today. None of these is extinct and all can be matched with modern species. In the case of the taxa recorded from the Don Valley site one beetle, *Anotylus gibbulus* Eppelsheim, is not known from North America. The same species is recorded as a fossil from Britain and France in Würmian Pleniglacial sediments (Ponel 1995), and is today believed to be extant only in the Caucasus Mountains (Hammond *et al.* 1979).

Moving on from demonstrably warm, last interglacial, sites there are at least 86 assemblages in North America attributed to the last, Wisconsinan, glacial stage. Approximately 31 of these are in the Great Lakes centred region. A detailed analysis of these sites is impractical in this review, but in general terms the warm faunas of the last interglacial are replaced by cold assemblages found today in the Boreal forest of North America, as well as in the Boreal montane zones of both the western and eastern United States. In specific time frames, and as the ice advanced toward certain sites, the Boreal insects are in turn replaced by species which are today found at the tundra-treeline transition, or in tundra areas of the Arctic or montane North America.

Nowhere is this better seen than in southern Ontario and the adjacent American states of Illinois, Indiana, Ohio and Pennsylvania, where several sites illustrate the transition of 'cool' to 'cold' climates. Analyses of sites close to Scarborough, Toronto, in Ontario (Morgan & Morgan 1980; Williams *et al.* 1981; Morgan 1989) illustrate that early Wisconsinan faunas contained both tree-line and tundra species. By late Wisconsinan time the Laurentide ice advance across Canada had replaced the treeline assemblages with true tundra species (Warner *et al.* 1988), and had forced the treeline – tundra transitional assemblages southward into the United States, eventually to a maximum ice marginal position from Illinois (Morgan & Morgan 1986), Indiana (Morgan *et al.* 1983c) through Cinncinatti, Ohio (Morgan & Pilny 1996) to north-central Pennsylvania (Figure 1b).

We believe that Coleoptera forced to this latitudinal position, which, in Europe, corresponds to southern France and northern Italy, were placed under extreme stress. As long as the Laurentide Ice Sheet continued to provide cold air flow which provided limited permafrost regimes in localised areas outside the ice margin, the fauna survived. However, changes to these conditions led to local extirpation. This happened to a number of the cold-adapted Coleoptera species commencing about 17,000 yr BP. By 14,500 yr BP most of the cold-adapted forms had vanished along much of the southern margin of the Laurentide Ice Sheet, surviving only in the higher areas of Appalachia and in the western mountains of the United States and Canada. These points are summarised in some detail by Morgan (1987, 1988). This localised extinction event will be discussed later.

Twenty-five sites are known in the Great Lakes region which date between 14,500 yr BP and the Holocene transition at 10,000 yr BP. None of these, with the exception of one site in the Appalachians, contains remains of the truly cold-adapted species which advanced south with the ice. There are many species which occupy the northern Boreal forests of North America and there are even a few species which range onto, but are not confined to, both Arctic and montane tundra regions. However, these are often mixed with species which are never found in tundra regions today, and even species which can be found in open areas of the Great Lakes today. The period from 14,500 to 12,000 yr BP was a period of non-analogue faunas – peculiar assemblages which mixed nothern Boreal, Prairie and Great Lakes taxa together. The last elements of these mixed assemblages did not disappear from southern Ontario until approximately 10,000 yr BP at the Holocene transition.

The Pleistocene – Holocene boundary in the Great Lakes region is not marked by any abrupt ecological changes in the Coleoptera fauna. It was a time of considerable change as water-bodies, such as Lake Algonquin, vanished, isostatic uplift accelerated, and new animals (including humans) and plant communities moved into the areas vacated by the retreating ice. The Coleoptera fauna gradually changed from Boreal-dominated species to those typically found in the region today (the Great Lakes – St. Lawrence Boreal – mixed hardwood elements). By early Holocene time a number of species are found in the region which today barely enter southernmost Canada, and this suggests a small thermal excursion (between about 8500 and 5000 yr BP) when temperatures may have been marginally above present. By 4000 yr BP the pre-European fauna was well-established throughout most of the Great Lakes region.

In the mountains of Appalachia, particularly in the New England States and Maine, as well as in southern Québec and New Brunswick, species are found which are remnants of the northern Boreal Forest and of open-ground faunas seen today in northernmost Manitoba, Ontario, Québec, Newfoundland and Labrador. As pointed out by Morgan and Morgan (1980) and Morgan (1988), these remnant faunas are good indicators that the climate during the Holocene maximum did not exceed the thermal tolerances of the species found in these refugia, and that these mountain-top assemblages served as potential 're-seeding' areas for Coleoptera invasion of lowland areas during any deteriorating climates. The same would apply to western montane refugial areas.

SUMMARY OF CHANGES IN THE COLEOPTERA FAUNA IN THE LATE QUATERNARY RECORD

The Coleoptera record of the Great Lakes region is long and complex; however, it can be summarised by the following observations:

- The climatic fluctuations created by ice advances and retreats in the region allowed different faunal assemblages to become established on many occasions. Similar climatic episodes (in terms of duration, temperature and humidities) allowed near-identical insect assemblages to be established in the various parts of the region. However, the consistent presence or absence of certain key species at certain time intervals does allow the assemblages to be used as stratigraphic indicators.

- In the course of the last interglacial – glacial – present interglacial cycle these faunal assemblages fall into at least six principal categories, namely;

  1. Mixed deciduous to southernmost Boreal assemblages   (Sangamonian/Holocene)
  2. Central Boreal assemblages   (Early, Mid and Late Wisconsinan)
  3. Boreal to northern treeline assemblages   (Early, Mid? and Late Wisconsinan)
  4. Northern treeline to tundra assemblages   (Early, Mid? and Late Wisconsinan)
  5. Openground (tundra) assemblages   (Mid? to Late Wisconsinan)
  6. Openground (non-tundra/non analogue) assemblages   (early-Late Wisconsinan)

- The Boreal-treeline to tundra assemblages which were well-established in Ontario and Québec during Early and Middle Wisconsinan time migrated south in front of the advancing Laurentide Ice Sheet.

- Throughout the last 100 ka there is only one apparent regional faunal 'extinction' which follows the start of the retreat of the Laurentide Ice Sheet from its maximum advance position. This broad extirpation event, which caused the demise of many species along the southern Laurentide ice margin, was probably created by a number of factors. These might have included winter rains on the ice margins, increasing solar insolation and summer warmth, complex landscape modification and movement of competing species (Morgan 1987). The Coleoptera fauna which followed the ice northward was missing many of the species which had moved southward during the ice advance. Obviously some adaptive strategies successfully utilised during most of Wisconsin time had been curtailed by the changed conditions which caused the rapid ice retreat.

- The faunal assemblages which migrated northward were in a state of flux for almost 5000 years, resulting in non-analogue faunas for part of this time. 'Openground' species filled the niche abandoned by the true tundra-adapted species.

- Modern Coleoptera assemblages became established in the region at least 8500 yr BP, although some species which indicate a slightly warmer climate than at present also are present in Early Holocene time.

FUTURE CLIMATE CHANGES AND THE COLEOPTERA FAUNA OF THE GREAT LAKES REGION.

Anthropogenic change, especially the build-up of greenhouse gases, will probably force some major climate changes in all parts of the world, and the Great Lakes region will be included. Current General Circulation Models suggest substantial modification of eastern North American climates (Figures 2a and b). The whole of the Great Lakes catchment basin, including the headwaters of adjacent catchment basins will likely experience a 4°C average temperature increase during summer months (June, July, August) under a doubled CO_2 scenario (expected by 2050 – 2100 AD). Average Winter (December, January, February) increases will be more extreme. The western half of the Great Lakes watershed, and adjacent catchments will have an 8°C increase; the eastern half of the Great Lakes basin and adjacent catchments a 7°C increase (Canadian Climate Modelling Group, pers. comm.).

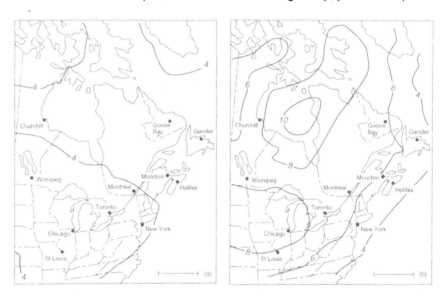

Figure 2. (a) Summer temperature increases from current norms in eastern North America under a doubled CO_2 atmosphere (Canadian Climate Modelling Group, pers. comm.) (b) Winter temperature increases from current norms in eastern North America under a doubled CO_2 atmosphere (Canadian Climate Modelling Group, pers. comm.). Scale bar = 500km.

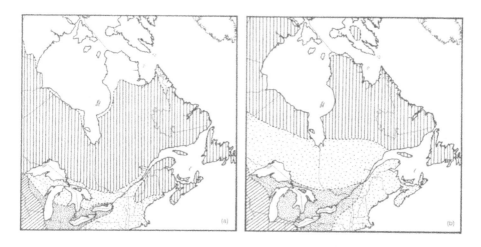

Figure 3. Potential ecotonal shifts in a doubled CO_2 climate in Canada and adjacent U.S.A. (a) Present vegetation zones: Dense Stipple – 'Carolinian' zone (hardwood forest area); Less-dense Stipple – St. Lawrence-Great Lakes mixed forest; Vertical Shading – Boreal Forest (northern edge represents treeline – Tundra boundary). Diagonal Shading – grassland (tall and short Prairie). (b) Future vegetation zones: Patterns as in (a). Potential changes under doubled CO_2 atmosphere (*ca.* 2100 AD). Note the northward shift of zonal boundaries.

The implications for such changes are remarkable, but unfortunately they are also relatively unpredictable. One potential scenario is shown in Figure 3. The degree of change is unprecedented over such a short time-frame. The closest comparison, although this was several magnitudes slower than the predicted future change, is with the conditions experienced at the Laurentide ice margin during the Late Pleistocene amelioration. At that time there was a widespread extirpation of certain species throughout the region. The twenty-first century warming will cause another extirpation event, but the prediction of what species may vanish is complicated by anthropogenic influences. Human land clearance, demands for water with draw-downs of aquifers, rivers and lakes, as well as regional flooding where new dams are constructed for water redistribution, will have significant impacts. Add to these the unknown consequences of genetic manipulation of species, or more conventional chemical and biochemical spraying to conserve food and forest resource crops, and the outcome is extremely unpredictable. Current government cutbacks in museum curation programmes, and the implied lack of importance attached to taxonomy and the training of new personnel engaged in nomenclatural classification, means that there is a very real chance that we will never know just what species have been extirpated or become extinct.

However, this anthropogenically-forced climatic event could allow us to establish how far, and how fast, species can move in the short-term warming experiment that awaits. Detailed baseline studies must be carried out as soon as possible with the aim of discovering what species are where today, and research councils must be committed to long-term collecting and curation programmes which must last throughout the twenty-first century. The following programme was verbally suggested at the Ontario Ecology–Etiology Conference at St. Catherines, Ontario (Morgan 1990b).

At least two collecting programmes should be created. A small and relatively cheap programme would involve a census of what Coleoptera species are present at specific elevations on selected mountains in the New England states and in Québec, New Brunswick and western Newfoundland, perhaps six mountains in all. This would establish a vertical thermal gradient for all the species recorded, coupled with ancillary ecological and climatic data. A second, larger, experiment would involve the establishment of 15 regional collecting areas, at 100 km separations, from southernmost (Point Pelee) to northernmost (Cape Henrietta Maria) Ontario. Here detailed collecting, conducted on perhaps a five-year interval basis, would keep track of Coleoptera across ecotonal boundaries in the province. This transect is practically south to north across Ontario. The ecotonal boundaries and isotherms tend to cross the province from east to west. Continuous, discontinuous and sporadic permafrost areas are represented on this transect. Northern Ontario represents the furthest south that lowland permafrost and tundra are found in North America. The southernmost location is well into the Carolinian zone, and the mixed hardwood and Boreal forests of eastern North America are also well represented in the central parts of the province. The logistics involved in both of these collecting programmes are outside the scope of this paper, but they are suggested as a realistic means of assessing the nature of future changes in the Coleoptera fauna. If Coleoptera are indeed good biological indicators of rapid climate change then details of their movements should help to point the way to what is happening in this portion of the mid-continent.

REFERENCES

Angus RB (1973) Pleistocene *Helophorus* (Coleoptera, Hydrophilidae) from Borislav and Starunia in the western Ukraine, with a reinterpretation of M. Lomnicki's species, description of new Siberian species, and comparison with British Weichselian faunas. Phil Trans Roy Soc B 265:299-326

Ashworth AC (1973) A Late-Glacial insect fauna from Red Moss, Lancashire, England. Ent Scand 3:211-224

Atkinson TC, Briffa KR, Coope GR, Joachim MJ, Perry DW (1986) Climatic calibration of Coleoptera data. *in* Berglund BE (ed) Handbook of Holocene Paleoecology and Paleohydrology, 851-859. John Wiley and Sons Ltd. Toronto

Atkinson TC, Briffa KR, Coope GR (1987) Seasonal temperatures in Britain during the last 22,000 years, reconstructed using beetle remains. Nature 325:587-592

Böcher J (1989) Boreal insects in northernmost Greenland: palaeoentomological evidence from the Kap Köbenhavn Formation (Plio-Pleistocene), Peary Land. Fauna Norv Ser B 36:37-43

Coope GR (1973) Tibetan species of dung beetle from Late Pleistocene deposits in England. Nature 245:335-336

Coope GR (1975) Mid-Weichselian climatic changes in western Europe, re-interpreted from coleopteran assemblages. *in* Suggate RP, Creswell MM (eds) Quaternary Studies, 101-108. Roy Soc New Zeal

Coope GR (1978) Constancy of insect species versus inconstancy of Quaternary environments. Diversity of Insect Faunas. Mound LA, Waloff N (eds) Roy Entomol Soc Lond Symposium 9:176-187

Coope GR, Brophy JA (1972) Late Glacial environmental changes indicated by a coleopteran succession from North Wales. Quaternary insects and their environments. Boreas 1:97-142

Coope GR, Morgan A, Osborne PJ (1971) Fossil Coleoptera as indicators of climatic fluctuations during the Last Glaciation in Britain. Paleogeog Paleoclim Paleoecol 10:87-101

Hammond P, Morgan A, Morgan AV (1979) On the *gibbulus* group of the genus *Anotylus* C. G. Thomson, and fossil occurrences of *Anotylus gibbulus* (Eppelsheim) (Coleoptera: Staphylinidae). Syst Entom 4:215-221

Matthews JV (1974) Quaternary environments at Cape Deceit (Seward Peninsula, Alaska): evolution of a tundra ecosystem. Geol Soc Amer Bull 85:1353-1384

Matthews JV (1979) Late Tertiary carabid fossils from Alaska and the Canadian Archipelago. *in* Erwin TL, Ball GE, Whitehead DR, and Halpern AL (eds) Carabid Beetles: Their Evolution, Natural History, and Classification, 425-446. Dr. W. Junk bv, The Hague

Miller BB, Graham RW, Morgan AV, Miller NG, McCoy WD, Palmer DF, Smith AJ, Pilny JJ (1994) A biota associated with Matuyama-age sediments in west-central Illinois. Quat Res 41:350-365

Morgan AV (1987) Late Wisconsin and early Holocene paleo-environments of east-central North America based on assemblages of fossil Coleoptera. *in* Ruddiman WF, H.E. Wright Jr (eds) The Geology of North America Volume K-3 North America and adjacent oceans during the last deglaciation, 353-370. Geol Soc of Amer Boulder CO

Morgan AV (1988) Late Pleistocene and Early Holocene Coleoptera in the Lower Great Lakes Region. *in* Laub RS, Miller NG, Steadman DW (eds) Late Pleistocene and Early Holocene Paleoecology and Archaeology of the Eastern Great Lakes Region, 195-206. Bull Buffalo Society of Nat Sci 33, Buffalo N.Y

Morgan AV (1990a) Prospects and Promises of Interglacials. Géographie physique et Quaternaire 44:251-256

Morgan, AV. (1990b). Coleoptera: Library of the Past, Key to the Future? Program and Abstracts, Ontario Ecology–Ethology Colloquium; Brock University, St. Catherines, Ontario:6

Morgan AV (1989) Late Pleistocene zoogeographic shifts and new collecting records for *Helophorus arcticus* Brown (Coleoptera: Hydrophilidae) in North America. Can J Zool 67:1171-1179

Morgan AV, Kuc M, Andrews JT (1993) Paleoecology and age of the Flitaway and Isortoq interglacial deposits, north-central Baffin Island, Northwest Territories, Canada. Can J Earth Sci 30:154-174

Morgan AV, Morgan A (1980) Faunal assemblages and species shifts during the Late Pleistocene in Canada and the northern United States. Can Ent 112:1105-1128

Morgan AV, Morgan A (1986) A preliminary note on fossil faunas from Central Illinois. Quaternary Records of Central and Northern Illinois: 81-83 XI AMQUA Field Guide II Geol. Surv Champaign II

Morgan AV, Morgan A, Ashworth A C, Matthews J V (1983a) Late Wisconsin fossil beetles in North America. *in* HE Wright Jr. (ed) Late Quaternary Environments of the United States, 354-363. Univ Minn Press Minneapolis Mn

Morgan AV, Morgan A, Miller R.F (1983b) Range extension and fossil occurrences of *Holoboreaphilus nordenskioeldi* (Mäklin) (Coleoptera: Staphylinidae) in North America. Can J Zool 62:463-467

Morgan AV, Morgan A, Miller RF (1983c) A preliminary report on two fossil insect assemblages from west central Indiana. *in* Interlobate Stratigraphy of the Wabash Valley, Indiana, 133-136. Midwest Friends of the Pleistocene Field Guide Ind Geol Surv Lafayette Ind

Morgan AV, Pilny JJ (1996) Fossil Coleoptera from a maximum Late Wisconsinan ice advance site, Cinncinatti, Ohio. (Submitted to: Paleogeog Paleoclim Paleoecol)

Pilny JJ, Morgan AV (1987) Paleoentomology and paleoecology of a possible Sangamon age site near Innerkip, Ontario. Quat Res 28:157-174

Ponel P (1995) Rissian, Eemian and Würmian Coleoptera assemblages from La Grand Pile (Vosges, France). Paleogeog Paleoclim Paleoecol 114:1-41

Warner BG, Morgan AV, Karrow PF (1988) A Wisconsinan Interstadial Arctic flora and fauna from Clarksburg, southwestern Ontario, Canada. Paleogeog Paleoclim Paleoecol 68:27-47

West RG (1968) Pleistocene Geology and Biology. Longman Group, London

Williams NE, Westgate JA, Williams DD, Morgan A, Morgan AV (1981) Invertebrate fossils (Insecta: Trichoptera, Diptera, Coleoptera) from the Scarborough Formation, Toronto, Ontario and their Paleoenvironmental Significance. Quat Res 16:146-166

The response of Coleoptera to late-Quaternary climate changes: Evidence from north-east France

Philippe Ponel

Laboratoire de Botanique historique et Palynologie (URA CNRS 1152)

Faculté des Sciences et Techniques de Saint-Jérôme

Avenue Escadrille-Normandie-Niémen

F - 13397 Marseille cedex 20

France

INTRODUCTION

Recent studies have emphasised the importance of the migrations of many arthropod species, especially of Coleoptera, as a result of Quaternary climatic changes. Such migrations have been described in north-western Europe (Coope 1975, 1994), Eastern Europe (Nazarov 1984), Siberia (Kiselyov 1973), North America (Morgan & Morgan 1980; Morgan *et al.* 1984) and South America (Hoganson & Ashworth 1991).

These migrations consist mainly of southwards movements by northern species during Glacial periods (Coope 1973) and of northwards movements by southern or even Mediterranean species during temperate or warm periods (Coope 1990). A recent palaeoentomological study carried out at la Grande Pile, in the Vosges piedmont (Département de Haute-Saône, France) (Figure 1), provided evidence demonstrating that the French arthropod fauna too was affected by these large amplitude migrations consequent upon climatic changes (Ponel 1995).

NATO ASI Series, Vol. I 47
Past and Future Rapid Environmental Changes:
The Spatial and Evolutionary Responses of
Terrestrial Biota
Edited by Brian Huntley et al.
© Springer-Verlag Berlin Heidelberg 1997

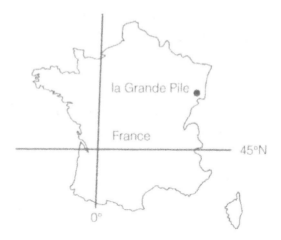

Figure 1. The geographical location of the site of la Grande Pile.

THE SITE OF LA GRANDE PILE AND ITS PALAEOECOLOGICAL INTEREST

As first shown by Woillard (1974), and more recently by Beaulieu and Reille (1992), the sedimentary sequence at la Grande Pile spans the last 140,000 years, from the end of the Rissian glaciation to the present, through the Eemian interglacial and the Würmian glaciation. It is one of the rare, easily accessible, French sediment sequences spanning the whole last climatic cycle. Coleoptera analysis of 41 samples has yielded remains of 394 taxa. Although the Holocene was not sampled, because of possible perturbation due to superficial peat exploitation, a continuous record of fossil beetle assemblages is available for the first time from the end of the penultimate glaciation to the end of the Würmian Pleniglacial. This record indicates clearly how insects (especially Coleoptera) responded to late-Quaternary climatic fluctuations.

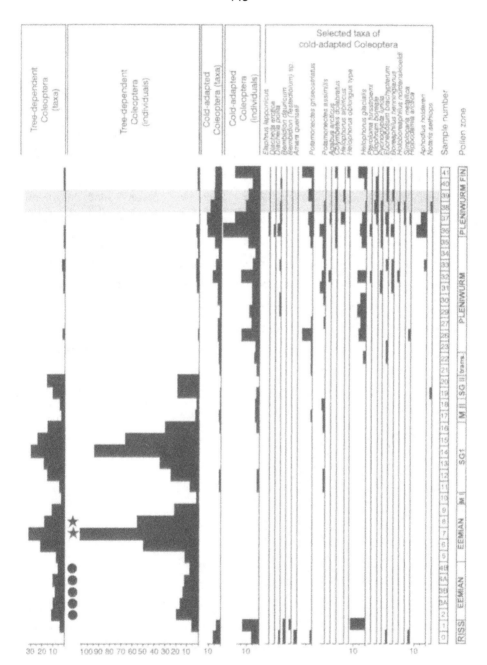

Figure 2. Occurrences of tree-dependent and cold-adapted arthropods in fossil assemblages at la Grande Pile. The histograms indicate the number of taxa or individuals. (MI: Mélisey I; MII: Mélisey II; SGI: St Germain I; SGII: St Germain II) Shaded area: high representation of *Lepidurus arcticus*. ●: presence of *Triodonta*; ★: high representation of *Platypus oxyurus;*

THE COLONISATION OF THE AREA BY AN ARCTIC–ALPINE FAUNA DURING THE RISSIAN AND WÜRMIAN GLACIAL PERIODS

Figure 2 shows occurrences of all the cold–adapted Coleoptera whose present-day distribution is restricted to northern Europe, the Arctic, high mountain regions of lower latitudes, or a combination of the latter two (Table 1, Figure 3). The cold periods are sharply delimited, especially the Riss and the Würmian Pleniglacial. They are marked by an invasion of Arctic species belonging to a great variety of families (Carabidae, Dytiscidae, Hydrophilidae, Silphidae, Staphylinidae, Byrrhidae, Coccinellidae, Scarabaeidae and Curculionidae). The Mélisey I and Mélisey II stadials, considered to be cold periods on the basis of pollen analysis (Beaulieu & Reille 1992), are characterised by a marked decrease of tree-dependent Coleoptera and hence undoubtedly appear as episodes of marked forest regression (Figure 2). Nevertheless, very few or even no Arctic–Alpine Coleoptera are recorded during these periods. The decrease in cold-adapted Coleoptera in samples from the middle of the Würmian Pleniglacial does not correspond to warmer climatic conditions; in fact this reduction coincides with a peak in abundance of *Lepidurus arcticus*, a voracious aquatic predator probably responsible for the disappearance of most of the other aquatic organisms. This species had a vigorous development over a very short period (Figure 2).

THE OCCURRENCE OF SOUTHERN ELEMENTS DURING THE EEMIAN INTERGLACIAL

Few southern (south of la Grande Pile) or mediterranean Coleoptera are recorded during the Eemian; only the histogram of tree-dependent taxa enables one to define the temperate episodes throughout the sequence: the Eemian, St Germain I and St Germain II are clearly indicated by great quantities of tree-dependent Coleoptera (Figure 2). However, most of the species that have been identified can be found today in the vicinity of la Grande Pile (Ponel 1995). This suggests for that region a climate not markedly different from the present. The only notable exceptions relate to the Scarabaeidae *Triodonta* sp. and to the Platypodidae *Platypus oxyurus* (Figure 2)

Figure 3. Modern European distributions of selected cold-adapted Coleoptera found in fossil assemblages at la Grande Pile.

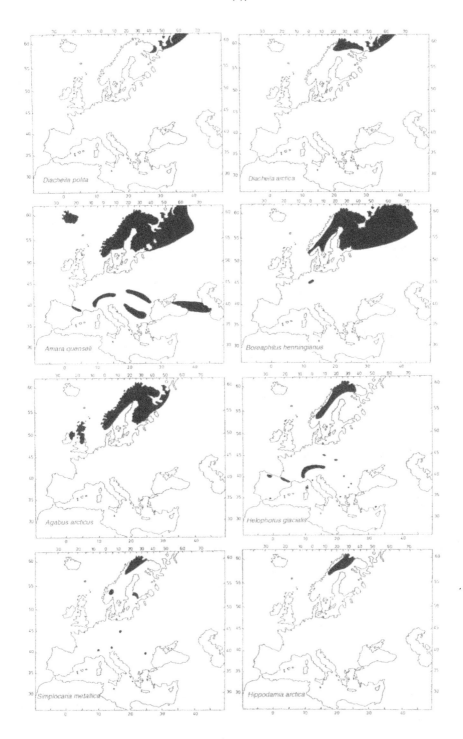

Table 1. Cold-adapted, Arctic–Alpine arthropods (Coleoptera; Crustacea – Notostraca) and southern Coleoptera recorded at la Grande Pile during the last climatic cycle.

COLD-ADAPTED AND ARCTIC–ALPINE COLEOPTERA

- *Diacheila arctica* (Gyllenhal) – Circumpolar (Lindroth 1985).
- *Diacheila polita* (Faldermann) – Northwestern North America and Eurasia (Lindroth 1985).
- *Elaphrus lapponicus* Gyllenhal – Circumpolar, isolated in the north of Britain (Lindroth 1985).
- *Bembidion dauricum* (Motschulsky) – Circumpolar (Lindroth 1985).
- *Bembidion* (*Testediolum*) sp. – Genus including many high altitude species (Jeannel 1939).
- *Amara quenseli* (Schönherr) – Circumpolar, boreoalpine species; Iceland, Scotland (Lindroth 1986).
- *Potamonectes griseostriatus* (DeGeer) – Boreoalpine species; northern Europe, mountains of central and southern Europe; Atlas; northern Asia and high mountains; boreal America (Guignot 1947).
- *Potamonectes assimilis* (Paykull) – North European and Siberian species, present in France only in the north-east: Alsace (Guignot 1947).
- *Agabus arcticus* (Paykull) – North of the British Isles and Fennoscandia (Balfour-Browne 1950).
- *Colymbetes dolabratus* (Paykull) – North of Fennoscandia and Russia, Siberia, North America, Greenland and Iceland (Coope 1968).
- *Helophorus sibiricus* (Motschulsky) – A Holarctic species, in the Palaearctic region ranging from northern Fennoscandia through Russia and Siberia to Mongolia and China; in the Nearctic region mentioned from Alaska and Canada (Hansen 1987).
- *Helophorus oblongus* type LeConte – North America and Siberia (Angus 1973).
- *Helophorus glacialis* Villa – A boreomontane species. Europe; besides Fennoscandia it occurs in the mountains further south, ranging from Spain to the Carpathians and the Balkans (Hansen 1987).
- *Pteroloma forsstroemi* (Gyllenhall) – Fennoscandia, mountains of central Europe (du Châtenet 1986).
- *Pycnoglypta lurida* (Gyllenhall) – Northern Europe, in the south toward Denmark and northern Germany; Siberia, North America (Zanetti 1987).
- *Olophrum boreale* (Paykull) – North of Fennoscandia, Russia, Siberia, North America (Campbell 1983).
- *Eucnecosum brachypterum* (Gravenhorst) – British Isles, northern Fennoscandia, central Europe from Germany to Russia, Alps, Transylvania, Bulgaria, Caucasus, Siberia, north Mongolia, North America (Zanetti 1987).
- *Boreaphilus henningianus* Sahlberg – Scandinavia, north Russia, Harz Mountains (Coope 1962).
- *Holoboreaphilus nordenskioeldi* (Mäklin) – Isolated localities in the north of North America and Asia (Coope 1975).
- *Simplocaria metallica* (Sturm) – Fennoscandia, mountains of central Europe (Coope 1961).
- *Hippodamia arctica* (Schneider) – North of Scandinavia, North America and Asia (Iablokoff-Khnzorian 1982).
- *Aphodius holdereri* Reitter – Tibetan endemic (Coope 1973).
- *Notaris aethiops* (Fabricius) – Boreal Europe and Asia; Lapland, Iceland; isolated localities in Germany, Switzerland and France: Massif Central (Hoffmann 1958).

COLD–ADAPTED CRUSTACEA

- *Lepidurus arcticus* Pallas – North of the Arctic Circle (Taylor & Coope 1985).

SOUTHERN COLEOPTERA

- *Triodonta* sp. – Genus including species with Mediterranean distribution (Baraud 1977).
- *Platypus oxyurus* (Dufour) – Pyrenees, Corsica, Calabria, Aegean islands, Turkey, Iran, India (Balachowsky 1949).

The species belonging to the genus *Triodonta* are today mainly confined to the western Mediterranean basin, one of them (*T. aquila*), however, reaches the Département of Côte d'Or (Baraud 1977), not very far away from the Vosges piedmont. The modern relictual distribution of *Platypus oxyurus* in southern Europe is indicated in Table 1. This bark-beetle is specific to the genus *Abies*, in the trunks of which it digs deep, ramified, galleries (Balachowsky 1949). The present-day restriction of its distribution suggests that some climatic factors may be involved, since *Abies* is today more widespread in Europe than *P. oxyurus*. In Britain it has been recorded in deposits that have been correlated with an earlier interglacial (Hoxnian) (Shotton & Osborne 1965; Coope 1990). *P. oxyurus* was also recently discovered in Upper Holsteinian deposits in the Vercors, Département of Isère, French Alps (P. Ponel, unpublished data).

CONCLUSIONS AND PERSPECTIVES

These data demonstrate that Quaternary climatic changes have deeply affected the arthropod fauna in north-eastern France, and that the response of this fauna to climatic fluctuations is large-scale shifts in geographical distribution. However many questions remain unanswered: how far southwards did the Arctic–Alpine elements migrate during glacial periods? To what extent was the Mediterranean fauna affected by these very cold periods? Did some regions in southern France constitute refuges for thermophilous arthropods during these episodes? How can the high endemism level in the Mediterranean mountains be interpreted in the light of palaeoclimatic data? Hopefully, further research will contribute to answering these fascinating questions.

ACKNOWLEDGEMENTS

I thank M. Pellet, B. Huntley and J.R.M. Allen who translated my text into English. A. Ashworth and A.V. Morgan for their comments. The Coleoptera diagram was greatly improved by J.R.M. Allen.

REFERENCES

Angus RB (1973) Pleistocene *Helophorus* from Borislav and Starunia in the Western Ukraine with a reinterpretation of M. Lomnicki's species. Phil Trans R Soc Lond B265:299-326

Balachowsky AS (1949) Faune de France, 50, Coléoptères Scolytidae. Lib Fac Sci, Paris

Balfour-Browne F (1950) British Water Beetles 2. Ray Society, London

Baraud J (1977) Coléoptères Scarabaeoidea. Suppl Nouv Rev Ent, Toulouse

Beaulieu J-L de, Reille M (1992) The last climatic cycle at la Grande Pile (Vosges, France): A new pollen profile. Quaternary Science Reviews 11:431-438

Campbell JM (1983) A revision of the North American Omaliinae (Coleoptera: Staphylinidae). The genus Olophrum Erichson. Can Ent 115:577-622

Châtenet G de (1986) Guide des Coléoptères d'Europe. Delachaux & Niestlé, Lausanne

Coope GR (1961) A Pleistocene coleopterous fauna with arctic affinities from Fladbury, Worcestershire. Quarterly Journal of the Geological Society of London 118:103-123

Coope GR (1962) Coleoptera from a Peat interbedded between two Boulder Clays at Burnhead near Airdrie. Trans Geol Soc Glasgow 24:279-286

Coope GR (1968) An insect fauna from Mid Weichselian deposits at Brandon, Warwickshire. Phil Trans R Soc Lond B254:425-456

Coope GR (1973) Tibetan Species of Dung Beetle from Late Pleistocene Deposits in England. Nature 245:335-336

Coope GR (1975) Mid-Weichselian Climatic Changes in Western Europe, Re-interpreted from Coleopteran Assemblages. *in* Suggate RP, Cresswell MM (eds) Quaternary Studies, 101-108. The Royal Society of New Zealand, Wellington

Coope GR (1990) The invasion of northern Europe during the Pleistocene by Mediterranean species of Coleoptera. *in* Di Castri F, Hansen AJ, Debussche M (eds) Biological Invasions in Europe and the Mediterranean Basin, 203-215. Kluwer, Dordrecht

Coope GR (1994) The response of insect faunas to glacial–interglacial climatic fluctuations. Phil Trans R Soc Lond B344:19-26

Guignot F (1947) Coléoptères Hydrocanthares. Lechevalier, Paris

Hansen M (1987) The Hydrophiloidea (Coleoptera) of Fennoscandia and Denmark. Brill/Scand. Science Press, Leiden-Copenhagen

Hoffmann A (1958) Coléoptères Curculionides (III). Lib Fac Sci, Paris

Hoganson JW, Ashworth AC (1991) Fossil beetle evidence for climatic change 18,000-10,000 years B.P. in southern Chile. Quat Res 36:2-17

Iablokoff-Khnzorian SM (1982) Les coccinelles. Boubée, Paris

Jeannel R (1939) Faune de France, 39, Coléoptères Carabiques (I). Lib Fac Sci, Paris

Kiselyov SV (1973) Late Pleistocene Coleoptera of Transuralia. Paleontological Journal 4:507-510

Lindroth CH (1985-1986) The Carabidae (Coleoptera) of Fennoscandia and Denmark. Brill/Scand. Science Press, Leiden-Copenhagen

Morgan AV, Morgan A (1980) Faunal assemblages and distributional shifts of Coleoptera during the Late Pleistocene in Canada and the northern United States. Can Ent 112:1105-1128

Morgan AV, Morgan A, Ashworth AC, Matthews JV (1984) Late Wisconsin Fossil Beetles in North America. Wright HE (ed) Late Quaternary Environments of the United States, 1, Porter SC (ed) The Late Pleistocene. Longman, London

Nazarov VI (1984) A reconstruction of the Anthopogene landscapes in the north-eastern part of Byelorussia according to palaeoentomological data. Nauka Press, Moscow

Ponel P (1995) Rissian, Eemian and Würmian Coleoptera assemblages from la Grande Pile (Vosges, France). Palaeogeogr Palaeoclimatol Palaeoecol 114:1-41

Shotton FW, Osborne PJ (1965) The fauna of the Hoxnian Interglacial deposits of Nechells, Birmingham. Phil Trans R Soc Lond B248:353-378

Taylor BJ, Coope GR (1985) Arthropods in the Quaternary of East Anglia – Their role as indices of local palaeoenvironments and regional palaeoclimates. Mod Geol 9:159-185

Woillard G (1974) Recherches sur le Pléistocene dans l'Est de la Belgique et dans les Vosges lorraines. Thèse, Univ Catholique de Louvain

Zanetti A (1987) Coleoptera Omaliinae. Calderini, Bologna

The spatial response of mammals to Quaternary climate changes

Russell W. Graham

Research and Collections Center

Illinois State Museum

1011 East Ash

Springfield, IL 62703

U.S.A.

INTRODUCTION

The Quaternary was characterized by rapid climatic fluctuations (Dansgaard *et al.* 1993) which had dramatic effects on terrestrial biotas, especially the mammalian fauna (Graham & Mead 1987). These climatic events controlled the composition and spatial distribution of continental mammal faunas by regulating intercontinental movements, by driving intracontinental shifts in the geographic distributions of species, and by creating environmental conditions for diversification and extinction. For this discussion, I will focus only on intercontinental and intracontinental movements of species. Diversification (evolution) and extinction are covered, respectively, by Lister (1996) and Sher (1996).

However, before I discuss the implications of species movements across landscapes, it is important to distinguish between the terms 'migration', 'immigration', 'dispersal' and 'geographic range shifts'. All of these terms have been used differently by biogeographers (e.g. Brown & Gibson 1983), population biologists (e.g. Chepko-Sade *et al.* 1987; Endler 1977), and palaeontologists (e.g. Koenigswald & Werdelin 1992). In this discussion, I will use 'dispersal' in the general sense of Tchernov (1992 pp 22 & 25) to describe historical processes of large scale changes, in both time and space, in the distribution of a taxon. In essence, this would be the same as geographic range shifts of a taxon. Whereas, immigration is the successful colonization of a new area (Tchernov 1992 p 25), migration refers to annual long distance movements made by large numbers of individuals in essentially the same

NATO ASI Series, Vol. I 47
Past and Future Rapid Environmental Changes:
The Spatial and Evolutionary Responses of
Terrestrial Biota
Edited by Brian Huntley et al.
© Springer-Verlag Berlin Heidelberg 1997

direction at basically the same time and usually there is a return migration (Endler 1977).

Immigrations and dispersals may occur at different time scales or respond to different magnitudes of climate change, but both processes have been important in shaping the composition of the mammal fauna of North America. Therefore, studying the effects of past climate events on mammal faunas may provide insight into the ways mammals may respond to future environmental changes like global warming (Graham 1992; Graham & Grimm 1990).

TRANSCONTINENTAL IMMIGRATIONS

The North American mammal fauna has been affected by biotic interchanges with both Eurasia, along the Bering Land Bridge, and South America, through the Isthmus of Panama and along the Central American Land Bridge. Throughout the Pleistocene, the emergence and submergence of land bridges fluctuated with sea level oscillations which were determined by the interdependence between the mass of continental ice sheets and fluctuations of global climates. The composition, and even direction of movement, of faunal immigrations have been determined by climatically sensitive environmental conditions (e.g. forest vs. grassland). Webb (1985b p 211) has noted that about 20 mammalian taxa immigrated to North America during the early Pleistocene and that 12 of these taxa were derived from Eurasia. For the late Pleistocene, Webb (1985b p 211) estimates that 27 immigrant mammalian genera entered North America and that 16, in addition to humans (*Homo sapiens*), were from Asia and only 9 came from South America.

The directional component of these immigrations has also varied through time. Marshall (1985) indicates that the overall number of genera entering South America from North America during the Pleistocene significantly outnumbers those genera moving in the opposite direction. Likewise, the number of invading mammals from Eurasia is greater than the number of North American mammals immigrating the other way. In essence, there appears to be a prevalence of immigration southwards away from cold, seasonal climates and towards warmer, less seasonal ones (Graham 1979; Vrba 1992). The Central American and Bering land bridges served to filter specific ecotypes. For instance, during the Pliocene and early Pleistocene, savanna-like environments on the Central American Land Bridge favoured the translocation of species adapted to open habitats, whereas during the late Pleistocene forest habitats favoured the exchange of arboreal mammals (Stehli & Webb 1985; Webb 1985a). Late Pleistocene steppe environments on the Bering Land Bridge selected for species generally adapted to grassland environments with few, if any, arboreal taxa (Graham 1979; Guthrie 1982; Kurtén & Anderson 1980).

It is also apparent that intercontinental immigrants were distributed differentially along environmental gradients (Graham 1985a). For instance, toxodonts were only able to invade Central America from South America, whereas other South American taxa (e.g. megathere sloths *Megatherium*, glyptodonts *Glyptotherium*, capybaras *Neochoerus*, *Hydrochoerus*, etc.) extended into the southeastern parts of the United States and some even reached more temperate (e.g., Shasta ground sloth *Nothrotheriops*, Harlan's ground sloth *Glossotherium*, armadillos *Dasypus*, etc.) and even arctic latitudes (eg. Jefferson's ground sloth *Megalonyx*) (Harington 1995; Kurtén & Anderson 1980). A similar, but latitudinally reversed pattern, is exhibited by Eurasian taxa that invaded North America, and the pattern is still apparent in the extant fauna of North America (Hoffmann 1986). *Saiga*, which no longer occurs in North America, has been found as Pleistocene fossils in Alaska, Yukon Territory, and Northwest Territories (Harington & Cinq-Mars 1995; Kurtén & Anderson 1980). Barren ground musk ox (*Ovibos*) and collared lemmings (*Dicrostonyx* spp.) extended farther south in the late Pleistocene (FAUNMAP Working Group 1994), but they did not extend as far south as other Eurasian immigrants like *Rangifer* and *Bootherium*. Finally, genera like *Microtus* (Repenning 1990) and *Bison* (McDonald 1981) ranged into Mexico and *Mammuthus* (Webb 1992) occurred as far south as Central America.

INTERCONTINENTAL DISPERSAL AND COMMUNITY RESPONSE TO ENVIRONMENTAL CHANGE

Theoretically, mammal communities can respond to environmental change in three distinctly different ways (Graham 1979) which have decidedly different implications for modeling future systems. Firstly, mammal communities can remain static and exhibit no response to environmental fluctuations regardless of the magnitude of this change. The palaeobiological record is full of examples that refute this position. Secondly, large groups of species (communities) can shift their geographic distributions as tightly-linked and highly-coevolved assemblages. If this hypothesis is correct, then life zone modelling (e.g. Harris & Cropper 1992) for future change is relatively easy because the location of entire communities, biomes, and ecosystems can be easily predicted for different climate scenarios. On the other hand, if species respond individually to environmental change, and they migrate in different directions, at different rates and at different times, then it is much more difficult to make predictions about the composition of biotic communities in response to environmental change (Graham 1992).

The fossil record can be used to distinguish between the life zone and individualistic models since they will predict different biogeographic patterns. To this end, an electronic relational database of *ca.* 2900 late-Quaternary mammal localities, FAUNMAP, has been compiled for the 48 contiguous states of the United States. It has been interfaced with a Geographic Information System (GIS) to construct distribution maps for individual species for seven different time periods (FAUNMAP Working Group 1994). Maps for the last 18 ka BP demonstrate that mammal species responded individualistically to late Quaternary climatic warming.

Figure 1 illustrates a clear example of this individualistic response. A late Pleistocene fossil locality, New Paris No. 4 (Guilday *et al.* 1964), is represented by a dot in southern Pennsylvania. The modern distributions of three different small mammal species are also shown in different shades. The remains of these three species have been found in the same deposits and excavation levels which suggests that they were probably contemporaneous in the past (Graham 1985b; Guilday *et al.* 1964). Therefore, an individualistic response is apparent in that the same late Pleistocene (10 – 18 ka BP) environmental fluctuations caused the thirteen-lined ground squirrel *Spermophilus tridecemlineatus* and the Hudson Bay collared lemming *Dicrostonyx hudsonius* to disperse westward and northward, respectively whereas the long-tailed shrew *Sorex dispar* still inhabits southern Pennsylvania.

Even though the responses of mammal species to late-Quaternary environmental changes were individualistic, an analysis of the FAUNMAP database indicates that there are groups of species that show similar directional patterns of dispersal (Table 1), although timing may vary for individual species. Some species dispersed northward rapidly. These northern species may have reached their present

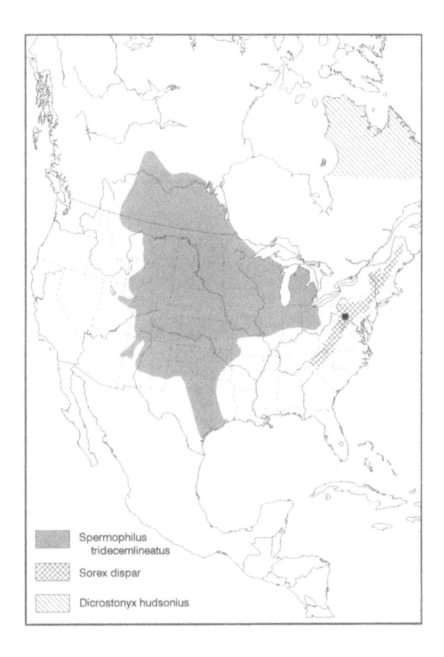

Figure 1: Location of New Paris No. 4 fossil site (Bedford County, Pennsylvania) and the modern distributions of *Spermophilus tridecemlineatus* (thrirteen-lined ground squirrel), *Sorex dispar* (long-tailed shrew), and *Dicrostonyx hudsonius* (Hudson Bay collared lemming).

Table 1: Categories of North American mammals exhibiting different directions and rates of dispersal in response to late-Quaternary (< 18 ka BP) global warming as evidenced from distribution maps compiled by the FAUNMAP Working Group (1994).

Dispersal	Species
Rapid Dispersal Northward	*Synaptomys cooperi* (southern bog lemming)
	Dicrostonyx spp. (collared lemming)
	Sorex cinereus (masked shrew)
	Sorex hoyi (pygmy shrew)
	Sorex palustris (water shrew)
Slower Dispersal Northward	*Clethrionomys gapperi* (red-backed vole)
	Microtus ochrogaster (prairie vole)
	Microtus pennsylvanicus (meadow vole)
Dispersal Eastward	*Cryptotis parva* (least shrew)
	Tamias striatus (eastern chipmunk)
	Blarina spp. (short-tailed shrew)
Dispersal Westward	*Thomomys talpoides* (northern plains pocket gopher)
	Microtus montanus (mountain vole)
	Microtus longicaudus (long-tailed vole)
	Cynomys ludovicianus (black-tailed prairie dog)
Substantially Unchanged Distributions	*Neotoma floridana* (eastern woodrat)
	Baiomys taylori (pygmy mouse)
	Neofiber alleni (Florida water rat)
	Scalopus aquaticus (eastern mole)

southern-most limits in the eastern United States by the end of the early Holocene, at the latest. Other species also dispersed northward, but their rate of movement was slower than the first group. Many of these species remained outside of their modern ranges through the middle Holocene and into the late Holocene. The distributions of other species shifted eastward as the climate changed during the late Quaternary, and some species became restricted to areas farther west in the Holocene. Finally, there were species that did not exhibit any substantial changes in their geographic distributions throughout the last 18 ka BP. Of particular interest are southern species like the eastern woodrat *Neotoma floridana* and pygmy mouse *Baiomys taylori* which maintained distributions as far north during the late Pleistocene as they did historically. In essence, these southern species were not displaced farther south into refugia during cold climatic phases. The general pattern of these distributional changes, like those of Quaternary plant dispersals (Davis 1976), best fits the pattern predicted by an individualistic model.

Even though the climate fluctuations of the Holocene were not as large as those of the late Pleistocene, some species exhibited significant changes in their distributions. For instance, the rice rat *Oryzomys palustris*, which inhabits the southeastern United States today, extended its distribution northward with climatic warming at the beginning of the Holocene. It occurred as far north as eastern Nebraska, southern Iowa, central Illinois, southern Indiana and southern Ohio by the late Holocene (FAUNMAP Working Group 1994 p 581). This species apparently retreated southward to its modern southeastern distribution some time later. Semken (1983) has summarized the various reasons proposed to explain the changes in the rice rat's distribution. It appears that climatically-driven environmental changes were significant. Other southern species like woodrats *Neotoma floridana* and *N. micropus* and cotton rat *Sigmodon hispidus* show similar northward extensions of their geographic distributions during the late Holocene. In fact, some species (e.g. nine-banded armadillo *Dasypus novemcinctus*, pygmy mouse and cotton rat) are still adjusting their distributions northward (Humphrey 1974; Stangl & Kasper 1987; Cameron & Spencer 1981). Frey (1992) has suggested that some northern species (e.g. meadow vole *Microtus pennsylvanicus*, northern jumping mouse *Zapus hudsonius*, and least weasel *Mustela nivalis*) are currently dispersing southward on the northern plains.

Mammals have been distributed and dispersed along environmental gradients throughout the Quaternary (Graham & Mead 1987). However, quantitative analyses of the FAUNMAP database have shown that faunal provinces can be identified for both the late Pleistocene and late Holocene (FAUNMAP Working Group 1996). However, because of the individualistic reorganization of communities, the species compositions of late-Pleistocene and late-Holocene provinces are markedly different. In fact, it is now apparent for mammals, as well as other biotic communities, that many of those of the late Pleistocene lack modern analogues (Graham 1985b). They contain combinations of species that do not geographically overlap today (e.g. Figure 1). Although the responses of mammal species to Holocene climatic changes were individualistic, they were not of the same magnitude as those of the late Pleistocene. Also, most of the changes occurred along environmentally sensitive ecotones like the prairie–forest border (McMillan & Klippel 1981). Therefore the species composition of Holocene mammal communities, especially the late Holocene ones, are quite similar to modern communities.

Even though the individualistic response of mammal species makes it more difficult to predict community response to environmental change, it will be important to consider in modelling the effects of future climate change. It is reasonable to assume that new communities may evolve with future climate changes if they are of sufficient magnitude. Clearly, static models of shifting communities, biomes and ecosystems as intact units over large latitudinal and longitudinal ranges are unrealistic.

ACKNOWLEDGEMENTS

I wish to thank two annonymous reviewers for their constructive comments on an earlier version of this paper. I am also indebted to Alan Morgan, Brian Huntley and Judy Allen for their editorial assistance and patience. Finally, I thank Alan Morgan, Brian Huntley, Honor Prentice, Wolfgang Cramer and NATO for the invitation to participate in the workshop. I thank Mary Ann Graham for producing Figure 1.

REFERENCES

Brown JH, Gibson AC (1983) Biogeography. C V Mosby Company St. Louis Toronto London
Cameron G, Spencer SR (1981) *Sigmodon hispidus*. Mammalian Species 158:1-9

Chepko-Sade BD, Shields WM, with Berger J, Halpin ZT, Jones WT, Rogers LL, Rood JP, Smith AT (1987) The effects of dispersal and social structure on effective population size. *in* Chepko-Sade BD, Halpin ZT (eds) Mammalian Dispersal Patterns - The Effects of Social Structure on Population Genetics, 287-321. Univ of Chicago Press, Chicago London

Dansgaard W, Johnsen SJ, Clausen HB, Dahl-Jensen D, Gundestrup NS, Hammer CU, Hvidberg CS, Steffensen JP, Sveinbjornsdottir AE, Jouzel J, Bond G (1993) Evidence for general instability of past climate from a 250-kyr ice-core record. Nature 364:218-220

Davis MB (1976) Pleistocene biogeography of temperate deciduous forests. Geosci and Man 13:13-26

Endler JA (1977) Geographic Variation, Speciation, and Clines. Princeton Univ Press, Princeton

FAUNMAP Working Group (1994) FAUNMAP: a database documenting late Quaternary distributions of mammal species in the United States. Ill St Mus Sci Pap 25, Nos. 1 and 2

FAUNMAP Working Group (1996) Spatial response of mammals to late Quaternary environmental fluctuations. Science 272:1601-1606

Frey JK (1992) Response of mammalian faunal element to climatic changes. Jour Mamm 73:43-50

Graham RW (1979) Paleoclimates and late Pleistocene faunal provinces in North America. *in* Humphrey RL, Stanford DJ (eds) Pre-Llano Culture of the Americas: Paradoxes and Possibilities, 49-69. Wash Anthro Soc Washington DC

Graham RW (1985a) Diversity and community structure of the late Pleistocene mammal fauna of North America. Acta Zool Fenn 170:181-192

Graham RW (1985b) Response of mammalian communities to environmental changes during the late Quaternary. *in* Diamond J, Case TJ (eds) Community Ecology, 300-313. Harper and Row, New York

Graham RW (1992) Late Pleistocene faunal changes as a guide to understanding effects of greenhouse warming on the mammalian fauna of North America. *in* Roberts RL, Lovejoy T (eds) Global Warming and Biological Diversity, 76-87. Yale University Press, New Haven London

Graham RW, Grimm EC (1990) Effects of global climatic change on the patterns of terrestrial biological communities. Trends Ecol and Evol 5:289-292

Graham RW, Mead JI (1987) Environmental fluctuations and evolution of mammalian faunas during the last deglaciation in North America. *in* Ruddiman WF, Wright Jr HE (eds) North America and Adjacent Oceans During the Last Deglaciation, 371-402. The Geology of North America, Geol Soc of Amer Vol K-3 Boulder

Guilday JE, Martin PS, McCrady AD (1964) New Paris No. 4: A late Pleistocene cave deposit in Bedford County, Pennsylvania. Nat Speleo Soc Bull 26:121-194

Guthrie RD (1982) Mammals of the mammoth steppe as paleoenvironmental indicators. *in* Hopkins DM, Matthews Jr JV, Schweger CE, Young SB (eds) Paleoecology of Beringia,307-326. Academic Press, Inc New York London Paris

Harington, CR (1995) Jefferson's ground sloth. Beringian Research Notes No 1

Harington CR, Cinq-Mars J (1995) Radiocarbon dates on saiga antelope (*Saiga tatarica*) fossils from Yukon and the Northwest Territories. Arctic 48:1-7

Harris LD, Cropper WP Jr (1992) Between the devil and the deep blue sea: implications of climate change for Florida's fauna. *in* Roberts RL, Lovejoy T (eds) Global Warming and Biological Diversity, 309-324. Yale University Press New Haven London

Hoffmann RS (1986) An ecological and zoogeographical analysis of animal migration across the Bering Land Bridge during the Quaternary Period. *in* Kontrimavichus VL (ed) Beringia in the Cenozoic Era, 464-481. AA Balkema, Rotterdam

Humphrey SR (1974) Zoogeography of the nine-banded armadillo (*Dasypus novemcinctus*) in the United States. BioSci 24:457-462

Koenigswald WV, Werdelin L (eds.) (1992) Mammalian Migration and Dispersal Events in the European Quaternary, Courier Forsch-Inst Senckenberg 153 Frankfurt

Kurtén B, Anderson, E (1980) Pleistocene Mammals of North America. Columbia Univ Press New York

Lister AM (1996) The evolutionary response of vertebrates to Quaternary environmental change. *in* Huntley B, Cramer W, Morgan AV, Prentice HC Allen JRM (eds) Past and future rapid environmental changes: The spatial and evolutionary responses of terrestrial biota, 287-302. Springer-Verlag, Berlin

Marshall LG (1985) Geochronology and land-mammal biochronology of the transamerican faunal interchange. *in* Stehli FG, Webb SD (eds) The Great American Biotic Interchange, 49-85. Topics In Geobiology 4, Plenum Press New York London

McDonald JN (1981) North American Bison - Their Classification and Evolution. Univ of California Press Berkeley Los Angeles London

McMillan RB, Klippel WE (1981) Post-glacial environmental change and hunting-gathering societies of the southern prairie peninsula. Jour Arch Sci 8:215-245

Repenning CA (1990) Of mice and ice in the late Pliocene of North America. Arctic 43:314-323

Semken HA Jr (1983) Holocene mammalian biogeography and climatic change in the eastern and central United States. *in* Wright, Jr HE (ed) Late Quaternary Environments of the United States, Volume 2: The Holocene, 182-207. Univ Minn Press, Minneapolis

Sher AV (1996) Late Quaternary extinction of large mammals in Northern Eurasia: A new look at the Siberian contribution. *in* Huntley B, Cramer W, Morgan AV, Prentice HC Allen JRM (eds) Past and future rapid environmental changes: The spatial and evolutionary responses of terrestrial biota, 319-340. Springer-Verlag, Berlin

Stangl FB Jr, Kasper S (1987) Evidence of communal nesting and winter-kill in a population of *Baiomys taylori* from north-central Texas. Tex Jour Sci 39:292-293

Stehli FG, Webb SD (1985) A kaleidoscope of plates, faunal and floral dispersals, and sea level changes. *in* Stehli FG, Webb SD (eds) The Great American Biotic Interchange, Topics In Geobiology 4, 3-16. Plenum Press New York London

Tchernov E (1992) Dispersal - a suggestion for a common usage of this term. *in* Koenigswald WV, Werdelin L (eds) Mammalian Migration and Dispersal Events in the European Quaternary, 21-26. Courier Forsch-Inst Senckenberg 153 Frankfurt

Vrba E (1992) Mammals as a key to evolutionary theory. Jour Mamm 73:1-28

Webb SD (1985a) Main pathways of mammalian diversification in North America. *in* Stehli FG, Webb SD (eds) The Great American Biotic Interchange, 201-217. Topics In Geobiology 4, Plenum Press New York London

Webb SD (1985b) Late Cenozoic mammal dispersals between the Americas. *in* Stehli FG, SD Webb (eds), The Great American Biotic Interchange, 357-386. Topics In Geobiology 4, Plenum Press New York London

Webb SD (1992) A brief history of New World Proboscidea with emphasis on their adaptations and interactions with man. *in* Fox JW, Smith CB, Wilkins KT (eds) Proboscidean and Paleoindian Interactions, 15-34. Baylor Univ Press Waco

The spatial response of non-marine Mollusca to past climate changes

Richard C. Preece
Department of Zoology
University of Cambridge
Downing Street
Cambridge CB2 3EJ
U.K.

There is a popular, even proverbial, belief that since snails move so slowly they are unlikely to disperse over any great distance and certainly not with any rapidity. In this review I will demonstrate that enormous distributional shifts have occurred in the geographical range of many species and that such movements can occur with surprising rapidity. The discussion will focus on four main topics. First, examples are given of some recent rapid distributional increases that should dispel the notion that snails are poor colonizers. Second, the diverse biogeographical composition of molluscan assemblages from the Middle Pleistocene of NW Europe are discussed, which show that huge distributional changes occurred after the major glaciations of the Anglian/Elsterian Stage. Third, the compositions of typical cold stage faunas are outlined and the reasons for the elimination of some arctic–alpine elements during the early Holocene discussed. Fourth, faunal successions during the Holocene itself are discussed and compared with vegetational records. The role of climate and other potential factors that may have caused the spatial reorganization of molluscan assemblages are also assessed.

MODERN INVASIONS

A direct way of establishing the rates at which non-marine molluscs can spread is to look at historical examples that have occurred during the last century or so.

The invasion of Europe by the ovoviviparous, parthenogenetic hydrobiid *Potamopyrgus jenkinsi*, was one of the most spectacular yet reported for any aquatic gastropod. It was first recorded from the River Thames in 1889, but was probably

NATO ASI Series, Vol. I 47
Past and Future Rapid Environmental Changes:
The Spatial and Evolutionary Responses of
Terrestrial Biota
Edited by Brian Huntley et al.
© Springer-Verlag Berlin Heidelberg 1997

introduced from New Zealand (where it is known as *P. antipodarum*) as early as 1859 (Hubendick 1950; Ponder 1988). It reached mainland Europe about 1900 and is now the dominant snail in many freshwater and some brackish habitats. Hubendick (1950) has published isochrone maps plotting its European invasion which show that its spread did not occur as an advancing front but was largely accomplished by 'jump dispersal' to specific centres and subsequent colonization from these.

Dispersal across watersheds is obviously the most serious problem facing a colonizing aquatic species. This is well illustrated by the case of the prosobranch *Lithoglyphus naticoides*. This is a Danubian species that was absent from the Rhine system during the Quaternary, despite the fact that the Danube and Rhine are neighbouring systems. Following the construction of the first man-made link between the two rivers in the last century, *L. naticoides* was quickly able to spread into the Rhine where it is now common. Other Ponto–Caspian species also gained access to the Rhine at the same time and such species now account for about 40% of the invertebrates in this river.

Bivalves belonging to the genus *Corbicula* are notorious colonizers and can occur in such numbers as to cause a serious nuisance. *C. fluminalis*, with a modern range in NE Africa and the Middle East, occurred in NW Europe during several interglacial periods but is unknown from any European Holocene deposit. *Corbicula* has recently invaded many Dutch rivers within only a four year period. There has been some confusion over the species involved but Kinzelbach (1991) has shown that two species, *C. fluminea* and *C. fluviatilis*, appeared in the Rhine in 1987 and that these now occur commonly throughout the lower and middle Rhine system. As there is no direct contact with Asia, Kinzelbach (1991) believed that these populations must have reached Europe in ships from the United States. Den Hartog *et al.* (1992) considered the reasons why this invasion of the Rhine by *Corbicula* has been so successful. They pointed out that both *Corbicula*, and the amphipod *Corophium curvispinum* which invaded the Rhine at the same time, are *r*-strategists and are consequently able to respond opportunistically when circumstances are favourable. The properties of rapid growth, production of several generations per year, early maturity and considerable fecundity are ideal attributes for a colonizing species. The success of these invasions must also be related to the present severely degraded state of the Rhine, which has resulted in a depletion of potential competitors.

Spectacular invasions are not confined just to aquatic taxa. The vermiform slug *Boettgerilla pallens*, a native of the Caucasus, has recently spread into NW Europe. Accurate details of this invasion are available for several countries including Britain. De Wilde *et al.* (1986) have published 10 km grid-maps of Belgium plotting records of *B. pallens* on a biennial or annual basis. The spread of this species is still occurring, aided perhaps by human horticultural activity. Why *Boettgerilla* should suddenly have expanded its range in this fashion remains uncertain. The situation is somewhat analagous to the expansion of the collared dove from the same region some years earlier.

MOLLUSCAN ASSEMBLAGES FROM THE INTERGLACIAL FOLLOWING THE ANGLIAN/ELSTERIAN GLACIATIONS OF NW EUROPE

The glaciations that occurred during the Anglian/Elsterian stage in the Middle Pleistocene are believed to have been the most extensive to have affected NW Europe. Conditions would have made it impossible for any thermophilous species to have survived *in situ*, and only a few hardy species of snail are known from this stage. All the woodland and warmth-loving taxa that were present during the ensuing interglacial must therefore have spread into NW Europe from refugia elsewhere. The precise location of such refugia has been the subject of much speculation (e.g. Holyoak 1989).

Surprisingly, some of the richest and most diverse molluscan faunas are known from the interglacial following this major cold stage. The most impressive faunas have been found in tufa deposits that formed in damp woodland. The most important sites include the tufas at Hörlis and Osterbuch in Germany; Arrest, Vernon and St Pierre-lès-Elbeuf, in France (Rousseau *et al.* 1992) and Hitchin (Kerney 1959) and Icklingham (Kerney 1976; Preece *et al.* 1991) in England. Stratigraphical and dating evidence suggest that most of these tufas formed during oxygen isotope stage 11 and are between 362 and 423 ka in age (Rousseau *et al.* 1992; Preece *et al.* 1991).

The molluscan faunas from these tufas are composed of several groups of species whose ranges never meet at the present day. The assemblages are rich in species of closed woodland and include several taxa which have modern central European ranges (*Platyla polita*, *Ruthenica filograna*, *Clausilia pumila* and *Macrogastra ventricosa*). One species from the British sites, originally listed as an extinct species (*Acicula diluviana*), has recently been shown to be conspecific with a species (*Platyla similis*) that now lives in SE Europe (Boeters *et al.* 1989). Another, *(Discus ruderatus)* has a modern boreo–alpine range (Kerney & Cameron 1979), yet this occurs with two taxa (*Leiostyla anglica* and *Perforatella subrufescens*) that have exclusively western/Atlantic modern ranges. Likewise, *Laminifera pauli*, which occurs in both the British sites and at St Pierre-lès-Elbeuf, is today restricted to forests in the western Pyrenees. The most remarkable species that occurs in all these tufas is another with an exclusively Atlantic modern range. *Lyrodiscus*, a subgenus of *Retinella*, today occurs only on the Canary Islands but members of this subgenus formally occurred in NW Europe during the Neogene and Middle Pleistocene (Rousseau & Puisségur 1990). *Retinella (Lyrodiscus) skertchlyi* is the species known from these Middle Pleistocene tufas.

These assemblages composed of such diverse biogeographical groups have no modern counterparts. Overall a warm, oceanic climate is suggested, but this must also have allowed the coexistence of some of the more continental taxa (e.g. *Discus ruderatus*). This is perhaps the most extreme case of a non-analogue situation exhibited by Quaternary land snail communities.

Contemporary Quaternary fluvial molluscan communities were equally distinctive. Fluvial deposits of this age are known from the Thames system at sites such as Swanscombe in Kent, and Clacton-on-Sea and East Hyde in Essex. A striking change in the molluscan faunas has been noted between the sediments that were thought to have accumulated early in the interglacial and those that were deposited later (Kerney 1971). In the later deposits species such as *Theodoxus danubialis* (= *T. serratiliniformis*), *Viviparus diluvianus*, *Valvata naticina* and *Corbicula fluminalis* occur. Since several of these taxa today have central European affinities, Kennard (1942) suggested that the appearance of this so called 'Rhenish fauna' resulted from the linkage of the Thames and Rhine systems.

The diverse composition of both the terrestrial communities from the tufas and the fluvial assemblages suggests unimpeded access to Britain at a time when it was not an island (Meijer & Preece 1995). This conclusion is supported by vertebrate evidence, particularly from the rich herpetofauna that existed in Britain at this time (Stuart 1995). Palaeogeographical considerations, such as these, must obviously be taken in account in any discussion of faunal composition and palaeoclimatic reconstruction.

COLD STAGE FAUNAS

In contrast to the faunas from interglacial stages, those from cold stages are far less diverse, despite the fact that Britain would have been joined to mainland Europe for virtually all of these periods. The fauna from the Last Cold Stage (Devensian/Weichselian) is now moderately well known, at least for its later part. Some thermophilous taxa did recolonize Britain during certain interstadials, but during stadials the fauna consisted of a few very hardy species (e.g. Holyoak 1982). There are very few obligate cold taxa known from British Pleistocene freshwater molluscan assemblages. One possible exception is the bivalve *Pisidium stewarti* (= *P. vincentianum*), which is today found in scattered localities in Asia, often at high elevation. Even in modern arctic regions there are no fluvial molluscs specifically confined to this biotope. Most molluscs require lime in order to construct their shells and the richest communties are consequently found in hard-water streams. Because of the scarcity of limestone in most northern areas, such suitable habitats are rare. It is consequently difficult to disentangle the effects of climate from those of geology, although excellent data exist for lacustrine species (Økland 1990).

Amongst the land snail faunas from cold stages some arctic-alpine elements, such as *Columella columella* and *Vertigo genesii*, were often present. The former is now extinct in Britain but the latter is known living from just two British sites, including Upper Teesdale, the site famous for its arctic–alpine flora (Coles & Colville 1980). Looking at arctic–alpine distributions such as these (Figure 1a, 1b), it is easy to conclude that climate is the most important factor controlling their ranges. This may not always be the case. Examination of the fossil record shows that both these species survived into the very early Holocene at several sites in the British Isles (Kerney *et al*. 1980; Preece *et al*. 1986), so their occurrence is not determined solely by low temperatures. At several of these British sites they were eliminated only when trees had spread into the local area. This can be established by directly comparing molluscan and vegetational records from identical profiles. It would appear that these species are intolerant of heavy shading and it is this factor, rather than a direct climatic one, that is most important. It is significant that in Upper Teesdale, one of the stations for *Vertigo genesii*, the pedigree of several of the arctic–alpine plants can be traced directly back to the Late-glacial. Their survival at this site has only been possible because at no time during the Holocene did a dense woodland canopy develop (Turner *et al.* 1973). The same situation holds for other species such as *Vertigo angustior*. This is not an arctic–alpine species and is unknown from the British Late-glacial, but it was widespread during the early to mid Holocene. Its dramatic reduction of range can also be directly linked to the spread of woodland during the early Holocene (e.g. Kerney *et al.* 1980; Preece & Day 1994).

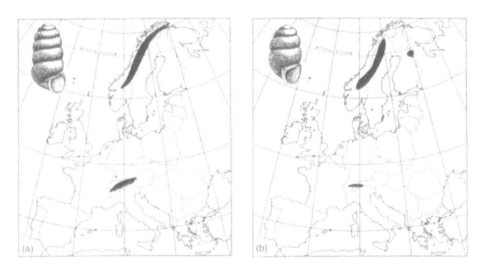

Figure 1. Modern distributions of (a) *Columella columella* and (b) *Vertigo genesii*, two land snails that were common in Britain during the Late-glacial and early Holocene.

Climatic oscillations may not only be responsible for changes in the overall geographical distribution of certain species but might also produce dramatic shifts in their altitudinal range. For example, in south-eastern France the helicelline land snail *Trochoidea geyeri* is today generally found in open environments only above about 1000m. Below this altitude a closely related species, *Candidula unifasciata*, occurs in similar open habitats. During cold stages of the Quaternary the range of *T. geyeri* extended down into the lowlands formerly occupied by *C. unifasciata* during warm stages. In areas where one of these species is absent, the range of the other expands to fill the unoccupied zone. Magnin (1993) interpreted this as evidence of possible interspecific competition between these species. The present disjunct distribution of *T. geyeri* therefore results partly from climatic history and partly from competitive interactions with other species.

It has been known for some time that some species from cold stage deposits are morphologically distinct from their temperate stage counterparts. The differences can be observed in specimens from the same stratigraphical sequence so that site specific factors can be excluded. The best known example is the case of *Pupilla muscorum*, which is often much larger, both taller and broader, in cold stage deposits. This pattern has been observed in various glacial–interglacial sequences and appears to be a widespread phenomenon across much of Europe (Rousseau 1989; Rousseau & Laurin 1984; Rousseau 1996). The large form of *Pupilla* is common in the British Late-glacial where it occurs in marshland contexts. The small dumpy form so common today is, on the other hand, primarily associated with dry grassland habitats. These morphological differences have therefore been attributed to ecophenotypic effects (Rousseau & Laurin 1984), although various shell forms may occur sympatrically, implying that other factors may be involved. It is still not entirely clear whether climatic factors alone can produce an increase in size in *Pupilla*.

HOLOCENE FAUNAL SUCCESSIONS

In recent years much information has been obtained on the Post-glacial colonization of various areas of the British Isles (e.g. Kerney *et al.* 1980; Preece 1980; Preece & Robinson 1984; Preece *et al.* 1984, 1986; Willing 1985; Preece & Day 1994). These investigations have mostly been based on detailed biostratigraphical studies of tufa deposits, which offer enormous advantages over other types of sediment (Preece 1991). Not only is it possible to recover successive fossil communities of land snails but, with the advent of AMS dating, detailed chronologies can be provided for each site. It is now possible to establish whether the appearance of a species occurred at the same time in different regions. Such data can provide the basis for estimating rates of spread and possible routes taken.

This approach has been utilized by palaeobotanists for some years. Using pollen-analytical data they have been able to present maps of vegetation at various scales for successive 1000 (or sometimes even 500) year intervals (e.g. Huntley & Birks 1983; Birks 1989). It will be some time before Quaternary malacologists achieve the necessary geographical coverage of well-dated sites. Nevertheless, dating of first appearance datums of certain critical species, such as *Discus rotundatus* (Table 1), does suggest rapid responses, equivalent to those seen in many pollen records.

Table 1. Dates of first appearance of *Discus rotundatus* at selected British sites.

Site	Date	Lab. Number	Reference
Wateringbury, Kent	8470 ± 190	Q-1425	Kerney *et al.* 1980
Holywell Coombe, Folkestone	8630 ± 120	OxA-2157	Preece 1993
Sidlings Copse, Oxford	8990 ± 90*	OxA-3859	Preece & Day 1994
Caerwys, North Wales	8100 ± 180	BM-1736R	Preece & Turner 1990
Newlands Cross, Co Dublin	8300 ± 90*	OxA-707	
	8930 ± 150*	OxA-708	Preece *et al.* 1986

* = shell dates

The date of the first appearance of a particular species is not necessarily linked directly with a change of climate. First, there is likely to have been a temporal lag between the climatic stimulus and the biological response. Second, the appearance of many species may have nothing whatever to do with climate, but may simply reflect a stage in the natural ecological succession. At a few sites that have remained permanently waterlogged it has been possible to recover pollen (and other organic fossils) as well as land snails. Comparisons of the faunal and vegetational successions at such sites has shown that the response of different groups is often out of phase (Preece, unpublished). At a tufa site near Oxford, some newly colonizing land snails appeared during periods of relative vegetational stability (Figure 2; Preece & Day 1994).

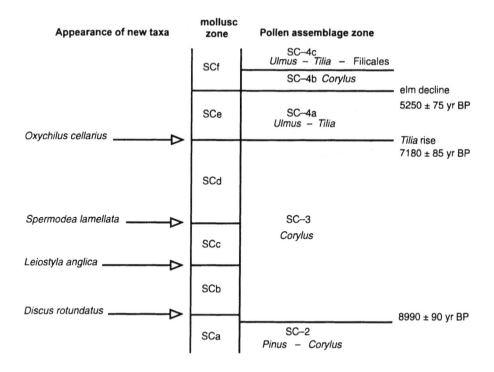

Figure 2. Comparison of molluscan and vegetational successions from Holocene tufa at Sidlings copse, Oxford (modified from Preece & Day 1994). Note the arrival of several species of land snail (shown on left) during the period of vegetational stability represented by pollen assemblage zone SC–3.

It is consequently difficult to appreciate the precise role that climate has in structuring the composition of molluscan communities, except in rather general terms. The diversity of land snail communities from Holocene tufas in southern Britain far exceeds that known from even the richest sites in the region today. Between 5 and 7 ka ago species such as *Leiostyla anglica*, *Spermodea lamellata*, *Vertigo alpestris* and *V. pusilla* were widespread in southern Britain. Today they have never been found together at any single site in this region and the isolated records for each are presumably relict from their much more extensive Holocene ranges. The important question to answer is why should the Holocene faunas have been so rich. Much of the early literature suggested that they were a reflection of a warmer and wetter climate that was thought to have characterised the thermal optimum of the Holocene. The decline or disappearance of many species with western distributions was then taken as evidence of climate change during the late Holocene. An alternative view (e.g. Sparks 1964) asserted that forest clearance and other anthropogenic modifications of the landscape during the Neolithic and Bronze Age were the main reason for the decline in diversity. The two causes are not, of course, mutually exclusive and both may have been contributory factors. Examination of some sequences (e.g. Kerney *et al.* 1980) does indicate that a few of these critical species declined before any obvious signs of human presence. This suggests that an anthropogenic explanation cannot be universally applied.

There is a general lack of detailed information on the precise way that climate may be exerting its effect on any given species. Many palaeoclimatic reconstructions are derived ultimately from modern distributional data of critical taxa, and these are assumed to reflect the climatic tolerances of the species concerned. In very few cases have these inferences been supported by independent laboratory studies or observations. Iversen (1944), in a now famous study, compared the meteorological conditions existing at sites in Denmark where ivy *Hedera helix* and holly *Ilex aquifolium* lived and where they did not occur. Broadly, he was able to demonstrate that these plants, which both have Atlantic-Mediterranean distributions, are markedly intolerant of winter frosts, and require minimum mean summer temperatures of about +14°C. He also showed that mistletoe *Viscum album* will withstand winter temperatures down to about -7°C, but needs rather higher mean summer temperatures of about +17°C.

Kerney (1968) noted the close similarity between the ranges of the land snails *Lauria cylindracea* and *Pomatias elegans* and the modern distributions of frost-sensitive taxa such as *Ilex* and *Hedera*. He concluded that the ranges of the molluscs are also likely to be governed by sensitivity to prolonged winter cold. Both these species had somewhat wider ranges during the mid Holocene, especially *Pomatias elegans*, which occurred at many sites beyond the eastern limit of its present range. Kerney (1968) suggested that the contraction of their ranges may plausibly be linked to a decline in winter temperatures since the thermal optimum. Similar distributional shrinkage is exhibited by *Ena montana*, although this species has a central European range quite different from those of the two species just discussed. In Britain *E. montana* is known from late Holocene deposits in Northamptonshire and Norfolk lying to the north of its present limit, which falls within the July isotherm for about +16·5°C. Kerney suggested that the northern distributional limit of *E. montana* may be determined by its demand for high summer temperatures and that the changes in range may again result from climatic change since the thermal optimum.

Such conclusions, although entirely reasonable, are largely inferential and have not yet been supported by detailed studies of the climatic tolerances of the species concerned. The same is true for most other organisms which have been used to infer changes of climate. Experimental information does exist for some of the larger species of land snail, particularly those belonging to the family Helicidae (e.g. Cameron 1970a, b). In a recent field survey in the neighbourhood of Basel, Switzerland, the helicid *Arianta arbustorum* had become extinct at 16 out of 29 sites between 1908 and 1991 (Baur & Baur 1993). At 8 of these sites its disappearance could be attributed to habitat destruction, but the remaining 8 sites appeared entirely suitable and still supported the associated molluscan fauna recorded in 1908. The only apparent difference between the sites retaining *A. arbustorum* and those which had lost it was altitude, implying that changes in the temperature regimes may be involved. Using remote sensing it was shown that the surface temperature of the vegetation in summer was significantly higher at those sites where *A. arbustorum* had become extinct and that this correlated with mean maximum temperatures recorded at potential oviposition sites of the snails. The satellite imagery showed that built-up areas do affect surrounding vegetation by emitting thermal radiation. This suggested that local climatic warming due to extensive urban development might have caused the local extinction of these *A. arbustorum* populations (Baur & Baur, 1993). Experimental work demonstrated that the likely controlling factor was the temperature at which the snail eggs developed; temperatures above a critical threshold significantly affected hatching success. Interspecific differences were also observed between *Arianta* and *Cepaea* which were consisitent with their contrasting geographical distributions.

DISCUSSION

The examples of recent distributional changes discussed above demonstrate that snails are capable of rapid dispersal and an examination of the fossil record shows that similar geographical shifts have occurred in the past. Most of the modern invasions are in response, not to climatic changes, but to opportunities that existed following anthropogenic dispersal to new regions (*P. jenkinsi, L. naticoides,*

Corbicula). Most of these taxa have important biological attributes that confer competitive advantages to the colonizing species. In the case of *Potamopyrgus* these include its wide ecological tolerance and its parthenogenetic reproduction that would allow a single individual to found a new colony.

Examination of the fossil record shows that important distributional changes have occurred without human intervention. The widespread occurrence of the '*Lyrodiscus* fauna' following the Anglian/Elsterian glaciation in NW Europe shows that, given suitable conditions, major changes of range can occur. In the Holocene it is possible to gain some idea of the rates of spread by reference to radiocarbon dates.

Climatic change is probably the single most important factor causing the spatial reorganizations of land snail communities during the Quaternary. Its effects, however, may not always be direct as it was in the case of the Swiss *Arianta*. More often its effects are likely to have been indirect, perhaps operating through its control of vegetation. The disappearance of several arctic–alpine species from Britain occurred, not as a result of the sharp thermal rise at the beginning of the Holocene, but as a result of the subsequent development of closed canopy woodland. The forest succession at a given site is governed by many factors, both physical (local edaphic and topographical conditions), as well as biological (competitive interactions between trees). The molluscan successions are also going to vary correspondingly and will to a large extent be dependent on the vegetational succession. Forest snails can obviously not occur without woodland. Distilling the purely climatic signal from such successions is clearly not an easy task.

Enormous changes of geographical range have occurred amongst the land snail faunas of NW Europe during the Late-glacial and throughout the Holocene. The most diverse assemblages occurred during the mid Holocene, a fact that has been taken as evidence of warmer and wetter conditions. Although human forest clearance and other anthropogenic activities undoubtedly contributed to the decline and local disappearance of several species, there are good reasons for believing that some of these losses were due to climatic deterioration since the thermal optimum. Much more experimental work is needed to support these conclusions.

During pre-Holocene interglacial stages even richer assemblages existed. The assumption is that the faunal successions during each interglacial stage reflect the development of forest, which in turn is driven ultimately by climate. The enormous differences in composition between the '*Lyrodiscus* fauna' and molluscan assemblages from Holocene tufas must to some extent reflect vegetational and climatic differences between the two stages. The species composition of Holocene tufa assemblages does not appear to have been severely affected by mesolithic activity, although several sites do show evidence of local disturbance (e.g. Preece *et al.* 1986). The pattern of colonization does differ between interglacial stages, which suggests that this includes an element of chance. There appears to be no clear reason why, for example, species such as *Corbicula fluminalis* and *Belgrandia marginata*, which were so common during most earlier interglacials, should not have returned to Britain during the Holocene. Similarly, it is not clear why in Britain *Discus ruderatus* should have occurred in the deciduous forest phases of interglacials, whereas it is confined to the beginning of the Holocene (cf. Kerney 1977). One fact is clear. Snails should not be regarded as the most pedestrian members of the animal kingdom. They can accomplish enormous distributional changes in comparatively brief periods and can even get to remote oceanic islands. Obtaining information on exactly how they achieve this is not easy, but there are now many documented cases of passive dispersal involving snails (e.g. Rees 1965).

REFERENCES

Baur B, Baur A (1993) Climatic warming due to thermal radiation from an urban area as possible cause for the local extinction of a land snail. J Appl Ecol 30:333-340
Birks HJB (1989) Holocene isochrone maps and patterns of tree-spreading in the British Isles. J Biogeogr 16:503-540
Boeters HD, Gittenberger E, Subai P (1989) Die Aciculidae (Mollusca: Gastropoda Prosobranchia). Zool Verh Leiden 252:1-234
Cameron RAD (1970a) The effect of temperature on activity of three species of helicid snail (Mollusca: Gastropoda). J Zool Lond 162:303-315
Cameron RAD (1970b) The survival, weight loss and behaviour of three species of land snail in conditions of low humidity. J Zool Lond 160:143-157
Coles B, Colville B (1980) A glacial relict mollusc. Nature 286:761
De Wilde JJ, van Goethem JL, Marquet R (1986) Distribution and dispersal of *Boettgerilla pallens* Simroth 1912 in Belgium (Gastropoda, Pulmonata, Boettgerillidae). Proc 8th Intern Malacol Congr (Budapest 1983):63-68
Den Hartog C, van den Brink FWB, van der Velde G (1992) Why was the invasion of the river Rhine by *Corophium curvispinum* and *Corbicula* species so successful? J Nat Hist 26:1121-1129
Holyoak DT (1982) Non-marine Mollusca of the Last Glacial Period (Devensian) in Britain. Malacologia 22:727-730
Holyoak DT (1989) The location of refugia for woodland snails during cold stages of the Quaternary. Quat Newsl 57:12-13

Hubendick B (1950) The effectiveness of passive dispersal in *Hydrobia jenkinsi*. Zool Bidr Upps 28:493-501

Huntley B, Birks HJB (1983) An atlas of past and present pollen maps for Europe 0 – 13,000 years ago. Cambridge University Press

Iversen J (1944) *Viscum, Hedera* and *Ilex* as climatic indicators. Geol För Stockh Förh 66:463-483

Kennard AS (1942) Faunas of the High Terrace at Swanscombe. Proc Geol Ass 53:105

Kerney MP (1959) An interglacial tufa near Hitchin, Hertfordshire. Proc Geol Ass 70:322-337

Kerney MP (1968) Britain's fauna of land Mollusca and its relation to the Post-glacial thermal optimum. Symp Zool Soc Lond 22:273-291

Kerney MP (1971) Interglacial deposits in Barnfield Pit, Swanscombe, and their molluscan fauna. Geol Soc Lond 127:69-93

Kerney MP (1976) Mollusca from an interglacial tufa in East Anglia, with the description of a new species of *Lyrodiscus* Pilsbry (Gastropoda: Zonitidae). J Conch Lond 29:47-50

Kerney MP (1977) British Quaternary non-marine Mollusca: a brief review. *in* Shotton FW (ed) British Quaternary Studies - recent advances, 32-42. Clarendon Press, Oxford

Kerney MP, Cameron RAD (1979) Land snails of Britain and North-west Europe. Collins, London

Kerney MP, Preece RC, Turner C (1980) Molluscan and plant biostratigraphy of some Late Devensian and Flandrian deposits in Kent. Phil Trans R Soc Lond B 291:1-43

Kinzelbach R (1991) Die Körbchenmuscheln *Corbicula fluminalis, Corbicula fluminea* und *Corbicula fluviatilis* in Europa (Bivalvia: Corbiculidae). Mainz Natuwiss Arch 29:215-228

Magnin F (1993) Competition between two land gastropods along altitudinal gradients in south-eastern France: neontological and palaeontological evidence. J Moll Stud 59:445-454

Meijer T, Preece RC (1995) Malacological evidence relating to the insularity of the British Isles during the Quaternary. *in* Preece RC (ed) Island Britain: a Quaternary perspective, 89-110. Geological Society Special Publication No 96

Økland J (1990) Lakes and snails. Universal Book Services / Dr W Backhuys, Oegstgeest

Ponder WF (1988) *Potamopyrgus antipodarum* - a molluscan coloniser of Europe and Australia. J Moll Stud 54:271-285

Preece RC (1980) The biostratigraphy and dating of the tufa deposit at the Mesolithic site at Blashenwell, Dorset, England. J Archaeol Sci 7:345-362

Preece RC (1991) Mapping snails in time: the prospect of elucidating the historical biogeography of the European malacofauna. Proc 10th Intern Malacol Congr (Tübingen 1989):477-479

Preece RC (1993) Late Glacial and Post-Glacial molluscan successions from the site of the Channel Tunnel in SE England. Scripta Geologica, Special Issue 2:387-395

Preece RC, Bennett KD, Robinson JE (1984) The biostratigraphy of an early Flandrian tufa at Inchrory, Glen Avon, Banffshire. Scott J Geol 20:143-159

Preece RC, Coxon P, Robinson JE (1986) New biostratigraphic evidence of the Post-glacial colonization of Ireland and for Mesolithic forest disturbance. J. Biogeogr 13:487-509

Preece RC, Day SP (1994) Comparison of Post-glacial molluscan and vegetational successions from a radiocarbon-dated tufa sequence in Oxfordshire. J Biogeogr 21:463-478

Preece RC, Lewis SG, Wymer JJ, Bridgland DR, Parfitt S (1991) Beeches Pit, West Stow (TL 798719). in Lewis SG, Whiteman, CA, Bridgland DR (eds) Central East Anglia and The Fen Basin, 94-104. Quaternary Research Association Field Guide, London

Preece RC, Robinson JE (1984) Late Devensian and Flandrian environmental history of the Ancholme Valley, Lincolnshire: molluscan and ostracod evidence. J Biogeogr 11:319-352

Preece RC, Turner C (1990) The tufas at Caerwys and Ddol. in Addison K, Watkins R (eds) North Wales, 162-166. Quaternary Research Association Field Guide, Coventry

Rees WJ (1965) The aerial dispersal of Mollusca. Proc Malac Soc Lond 36:269-282

Rousseau D-D (1989) Réponses des malacofaunes terrestres quaternaires aux contraintes climatiques en Europe septentrionale. Palaeogeogr Palaeoclimatol Palaeoecol 69:113-134

Rousseau D-D (1996) The weight of internal annd external constraints on *Pupilla muscorum* L. (Gastropoda: Stylommatophora) during the Quaternary in Europe. in Huntley B, Cramer W, Morgan AV, Prentice HC Allen JRM (eds) Past and future rapid environmental changes: The spatial and evolutionary responses of terrestrial biota, 303-318. Springer-Verlag, Berlin

Rousseau D-D, Laurin B (1984) Variations de *Pupilla muscorum* L. (Gastropoda) dans le Quaternaire d'Achenheim (Alsace): une analyse de l'interaction entre espèce et milieu. Geobios Mém Spécial 8:349-355

Rousseau D-D, Puisségur J-J, Lécolle F (1992) West-European terrestrial molluscs assemblages of isotopic stage 11 (Middle Pleistocene): climatic implications. Palaeogeogr Palaeoclimatol Palaeoecol 92:15-29

Rousseau D-D, Puisségur J-J (1990) Phylogénèse et biogéographie de *Retinella* (*Lyrodiscus*) Pilsbry (Gasteropoda: Zonitidae). Géobios 23:57-70

Sparks BW (1964) Non-marine Mollusca and Quaternary ecology. J Anim Ecol 33 (Suppl):87-98

Stuart AJ (1995) Insularity and Quaternary vertebrate faunas in Britain and Ireland. in Preece RC (ed) Island Britain: a Quaternary perspective, 111-125. Geological Society Special Publication No 96

Turner J, Hewetson VP, Hibbert FA, Lowry KH, Chambers C (1973) The history of the vegetation and flora of Widdybank Fell and the Cow Green reservoir basin, Upper Teesdale. Phil Trans R Soc Lond B 265:327-408

Willing MJ (1985) The biostratigraphy of Flandrian tufas in the Cotswold and Mendip districts. PhD thesis, University of Sussex

Section 3

Mechanisms enabling spatial responses

Reinterpreting the fossil pollen record of Holocene tree migration

Margaret B. Davis and Shinya Sugita
Department of Ecology, Evolution and Behavior
1987 Upper Buford Circle
University of Minnesota
St. Paul, MN 55108
U. S. A.

INTRODUCTION

The potential of plant and animal populations to spread over the landscape must be better understood in order to predict biotic responses to Global Change. The fossil record has potential for providing a record of indigenous species as they have shifted ranges in response to past changes of climate (Davis 1976, 1981; Huntley & Birks 1983; Prentice *et al.* 1991). Our purpose in this paper is to examine whether fossil pollen can provide a clear record of range shifts. We use a model of pollen dispersal to ask how the sizes and locations of lakes affect the way fossil pollen records an approaching population. Can small populations established in advance of the species front be detected? Do changes in pollen deposition give accurate estimates of the intrinsic growth rate of populations of invading species? We are using a simulation model, POLLSCAPE (Sugita 1994), that simulates heterogeneous vegetation on a landscape and calculates pollen dispersal to lakes. The results of the experiments provide guidelines for interpretation of fossil pollen records, and suggest how future studies can be designed to maximize information on past range shifts in response to changing climate.

DESCRIPTION OF SIMULATION EXPERIMENTS

POLLSCAPE is a spatial model that simulates pollen dispersal and deposition within a landscape with patchy vegetation. We specified two kinds of species patches,

NATO ASI Series, Vol. I 47
Past and Future Rapid Environmental Changes:
The Spatial and Evolutionary Responses of
Terrestrial Biota
Edited by Brian Huntley et al.
© Springer-Verlag Berlin Heidelberg 1997

superimposed on a background occupied by a third species. This patchy vegetation occupied a 50 x 50 km square which was located 500 km from the eastern edge of a large landscape 1000 by 2000 km in size (Figure 1). The landscape outside the 50 x 50 km square was homogeneous, because pollen travelling such a large distance to a lake does not sense patches of the sizes we used (Sugita 1994), but the overall species abundances were the same as within the heterogeneous square. A lake of a specified size was placed randomly within a 4 x 4 km area at the centre of the heterogeneous square. The model then scanned the landscape at a series of distances out to the edge of the entire landscape and calculated the proportion of the area at each distance occupied by each tree species. Pollen dispersal to the lake from that distance was then calculated using Sugita's (1993) model for the source area of pollen deposited on the surface of a lake. Wind speed, pollen fall speed, pollen productivity and lake size are all parameters that affect the calculation of pollen loading on the lake; values were taken from the literature and are specified in Table 1. Simulations were repeated thirty times for each experiment in order to pick up the effects of lake locations within various kinds of vegetation patches. Lake sizes were varied in different experiments.

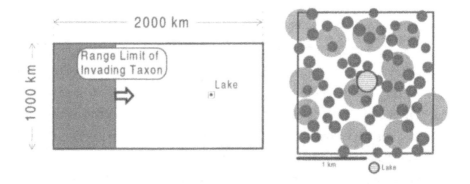

Figure 1. The landscape used in the simulations is shown in the figure to the left. To the right is an enlargement of part of the 50 x 50 km area within which the sampling sites were located (Sugita 1994). In the initial experiment, the large patches represent 15 ha areas dominated by sugar maple *Acer saccharum*, the small patches 2 ha areas dominated by birch *Betula alleghaniensis*, and the matrix is dominated by pine *Pinus strobus*. The range limit of the invading taxon, hemlock *Tsuga canadensis* in this experiment, moved across the landscape at a steady rate; within the range limit hemlock replaced 90% of the pine and 10% of the sugar maple and yellow birch.

The migrational front of an invader was simulated by a line 1000 km in length oriented north-south, which moved across the area from west to east (Figure 1). Within this range limit, the invading species replaced the existing taxa with probabilities equal to 90% for the background species and 10% for taxa in patches. This resulted in a patchy density for the invading species, similar to the primaeval forest containing hemlock (Davis *et al.* 1994).

Table 1. Parameters used in model

Wind speed = 3 m sec^{-1}, dispersing pollen equally in all directions

Lake radius – varied in different experiments

Taxon	Fall speed (m sec^{-1})	Reference cited	Productivity (relative to birch)	Reference cited
hemlock *Tsuga canadensis*	0·071	Eisenhut (1961)	0·55	Sugita et al. (in press); Sugita and Calcote (in prep)
pine *Pinus strobus*	0·031	"	1·60	"
Birch *Betula* spp.	0·024	"	1·0	"
spruce *Picea* spp.	0·056	"	0·69	"
aspen *Populus* spp.	0·025	"	0·02	"
maple *Acer* spp.	0·056	Heathcote (1978); Sugita (1993)	0·15	"

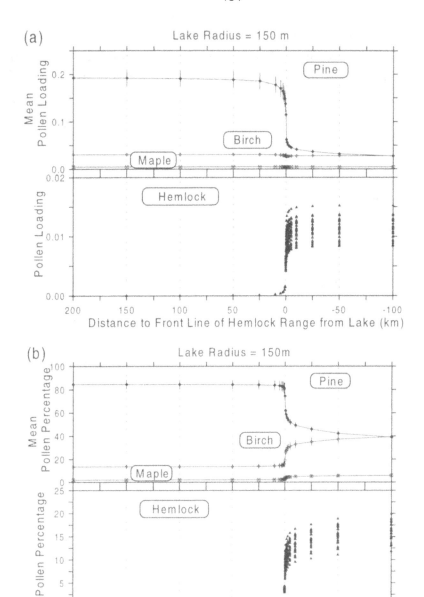

Figure 2-a. Simulated pollen loading (grains cm^{-2} yr^{-1}) on the surface of a 7 ha lake when the hemlock range limit was various distances away from the lake. The experiment was repeated 30 times with the lake in different locations within a 4 x 4 km area. The results for hemlock pollen are shown as a scatter diagram, while mean values and standard deviations are shown for pine, birch and maple pollen.
Figure 2-b. The same experiment with output displayed as percentages.

RESULTS AND DISCUSSION

Experiments and Empirical Data on Hemlock Invasion

Simulation

In our simulations, hemlock invaded a landscape dominated by pine, with patches of sugar maple and yellow birch. The frontier of the hemlock population moved at a constant rate, replacing 90% of the pine but only 10% of the birch and maple, thus forming a forest mosaic of stands dominated by hemlock and stands dominated by hardwoods (Davis *et al.* 1994).

Pollen loadings were calculated for 30 lakes of 150 m radius (7 ha) located at different locations within the heterogeneous landscape as the hemlock front advanced toward the lakes and passed beyond them toward the far border of the landscape (Figure 2-a). As the simulated species front reached each lake, pollen loadings increased steeply, quickly reaching a plateau as the front moved a few kilometres beyond each lake. The pollen values of the plateau varied among experiments, depending on the lake location within a maple-dominated patch or within the hemlock matrix. As the migration front continued across the landscape, pollen loading for hemlock increased very slightly as the regional pollen rain continued to change. However, because hemlock pollen is poorly dispersed, this effect is small. The effect of the invasion is more striking when pollen percentages are considered (Figure 2-b). As the hemlock front reached each lake, pine pollen percentages declined dramatically, although not by 90%. The decline continued as the front continued to move past the lake, with hemlock replacing pine in the regional vegetation. Birch and maple pollen percentages both showed sharp increases as the hemlock front reached the lake – potentially misleading changes that result from the decline in total pollen production by the vegetation.

Comparison with empirical data

Hemlock pollen percentages from lake sediments in Michigan, Wisconsin and Minnesota dating from the early 19th century, just before the landscape was logged and settled, have been displayed in a scatter diagram, plotted against the distance of each lake from the hemlock range limit (Figure 5 in Davis *et al.* 1991). As in the simulations, pollen values begin to rise 10-20 km outside the range limit, rising steeply at the range limit itself. Within the hemlock range, percentages of hemlock pollen vary from zero to 33%, reflecting local patchiness in the vegetation. The simulation result is further validated by fossil hemlock pollen deposited over the last 6000 years (Figure 18.5 in Davis 1987). As hemlock migrated from east to west across Michigan and Wisconsin, the profile of hemlock pollen percentages relative to the position of the advancing frontier remained very steep, except when population densities were low during the initial colonization, 6000 yr BP, and just after the hemlock decline 4800 yr BP. A steep increase in hemlock pollen percentages at the species' range limit has been characteristic in the Great Lakes region at all times when hemlock has been a dominant species, regardless of where the range limit was located.

Comparison with logistic population model

Sigmoid pollen increase

A significant increase in pollen deposition rates and percentages soon after pollen grains of a taxon first appear in the sediment has been observed in Holocene pollen diagrams many times and for many different taxa. The marked change in pollen abundance has been interpreted to represent the arrival of a tree taxon and its entry into forest communities. The resemblance of the sigmoid pollen increase (Figure 2-a and 2-b) to a logistic population growth curve has led investigators to use the pollen changes to estimate population growth rates (Tsukada 1981; Tsukada & Sugita 1982; Bennett 1983, 1988; Delcourt & Delcourt 1987; Walker & Chen 1987; MacDonald & Cwynar 1991). Our simulation makes it clear, however, that the sigmoid increase in pollen loading at the sampling site was caused by a *linear* increase in the landscape area occupied by hemlock, not by a local increase in population density. The sigmoid form to changes in pollen loading at the sampling site results from changes in the area occupied by hemlock within the relevant pollen source area. The relevant source area, which reflects the scale at which local variation in vegetation can be detected by pollen (Sugita 1994), is circular, and the proportion of its area that is occupied by hemlock changes as the species front advances across the landscape. As hemlock advances, the area behind the species front within the relevant source area increases by small increments per unit time at first, then larger increments as the frontier extends the full diameter of the circle, and then smaller increments again as the front passes to the far edge of the relevant area. Beyond this point the increase in pollen loading is small because relatively few hemlock pollen grains, which are poorly dispersed, come from the regional pollen rain. Thus once the species front has passed beyond the relevant source area, the pollen loading appears more or less constant, mimicking constant population densities at carrying capacity in a logistic curve. The simulation makes it clear, however, that the sigmoid shape of the increase in pollen loading can be explained as a function of pollen dispersal and migration rate, rather than local population growth.

We are not arguing that population growth (rather than the instantaneous replacement in our model) is not occurring as a species invades a landscape, but only that the Logistic Model was designed to describe population growth in a single population, not population growth at the landscape scale, merging together many different rates as migrating species invade a heterogeneous landscape. Pollen accumulating in a 7 ha lake records vegetation patches within a relevant source area approximately 1 km in diameter. Such a large area will include many individual forest stands where hemlock populations are growing rapidly, as well as many patches where hemlock population growth is slow or is not occurring at all (Davis *et al.* 1994). The pollen from these patches will be superimposed on regional background which will also be changing as hemlock replaces other species within the region. The pollen record is thus at the wrong spatial scale to record the behavior of an individual population. The process of invasion and subsequent population growth is better studied at the spatial scale of individual forest stands using more local pollen collection sites, such as small forest hollows (Sugita 1994; Calcote 1995).

Sensitivity experiments

Migration rate affects the shape of the sigmoid curve and thus it will affect estimates of the intrinsic rate of population increase – r in the Logistic Model – that are based on the slope of the curve. We repeated the hemlock migration simulation shown in Figures 2-a and 2-b, varying the migration rate by a factor of ten. The results are shown in Figure 3-a, plotted as average hemlock loading against time. The trajectories of changes in pollen loadings in the case of slow migration rate *versus* rapid migration rate were strikingly different. This result is supported by the contrast between the sigmoid rise in hemlock pollen percentages at three Michigan lakes invaded when hemlock migration rates were rapid, and two lakes invaded when hemlock migration rates were slow (Figure 4). Several investigators have noted differences in the slope of the sigmoid increase in various parts of a species range, interpreting them as differences in r (MacDonald & Cwynar 1991; Bennett 1988; Delcourt & Delcourt 1987). It seems likely to us that these differences reflect differences in migration rates, caused either by variation in the rate of climatic change at different times, or by changes in climatic limitation near the centre of a species' range *versus* near its edge.

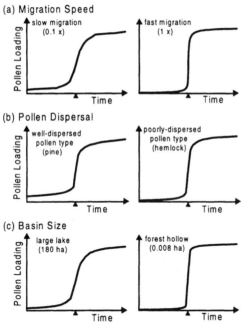

Figure 3. Pollen loading plotted against time as a population invades a simulated landscape, the population front moving at a constant rate. The shape of the sigmoid increase in pollen loading and its rate of change depends on (a) the rate of migration, (b) the dispersal-ability of pollen, and (c) lake size. Each solid line represents the mean of 30 simulations. (▲ indicates arrival of the invasion front at the lake)

Pollen dispersal also affects the shape of the sigmoid pollen loading curve as the species front enters the relevant pollen source area surrounding the sampling site. We repeated the simulation with a landscape in which large patches of aspen, small patches of birch, and a matrix of spruce were invaded by pine. As it invaded, pine replaced 90% of the spruce and 10% each of aspen and birch. The changes in pollen loading and percentages were very different from the previous case, due to stronger inputs from trees at a distance (Figure 3-b). Percentages of pine rose to 10% when the species front was 100 km from the site and continued to rise as the population approached. However in this case as well, the steepest pollen increase occurred as the front passed over the lake. After this event, pine pollen loading and percentages continued to rise slowly, recording the increase in pine within the regional vegetation. In the case of this well-dispersed pollen type, regional vegetation beyond the relevant source area contributed appreciable quantities of pine pollen to the lake. Slow migration by a poorly-dispersed species gives much the same shape to the sigmoid curve as fast migration by a well dispersed pollen type.

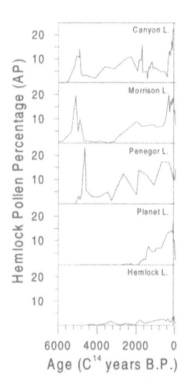

Figure 4. Hemlock pollen percentages at five lakes in northern Michigan. The first three were invaded by hemlock about 5000 years ago, when the hemlock frontier was migrating westward rapidly. Planet and Hemlock Lakes, in contrast, were invaded during the Late Holocene, when hemlock was extending its range very slowly toward the distribution limit recorded by the land surveys in the early 19th century.

Lake size will also affect the pattern of pollen increase. Given a constant rate of advance of a species front, large lakes will show a pattern similar to the trajectory for a well-dispersed pollen type or for a slow migration rate, while small lakes and forest hollows will show the steep rise characteristic of rapid migration or a poorly dispersed pollen type (Figure 3-c). However, the model is not particularly sensitive to lake areas within the same order of magnitude. The results shown in Figure 3-c compare lake areas that differ by 5 orders of magnitude.

Large differences in population growth rates among landscape patches of various sizes at the species frontier might also affect the shape of the sigmoid pollen increase. Simulations that include more complex invasion scenarios than are presented here are in progress to test this hypothesis.

Jump-dispersal

Population biologists have made a distinction between range extension by a continuous population front ('population diffusion') and range extension through establishment of advance colonies and later infilling of the landscape ('jump-dispersal') (Pielou 1979; MacDonald & Cwynar 1985, 1991; Hengeveld 1989). The distinction is a matter of scale (Woods & Davis 1989), but when advance colonies are established tens of kilometres in advance of the front they can result in a much faster rate of invasion of large areas of landscape. If the advance colonies are outside the range of gene exchange with the parent population, evolutionary changes may be facilitated (Cwynar & MacDonald 1987). Consequently it will be important to identify large scale jump-dispersal events from the fossil record.

Our simulations emphasize the difficulty palynologists will have in distinguishing between these alternatives. The relevant source area (the scale at which local variations in vegetation are detectable) for a poorly dispersed pollen type such as hemlock is so small – about 1 km in diameter for a 7 ha lake with northern hardwood vegetation in patch configuration like the model – that it becomes impractical to study enough lakes to create a grid of sites dense enough to monitor the entire landscape. Even in the case of pine, the diameter of the relevant source area for the landscape we simulated would be about 2 km for 7 ha lakes, making a grid of sites impractical. Larger lakes monitor a larger area, but the outlying patch must be quite large (about as large as the lake) to be detectable (Sugita *et al.* in press)

With luck, a fossil sampling site might happen to be located within an ancient outlying population, in which case pollen percentages will be higher than the background pollen observed at surrounding sites (Davis *et al.* 1991). Thus our theoretical result corroborates the contention by Davis *et al.*, based on empirical data, that it may be possible to demonstrate the existence of outlying colonies, but virtually impossible to demonstrate their absence.

CONCLUSIONS

The simulation model, validated through comparison with pollen records from Michigan lake sediments, adds to the insights on plant range shifts that have been derived from the Holocene fossil pollen record, suggesting:

1. The small quantities of fossil pollen deposited just prior to the first marked increase in deposition may represent pollen blown in from an advancing population front, rather than the local establishment of a small population. Actual invasion in all of our simulations was recorded by a steep increase in pollen loading and pollen percentages. Our simulation was realistic for species like hemlock and American beech *Fagus grandifolia*, which prior to logging grew in high densities within 5 km of the species front (Davis *et al.* 1991). Even if opinions differ on the timing of invasions at individual lakes, however, the overall record of tree migrations during the Holocene, as well as estimates of rates of migration, are supported by our simulations, even where lakes of different sizes have been compared.

2. Small advance colonies are hard to detect in the fossil record, making it difficult to distinguish jump-dispersal from a continuous front.

3. Differences in the rates of increase of fossil pollen from one region to another, and differences in rates of increase among taxa should be re-examined. Differences in the shape of the sigmoid curve that records invasion may result from differences in pollen dispersal, differences in the rate of advance of migrating species, and in extreme cases, on basin size. It will be important to detect accurately differences in rates of migration, because they may provide information on rates of climate change, or on ecological conditions that inhibited or promoted invasions.

4. The population phenomena that accompany invasion remain poorly understood. Small scale studies using forest hollows or humus profiles should be used for this

purpose, to better understand how invasion of individual vegetation communities occurs in response to changes in climate.

ACKNOWLEDGEMENTS

This work has been supported by NSF Grant DEB 9221375 and by the Mellon Foundation. We gratefully acknowledge discussion and critical review of the manuscript by RR Calcote, as well as helpful comments from anonymous reviewers.

REFERENCES

Bennett KD (1983) Postglacial population expansion of forest trees in Norfolk, UK. Nature 303:164-167

Bennett KD (1988) Holocene geographic spread and population expansion of *Fagus grandifolia* in Ontario, Canada. Journal of Ecology 76: 547-557.

Calcote R (1995) Pollen source area and pollen productivity: evidence from forest hollows. Journal of Ecology 83:591-602

Cwynar LC, MacDonald GM (1987) Geographical variation of lodgepole pine in relation to population history. American Naturalist 129:463-469

Davis MB (1976) Pleistocene biogeography of temperate deciduous forests. Geoscience and Man 13:13-26

Davis MB (1981) Quaternary history and the stability of forest communities. *in* West DC, Shugart HH, Botkin DB (eds) Forest Succession: Concepts and Application, 132-153. Springer-Verlag, New York

Davis MB (1987) Invasions of forest communities during the Holocene: Beech and hemlock in the Great Lakes region. *in* Gray AJ, Crawley MJ, Edwards PJ (eds) Colonization, Succession and Stability, 373-393. Blackwell Scientific Publications, Oxford

Davis MB, Sugita S, Calcote RR, Ferrari JB, Frelich LE (1994) Historical development of alternate communities in a hemlock-hardwood forest in northern Michigan, USA. *in* Edwards PJ, May R, Webb NR (eds) Large-scale Ecology and Conservation Biology, 19-39. Blackwell Scientific Publications, Oxford

Davis MB, Schwartz MW, Woods K (1991) Detecting a species limit from pollen in sediments. Journal of Biogeography 18:653-668

Delcourt PA, Delcourt HR (1987) Long-term Forest Dynamics of the Temperate Zone. Springer-Verlag, New York

Eisenhut G (1961) Untersuchungen über dei Morphologie und Ökologie der Pollenkörner heimischer und fremdländischer Waldbäume. Paul Parey, Hamburg (Translated to English by Jackson ST, Jaumann P, 1989)

Heathcote IW (1978) Differential pollen deposition and water circulation in small Minnesota lakes. Unpublished Ph.D. dissertation Yale University

Hengeveld R (1989) Dynamics of Biological Invasions. Chapman and Hall London

Huntley B, Birks HJB (1983) An Atlas of Past and Present Pollen Maps for Europe: 0-13000 Years Ago. Cambridge University Press, Cambridge

MacDonald GM, Cwynar LC (1985) A fossil pollen-based reconstruction of the late Quaternary history of lodgepole pine (*Pinus contorta* ssp. *latifolia*) in the western interior of Canada. Canadian Journal of Forest Research 15:1039-1044

MacDonald GM, Cwynar LC (1991) Postglacial population history of *Pinus contorta* *ssp. latifolia* in the western interior of Canada. Journal of Ecology 79:417-429

Pielou EC (1979) Biogeography. Wiley, New York

Prentice IC, Bartlein PJ, Webb T III (1991) Vegetation and climate change in eastern North America since the Last Glacial Maximum. Ecology 72:2038-2056

Sugita S (1993) A model of pollen source area for an entire lake surface. Quaternary Research 39:239-244

Sugita S (1994) Pollen representation of vegetation in Quaternary sediments: theory and method in patchy vegetation. Journal of Ecology 82:881-897

Sugita S, Calcote RR (in prep) Estimating pollen productivity: A new method with application for Quaternary palynology.

Sugita S, MacDonald GM, Larsen CPS (in press) Reconstruction of fire disturbance and forest succession from fossil pollen in lake sediments: potential and limitations. *in* Clark JS, Cachier H, Goldammer JG, Stocks BJ (eds) Sediment Records of Biomass Burning and Global Change. Springer-Verlag, Berlin

Tsukada M (1981) *Cryptomeria japonica* D.Don. I. Pollen dispersal and logistic forest expansion. Japanese Journal of Ecology 31:371-383

Tsukada M, Sugita S (1982) Late Quaternary dynamics of pollen influx at Mineral Lake, Washington. Botanical Magazine, Tokyo 95:401-418

Walker D, Chen Y (1987) Palynological light on rainforest dynamics. Quaternary Science Review 6:77-92

Woods KD, Davis MB (1989) Paleoecology of range limits: beech in the Upper Peninsula of Michigan. Ecology 70:681-696

Mechanisms of vegetation response to climate change

Philip Grime

NERC Unit of Comparative Plant Ecology

Department of Animal and Plant Sciences

The University of Sheffield

Sheffield S10 2TN

U.K.

INTRODUCTION

It is deceptively easy for plant ecologists to devise predictions of how vegetation might respond to climate change. Future shifts in structure and species composition can be suggested on the basis of correlations between present day floristics and climate (Raunkiaer 1934; Holdridge 1947; Box 1981; Woodward 1992) or by reference to the palaeoecological record of vegetation response to past changes in climate (Huntley & Birks 1983, Davis *et al.* 1986). However, there are two main reasons to suspect that both technical and philosophical developments will be necessary before this area of research graduates from speculation to science. In the first place, it is clear that our capacity to devise relevant predictions on the basis of either present correlations or past events is limited by:

a) the unprecedented rate and specific nature of current and predicted anthropogenically-driven climate changes; and

b) the need for predictions of climate change impacts to take account of a more potent and pervasive agent of change – the increasingly disruptive use of land by man.

This paper illustrates some of these complications, but the main objective is to suggest a protocol whereby hypotheses related to impacts of climate change on vegetation may be tested and refined.

NATO ASI Series, Vol. I 47
Past and Future Rapid Environmental Changes:
The Spatial and Evolutionary Responses of
Terrestrial Biota
Edited by Brian Huntley et al.
© Springer-Verlag Berlin Heidelberg 1997

A RESEARCH PROTOCOL

Figure 1 illustrates a set of procedures which appears to be gaining acceptance as an approach to understanding the role of plants in ecosystem responses to environmental change. An important step in devising the protocol has been the development of standardised screening procedures in which comparatively large numbers of plants of contrasted ecology and drawn from regional floras in various parts of the world have been grown under controlled conditions and their responses to individual factors measured. Examples of such experiments are available from Canada (Keddy 1992), Australia (Jurado *et al.* 1991), Japan (Washitani & Masuda 1990) and Northern England (Grime & Hunt 1975; Grime *et al.* 1981; Hunt *et al.* 1993), and recently a manual of screening methods has been published (Hendry & Grime, 1993). The purpose of such screening is to allow objective review of variation in basic attributes of plant morphology, physiology and biochemistry in order to recognise recurring functional types. The perspective emerging is that despite the apparently infinite variety in plant design, evolutionary and ecological specialisation in plants throughout the world is tightly constrained with respect to certain key attributes such as seed size (Hodgson & Mackey 1986; Jurado *et al.* 1991), the apportionment of captured resources between growth, storage and reproduction (Grime 1977), genome size (Grime & Mowforth 1982), leaf longevity (Reich *et al.* 1992) and root penetration (Reader *et al.* 1992). It appears that certain of the attributes used singly (Noble & Slatyer 1979; van der Valk 1981) or as sets (Grime *et al.* 1987a) can provide a basis for predicting plant responses to specific changes in climate, soils or management. Where sufficient data are available to classify all component species into functional types it may be possible to predict vegetation responses to specific scenarios of changed climate or landuse.

The recognition of plant functional types and the development of predictions of vegetation dynamics relevant to future environments are rapidly expanding research activities. However, both are limited at present by lack of screening data and by the shortage of adequate tests of the predictions. In the protocol of Figure 1 it is suggested that there are two alternative methods by which hypotheses concerning community and ecosystem function may be tested and refined. The first, illustrated in the left-hand side of the figure, involves comparison of model predictions against data collected by monitoring. With respect to both land-use and climate-change this represents an efficient mechanism for recognising the imperfections of models in that, as indicated by the arrows in Figure 1, discrepancies can stimulate not only changes in the model but also, where necessary, further data inputs from new screening procedures. Unfortunately, monitoring studies on communities and ecosystems are an exceedingly scarce resource and, for many purposes, alternative mechanisms of hypothesis testing (illustrated on the right-hand side of Figure 1) must be sought. These follow a logical pathway similar to that involving monitoring but rely upon manipulative experiments. Some of these experiments can be conducted on a small scale and involve synthesis of vegetation under controlled conditions (e.g. Grime *et al.* 1987b, Lawton *et al.* 1993). For many purposes, however, it is necessary to perform replicated manipulations of climate and land-use at a scale that often necessitates an alliance between ecologists and engineers.

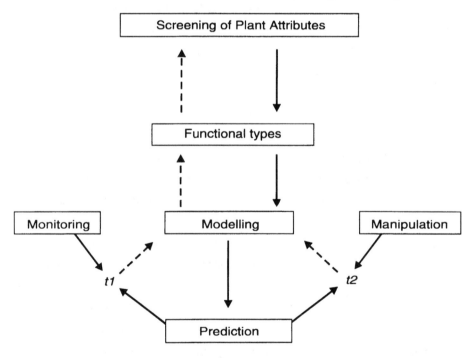

Figure 1. Protocol for development and testing of predictions of vegetation responses to climate change. Discrepancies revealed in the tests at *t1* and *t2* initiate further modelling cycles, each of which may necessitate refinement of the functional types or even additional screening.

The remainder of this paper provides extremely condensed accounts of three recent examples of tests of predictions of vegetation responses following one or other of the two pathways of Figure 1. The examples refer to processes already implicated in vegetation responses to climate and the first is a particularly good example of the need to consider interactions between land-use and climate change.

TESTING MECHANISTIC THEORIES OF VEGETATION RESPONSE TO CLIMATE CHANGE

*Primary plant strategies (*sensu *Grime 1974) and vegetation responses to CO_2 enrichment*

The most clearly documented symptom of climate change is the continuous rise in atmospheric carbon dioxide concentration from the onset of the Industrial Age to the present day (Houghton *et al.* 1990). In view of the sensitivity of plant growth, water relations and soil microbiology (Diaz *et al.* 1993) to CO_2 it is important to establish what effects, if any, have already taken place in various parts of the world as a consequence of this global elevation of carbon dioxide. Recently, following the protocol of Figure 1, a comprehensive screening of herbaceous plant responses to elevated CO_2 has been conducted under non-limiting conditions of mineral nutrient supply and temperature, in natural summer daylight. Growth of seedlings for 6 – 8 weeks at each of four levels of CO_2 (ambient, 500, 650 and 800 ppmv) has allowed curve-fitting procedures to be used to characterise the response of each species. Experimental methods and results for forty native C3 species are described by Hunt *et al.* (1991, 1993). Large species-specific differences in response to elevated CO_2 varying from zero in some ephemerals and slow-growing perennials to very large increases in dry-matter production in some species were detected. Among the latter, potentially large, fast-growing, clonal perennials of productive habitats, C-strategists in the terminology of Grime (1974), are conspicuous, prompting the hypothesis that responsiveness to elevated CO_2 may be dependent upon the coincidence of high relative growth rates and large carbon sinks.

One way in which to attempt to begin to test the hypothesis that C-strategists have responded differentially to elevating carbon dioxide is to consult the records of increasing and decreasing species now available from those European countries which have maintained national recording schemes over recent decades. In a recent paper (Thompson 1994) a discriminant analysis was conducted to compare the functional attributes of increasing and decreasing species in each of six countries in western Europe. The results show that the importance of particular discriminating variables varied between countries and there was no consistent evidence of a general expansion in abundance in the species classified by Grime *et al.* (1988) as C-strategists. However, a striking feature of the results was the extent to which expanding and declining species in the Netherlands and England could be classified by reference to S-radius, a five-point scale in which high scores indicate plants with stress-tolerant traits (e.g. slow growth rates, long life-spans and low rates of tissue turnover). In the next stage of analysis for each of the six countries the mean S-radius was calculated for two groups of species composed, respectively, of plants currently expanding and currently declining in abundance in the country concerned. In the four countries of low population density mean S-radius differed only slightly between expanding and declining species but in the two most populous countries (Netherlands and England) the S-radius of decreasing species was almost twice that of the increasing species. This analysis by Thompson thus provides strong circumstantial support for the contention (Hodgson 1986; Grime *et al.* 1988) that above a critical threshold of human population density major impacts of eutrophication and disruptive land-use are inevitable, causing contemporary vascular plant floras to polarise into 'winning' and 'losing' components, each with distinctive and predictable biologies. Support for this interpretation is also evident in a recent unpublished analysis in which the same floristic data are examined with respect to Ellenberg Nitrogen Numbers (Ellenberg *et al.* 1992). The dichotomy between a declining element of the flora with low growth rates and low nitrogen requirements and an expanding sector of more nitrogen-demanding, fast-growing species was again found to be peculiar to the Netherlands and England.

These results confirm that, in Europe at least, land-use is the most powerful contemporary force modifying vegetation cover. Does this mean that we can eliminate CO_2 enrichment as a major influence at the present time? I believe that to do so would involve risk of serious error. Robust, fast-growing perennial herbs of low S-radius and high Ellenberg N number are included in the expanding sectors of the floras of the Netherlands and England and have been shown to have a strong potential to respond to CO_2 enrichment (Hunt *et al.* 1991, 1993). The possibility remains, therefore, that some of the current species expansions are driven by a synergistic interaction between changing land-use (particularly, N and P eutrophication) and elevating CO_2. Further clarification of this issue appears to lie beyond the reach of the monitoring approach; we need long-term, outdoor experimental manipulations involving factorial combinations between CO_2 enrichment and variables associated with landuse.

Nuclear DNA amounts and plant responses to climate

From large-scale screening operations (eg Bennett & Smith 1976) and various other investigations (Hartsema 1961; Bennett 1971, 1976; Grime & Mowforth 1982; Grime *et al.* 1985), it has been established that there is more than a thousand-fold variation in nuclear DNA amount in vascular plants and that differences in DNA amount in cool temperate regions such as the British Isles coincide with differences in the timing of shoot growth. A mechanism interpreting variation in DNA amount, cell size and the length of the cell cycle (the three attributes are inextricably linked) as a consequence of climatic selection has been proposed (Grime & Mowforth 1982) suggesting that the delayed phenology of many British plants is imposed by the greater sensitivity of small-celled, low-DNA species to the inhibitory effects of low temperature on cell division. If this hypothesis is correct, we may expect that a rise in temperature would confer an advantage on low-DNA plants by differentially lengthening their growing seasons. Recently, this prediction has been tested by reference to the results of a 35-year monitoring study conducted on roadside vegetation at Bibury in Gloucestershire. The results of this analysis (Grime *et al.* 1994) expressed as coefficients of variation in plants grouped according to genome size confirm that plants of low-DNA amount are differentially sensitive to year to year variation in climate.

Plant relative growth rates and community responses to extreme events.

Events such as severe frost, droughts and fire have impacts that are large relative to their duration. Through effects on extinctions and invasions, extreme events are therefore likely to be important determinants of the rate and direction of vegetation responses to climate change (Walker 1991). Following the proposal of Westman (1978) it is useful to distinguish between two aspects of the response of vegetation to an extreme event; resistance is the ability of plant biomass to avoid displacement from control levels whilst resilience is the speed of recovery to control levels. Leps *et al.* (1982) built on earlier speculation by Levitt (1980) and Grime (1974) in suggesting that particular variable traits of life history and resource allocation are of universal value as predictors of resistance and resilience. They argue that traits which promote the tolerance of mineral nutrient stress (long lived organs, low rates of nutrient turnover) lead to correlated resistance to other forms of damage. It was further predicted that the existence of within-plant trade-offs leads to an association of these traits with low growth rates and therefore with low rates of recovery after damage (low resilience). A recent investigation closely adhering to the protocol of Figure 1 has tested the ability of species traits related to stress tolerance to predict resistance and resilience of five types of herbaceous vegetation subjected to three types of extreme event (frost, drought and fire). Essential details of the laboratory screening procedures employed to classify species and of the methods used to apply the extreme events are provided in MacGillivray *et al.* (1995). The results provide strong support for the hypothesis that the syndrome of plant traits (including low relative growth rates) associated with nutrient stress tolerance is positively correlated with resistance to extreme events and has a negative relationship with resilience. This suggests that resistance and resilience at the community level are mainly functions of traits possessed by the component species. A contrasting conclusion is drawn in the recent study by Tilman and Downing (1994) which emphasised the importance of biodiversity, a community level property, in determining resistance and resilience.

CONCLUSIONS

In order to place predictions of the impact of climate change on firm scientific ground it is essential to incorporate them into an iterative framework of hypothesis testing and hypothesis refinement. Monitoring studies at various geographical scales from national floras to small field plots can assist in this task but for some purposes measurements of vegetation response to experimental simulations of climate change are urgently required.

REFERENCES

Bennett MD (1971) The duration of meiosis. Proceedings of the Royal Society of London B 178:277-299

Bennett MD (1976) DNA amount, latitude and crop plant distribution. Environmental and Experimental Botany 16:93-108

Bennett MD, Smith JP (1976) Nuclear DNA amounts in angiosperms. Philosophical Transactions of the Royal Society B 274:227-274

Box EO (1981) Macroclimate and plant forms; an introduction to predictive modelling in phytogeography. Junk, The Hague

Davis MB, Woods KD, Webb SL, Futyma RB (1986) Dispersal versus climate: Expansion of Fagus and Tsuga into the upper Great Lakes region. Vegetatio 67:93-103

Diaz S, Grime JP, Harris J, McPherson E (1993) Evidence of a feedback mechanism limiting plant response to elevated carbon dioxide. Nature 364:616-617

Ellenberg H, Weber HE, Düll R, Wirth V, Werner W, Paulißen D (1992) Zeigwerte von Pflanzen in Mitteleuropa. Scripta Geobotanica 18, Erich Goltze, Göttingen

Grime JP (1974) Vegetation classification by reference to strategies. Nature 250:26-31

Grime JP (1977) Evidence for the existence of three primary strategies in plants and its relevance to ecological and evolutionary theory. American Naturalist 11:1169-1194

Grime JP, Hodgson JG, Hunt R (1988) Comparative Plant Ecology: A Functional approach to common British Species. Unwin Hyman, London

Grime JP, Hunt R (1975) Relative growth rate; its range and adaptive significance in local flora. Journal of Ecology 63:393-422

Grime JP, Hunt R, Krzanowski WJ (1987a) Evolutionary physiological ecology of plants. in Calow P (ed) Evolutionary Physiological Ecology, 105-126. Cambridge University Press, Cambridge

Grime JP, Mackey JML, Hillier SH, Read DJ (1987b) Floristic diversity in a model system using experimental microcosms. Nature 328:420-422

Grime JP, Mason G, Curtis AV, Rodman J, Band SR, Mowforth MA, Neal AM, Shaw SC (1981) A comparative study of germination characteristics in a local flora. Journal of Ecology 69:1017-1059

Grime JP, Mowforth MA (1982) Variation in genome size - an ecological interpretation. Nature 299:151-153

Grime JP, Shacklock JML, Band SR (1985) Nuclear DNA contents, shoot phenology and species coexistence in a limestone grassland community. New Phytologist 100:435-444

Grime JP, Willis AJ, Hunt R, Dunnett NP (1994) Climate-vegetation relationships in the Bibury road verge experiments. in Leigh RA, Johnston AE, (eds) Insight from Foresight: Long-term experiments in agricultural and ecological sciences, 271-285. CAB International, Wallingford

Hartsema AM (1961) Influence of temperature on flower formation and flowering of bulbous and tuberous plants. in Ruhland W (ed) Handbuch der Pflanzenphysiologie. 16 Ansenfaktoren in Wachstum und Entwicklung, 123-167. Springer, Berlin

Hendry GAF, Grime JP (1993) The Comparative Plant Ecology: A Laboratory Manual. Chapman and Hall ,London

Hodgson JG (1986) Commonness and rarity in plants with special reference to the Sheffield flora. II The relative importance of climate, soils and landuse. Biological Conservation 36:253-274

Hodgson JG, Mackey JML (1986) The ecological specialisation of dicotyledonous families within a local flora: some factors constraining optimisation of seed size and their possible evolutionary significance. New Phytologist 104:479-515

Holdridge LR (1947) Determination of world formations from simple climatic data. Science 105:367-368

Houghton JT, Jenkins GJ, Ephraums JJ. (eds) (1990) Climate Change: The IPCC Scientific Assessment. Cambridge University Press, Cambridge

Hunt R, Hand DW, Hannah MA, Neal AM (1991) Response to CO_2 enrichment in 27 herbaceous species. Functional Ecology 5:410-421

Hunt R, Hand DW, Hannah MA, Neal AM (1993) Further responses to CO_2 enrichment in British herbaceous species. Functional Ecology 7:661-668

Huntley B, Birks HJB (1983) An Atlas of past and present maps for Europe: 0-13,000 Years Ago. Cambridge University Press, Cambridge

Jurado E, Westoby M, Nelson D (1991) Diaspore weight, dispersal, growth form and perenniality of central Australian plants. Journal of Ecology 79:811-828

Keddy P (1992) A pragmatic approach to functional ecology. Functional Ecology 6:621-626

Lawton JH, Naeem S, Woodfin RM, Brown VK, Gange A, Godray HJC, Heads PA, Lawler S, Magda D, Thomas CD, Thompson LJ, Young S (1993) The Ecotron: a controlled environmental facility for the investigation of population and ecosystem processes. Philosophical Transactions of the Royal Society of London B 341:181-194

Leps J, Osbornova-Kosinova J, Rejmanek M (1982) Community stability, complexity and species life-history strategies. Vegetatio 50:53-63

Levitt J (1980) Responses of Plants to Environmental Stresses. Volume II. Water, Radiation, Salt and Other Stresses. Academic Press, New York

MacGillivray CW, Grime JP and the Integrated Screening Programme (ISP) team (1995) Testing predictions of the resistance and resilience of vegetation subjected to extreme events. Functional Ecology 9:640-649

Noble IR, Slatyer RO (1979) The use of vital attributes to predict successional changes in plant communities subject to recurrent disturbances. Vegetatio 43:5-21

Raunkiaer C (1934) The Life Forms of Plants and Statistical Plant Geography; being the collected papers of C. Raunkiaer. Translated into English by Carter HG, Tansley AG & Miss Fansboll. Clarendon Press, Oxford

Reader RJ, Jalili A, Grime JP, Spencer RE, Matthews N (1992) A comparative study of plasticity in seedling rooting depth in drying soil. Journal of Ecology 81:543-550

Reich PB, Walters MB, Ellsworth DS (1992) Leaf life-span in relation to leaf, plant and stand characteristics among diverse ecosystems. Ecological Monographs 62:365-392

Thompson K (1994) Predicting the fate of temperate species in response to human disturbance and global change. in Boyle TJB, Boyle CEB, (eds) NATO Advanced Research Workshop on Biodiversity, Temperate Ecosystems and Global Change, 61-76. Springer-Verlag, Berlin

Tilman D, Downing JA (1994) Biodiversity and Stability in grasslands. Nature 367:363-365 van der Valk AG (1981) Succession in wetlands: a Gleasonian approach. Ecology 62:688-696

Walker BH (1991) Ecological consequences of atmospheric and climate change. Climatic Change 18:301-316

Washitani I, Masuda M (1990) A comparative study of the germination characteristics of seeds from a moist tall grassland community. Functional Ecology 4:543-557

Westman WE (1978) Measuring the inertia and resilience of ecosystems. Bioscience 28:705-710

Woodward FI (1992) Predicting plant responses to global environmental change. New Phytologist 122:239-251

Plant invasions: Early and continuing expressions of global change

Richard N. Mack
Department of Botany
Washington State University
Pullman, WA 99164
U.S.A.

INTRODUCTION

Through prolific, current use and tacit agreement 'global change' refers to permanent alteration of the Biosphere as caused by humans. In practice, most emphasis in predicting global change stems from the on-going rise in levels of greenhouse gases in the atmosphere and their myriad environmental consequences (National Academy of Sciences 1988). So defined and practised, it has been treated as a subject with immensely important present and future implications but little past; parts of the Biosphere have been drastically and repeatedly altered, but humans' historic role has been largely viewed as restricted to non-global alterations (e.g. forest-clearing and burning, diversion of rivers, land-filling and flooding) (Turner et al. 1990). I contend that this conventional view, whether explicitly or implicitly stated, overlooks the role humans have already played in at least one important arena of global change - plant invasions (Drake et al. 1989). Unlike other forms of current or projected global change, humans have altered the composition of vegetation for several thousand years through their deliberate or accidental dispersal of plants beyond their native ranges (di Castri 1989). As I hope to demonstrate, there are important lessons to be gleaned from plant invasions, both as examples of global change in their own right and as phenomena that will complicate (and confound) predictions about the consequences of other forms of global change, including those stemming from increases in levels of greenhouse gases. I concentrate here on four such topics; in my estimation, all are either under- or unappreciated in the current concern about global change, although D'Antonio and Vitousek (1992) provide a related assessment of the role of alien grasses as agents of global change.

NATO ASI Series, Vol. I 47
Past and Future Rapid Environmental Changes:
The Spatial and Evolutionary Responses of
Terrestrial Biota
Edited by Brian Huntley et al.
© Springer-Verlag Berlin Heidelberg 1997

Definitions in invasion biology

First, a few definitions are necessary because our understanding of the geographical spread of plants at almost any scale has been hampered by conflicting definitions (cf. Baker and Stebbins 1965; Lodge 1993). Plant *migration* is used frequently in its broadest meaning: large-scale changes in the spatial distribution of species. Thus, a migration can refer to an expansion, a contraction, or even a shift in range, e.g. the radical geographical shifts experienced by species in the late Quaternary (Huntley & Birks 1983).

By giving migration such a broad definition, the term '*invasion*' can be reserved for that which I believe is its original intent - entry into a new range by members of a species that has detrimental consequences. While a migration does not necessarily denote any detrimental effect, the term 'invasion' always does. For plant invasions, these effects can be biotic or abiotic (Cross 1981; Vitousek 1990). Left unstated is whether the impetus for the invasion comes through change in the physical environment. In a strict sense, an invasion can occur without human intervention. But humans have so increasingly influenced plant dispersal and establishment, especially in the last 500 years, that I will further restrict my comments to invasions initiated or later influenced, or both, by humans. Such a restriction is appropriate, given the obvious role humans will play in the future in all manner of global change.

Implicit in describing an invasion are two other phenomena that must precede it. The first step in what may become an invasion is the *immigration* of members of the alien species (usually a small, genetically-impoverished group) to the new range. Immigration refers to nothing more than this entry. No statement (explicit or implicit) is made as to the fate of these immigrants or their descendants or even whether they will have descendants. The persistence of such descendants is a '*naturalization.*' For plants, such persistence occurs without cultivation. How long an alien species must reside in the new range before it is considered naturalized is equivocal (Webb 1985). Even though continuous human intervention is not necessary for the maintenance of the alien species, humans often do play a continuing role (e.g. subsequent dispersal within the new range). So, all invasions stem from naturalization, but a naturalization does not necessarily constitute an invasion, i.e. naturalizations are not by definition detrimental. Deciding what is 'detrimental' admittedly varies with the perspective, e.g. the mere presence of an alien species in a nature reserve could be considered detrimental.

MANY PLANT INVASIONS PRE-DATE THE INDUSTRIAL REVOLUTION

An awareness of the role humans play in affecting the environment emerged more than 130 years ago as illustrated in George P. Marsh's, "Man and Nature; or, Physical Geography as Modified by Human Action." In this 1865 treatise and its expanded version, "The Earth as Modified by Human Action" (1874), Marsh outlined his tally of the environmental consequences of human activity on the planet. His books cover diverse topics, both ludicrous (trees as protection against malaria, the electrical action of trees), and insightful (the consequences of mixing marine biota via the recently opened Suez Canal). Not surprisingly, he made no mention of increasing levels of CO_2 in the atmosphere. He did however discuss at length what has since become recognized as a partial explanation for these rising levels: the massive destruction of the earth's forests.

Marsh devoted much of one chapter to the spread of organisms beyond their native ranges by humans, including plant invasions. He cited contemporary examples of both the deliberate and accidental spread and establishment of plant species (e.g. the introduction of crops and ornamental plants worldwide, the accidental introduction of hundreds of species to St. Helena). He was clearly aware of the potential negative consequences of alien plants.

Most of the great upheavals in plant distribution caused by humans were however just beginning to unfold in the mid-19th century, as based on the historic growth of naturalized floras world-wide (e.g. Wells *et al.* 1986; Kloot 1987). These events and their frequency were tied obviously to the growth of transoceanic commerce and the spread of human settlement. It was these plant naturalizations and invasions that drew Marsh's attention, e.g. the proliferation of thistles into the Pampas, an invasion that Darwin (1898) witnessed. Initiation of these events was not directly tied to any physical alteration stemming from the Industrial Revolution.

OFTEN WITH MINIMAL INPUT BY HUMANS, PLANT INVASIONS HAVE ALREADY
PERMANENTLY ALTERED MUCH OF THE EARTH'S VEGETATION

Not only does the history of plant invasions span several millennia of human
transport and plant cultivation (di Castri 1989), their collective impact has also been
truly global. In the current discussion on the magnitude of any environmental
consequences that could stem from increasing levels of greenhouse gases, global
change is considered a collection of phenomena that will *likely* occur and have
permanent consequences for biota. In contrast, plant invasions as induced by
human activity have *already* occurred and have already wrought permanent
consequences.

The current status of the earth's vegetation reveals that much terrestrial vegetation
has already been altered to an extent that rivals predictions of biogeographic
changes from increasing levels of greenhouse gases (Emanuel *et al.* 1985; Solomon
et al. 1993). To estimate the scale of this global change, I tallied those land areas
for which I could conservatively determine that alien plants are now dominant. Some
of the larger regions thus radically transformed include the Intermountain West in
North America, the Central Valley of California, much of southern Florida, the fynbos
in South Africa, and parts of New Zealand (Ewel 1986; Healy 1974; Macdonald 1984;
Mack 1989) (Figure 1). Even though parts of these regions are now devoted to
crops, their susceptibility to plant invaders has been historically quite high (e.g. the
cereal-growing districts in the Pampas and the Columbia Plateau in western North
America) (Mack 1989).

The most extensive transformations have probably occurred since 1900 within
Neotropical biomes (Parsons 1970; Baker 1978). Janzen (1988) considers that less
than 2% of the 550,000 km^2 of Tropical Dry Forest that once stretched along the
Pacific Coast of Mesoamerica remains intact. Burned, cleared and converted largely
to pasture, these forest habitats now support mainly naturalized grasses, such as
Hyparrhenia rufa (Pohl 1983; G. Hartshorn, pers. comm.). The Cerrado and
Caatinga, two huge Brazilian vegetation regions with distinct dry seasons, have had
or will likely soon face similar fates. The southern half of the Cerrado has already
been invaded by *Melinis minutiflora*, an African grass, along with the less prominent
H. rufa and *Rhynchelytrum repens* (Eiten & Goodland 1979); thus, as much as
845,000 km^2 of this region may now be dominated by alien grasses. *M. minutiflora*

and *H. rufa* have also transformed much smaller areas in the savannas of Colombia and Venezuela (Blydenstein 1967; Baruch *et al.* 1985). The list of islands that have been extensively altered is legion but includes such well-known groups as the Hawaiian Islands and the Channel Islands of California, *plus* Mauritius, New Caledonia, and Singapore (Corlett 1992; Halvorson 1992; Lorence & Sussman 1986; Barrau 1953; Stone *et al.* 1992). The grand total for these regions is $4 \cdot 02 \times 10^6$ km^2 or $3 \cdot 1$ % of the earth's ice-free land surface ($1 \cdot 3 \times 10^8$ km^2) – approximately equivalent to the area of Europe, excluding the former Soviet Union and Scandinavia.

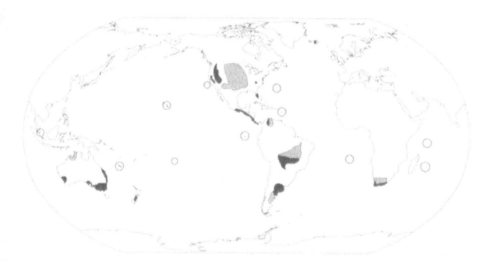

Figure1. Estimate of the earth's terrestrial surface in which the vegetation is now dominated by alien species (blackened) and those regions in which alien plants are prevalent but not yet dominant (stippled). Representative islands for which the vegetation is now dominated by alien plants are circled. (References include Healy 1974; Mack 1981, 1989; Macdonald 1984; and my own field estimates)

Both the total area estimated as well as the percentage of the earth that is subject to invasion were conservatively computed. First, I tallied invaded areas that are each > 400 km^2. In addition to excluding many, small oceanic islands that are dominated by alien plants (Fosberg 1957), this minimum area also excludes many continental invasions (e.g. *Spartina anglica* in Britain (Charman 1990)). Thus, Europe and most of eastern North America appear in Figure 1 as unaltered by plant invasions; yet both contain many localized invasions (di Castri *et al.* 1990). Even less quantifiable information is available on areas for which aliens are prevalent, although not dominant (e.g. rangelands in the North American Great Plains). Except as noted above, the figure does not include any of the immense area under cultivation (approx. 15×10^6 km^2, Richards 1990). Consequently, the tally does not generally include agricultural weeds that may have little ability to spread beyond cultivated land.

This global change in vegetation is permanent; there is no indication the native species would resume their former roles even if all human immigration ceased. In this regard, plant invasions impose a level of permanency not found with other early expressions of change wrought by humans (e.g. the effects of forest-clearing can often be reversed through subsequent succession). Any temporary character that these new communities may possess is usually due to the likelihood that other immigrants will arrive and supplant the current alien residents. For example, the alien species that now dominate the grasslands of California supplanted earlier-arriving alien immigrants (Mack 1989).

THE DYNAMICS OF PLANT INVASIONS WILL CONFOUND PREDICTIONS ON PLANTS' GEOGRAPHIC RESPONSES TO RISING LEVELS OF GREENHOUSE GASES

A striking feature of post-Columbian terrestrial plant invasions has been the swiftness with which these events have unfolded. Often the immigrants consist of a small, inconspicuous population of founders. If they or their immediate descendants survive at all – an unlikely circumstance – these populations have little net growth and instead undergo large swings in size caused by environmental stochasticity (Mack 1995). This so-called 'lag phase' is initially indistinguishable from the dynamics in other immigrant populations that lead to their local extinction. But for those rare alien species that become invaders, their range expansion and numerical growth (a 'log phase') occur quickly; the majority of the new range is occupied in less time than the whole lag phase (Mack 1985). For instance, the Eurasian annual *Bromus tectorum* occupied its new range in the Intermountain West of North America from *ca.* 1890 – 1930. It remained mostly in localized populations until about 1910. By 1915 it was attracting attention as an invader at the scale of counties, but by 1930 it was a dominant on more than 200,000 km^2 (Mack 1981). Although the time courses for plant invasions have certainly varied, the fundamental property of logistic range expansion has been a consistent feature (Mack 1985).

For plant invasions, the threshold level of change in the new environment to bring about change may be modest. Disturbance has long been cited as a prerequisite (Harper 1965), but the scale and character of disturbance among invasions appear to vary enormously (Mack 1985). For *B. tectorum* in western North America, the speed of the invasion was certainly assisted by rapid conversion of these grasslands to use for crops and livestock. But in other cases, dispersal alone may be all that is necessary to trigger an invasion (Ewel *et al.* 1982; Huenneke & Vitousek 1990).

Future change in plant distributions will likely result from the interaction of two forces that have not previously operated simultaneously: a changing climate (as illustrated by the Quaternary record of plant migration) and ubiquitous human influence on both the volume and character of plant dispersal and subsequent naturalization. Each is clearly capable of affecting global changes in plant distributions. Furthermore, plant invasions will continue, regardless of the consequences of increasing levels of greenhouse gases. The challenge, so far largely unappreciated (but see Sasek & Strain 1990; Beerling 1994), will be to predict how these forces will operate simultaneously. Although each could dampen the magnitude of the other (such as rising sea-levels flooding coastal sites that were once suitable for terrestrial plant entry), it is at least as likely that their effect will be synergistic – climatic change expanding the potential ranges for some species, while an ever-increasing level of human activity enhances the opportunities for plant immigration to reach these new ranges.

AS EXEMPLIFIED BY SOCIETY'S REACTION TO PLANT INVASIONS, IT MAY NOT ACTIVELY RESPOND TO OTHER FORMS OF GLOBAL CHANGE IN THE FUTURE

As perhaps the earliest examples of global change induced by humans, plant invasions provide another lesson quite aside from any arising from their nature and extent. Observations and reports of plant invasions made by Darwin, Marsh and others were disseminated widely enough that society was, in a sense, forewarned. Yet then, as now, the response was slow. Broad federal legislation to combat the importation and subsequent spread of alien plant species did not appear in the United States until well into the 20th century (U.S. Congress, Office of Technology Assessment 1993). In the United States and elsewhere society (other than perhaps the parties immediately affected) often views these changes with seeming indifference or even acquiescence and acts only in the most extreme cases.

Aldo Leopold (1966) illustrated this attitude to environmental change in his essay, "Cheat takes over." Cheat or cheatgrass (the aforementioned *Bromus tectorum*) has both an immense native range in arid Eurasia as well as large, far-flung new ranges in New Zealand, Argentina, Chile and most prominently, western North America (Pierson & Mack 1990). Leopold described how most people, including the California pest quarantine officer in his story, viewed hills in the Intermountain West blanketed with cheatgrass, not knowing how much this scene had changed from the caespitose grass-dominated landscape of less than a century earlier. He also described the attitude of those who attempted to raise livestock on rangelands now dominated by this alien grass. Instead of any concerted attempt to reverse the situation and restore the native vegetation (or at least eliminate the alien Cheat), Leopold found a "hopeless attitude" that was almost universal.

The public does not always however acquiesce to a plant invasion. *Eichhornia crassipes* (Water Hyacinth) probably escaped in the United States in the 1880's. Within a decade it had emerged from being a horticultural curiosity to become a pernicious pest that clogged otherwise navigable rivers in Florida. By 1899 the U.S. Congress had provided funds to combat it – the first such legislative action in the United States (Mack 1991). Several decades later the alien shrub *Berberis vulgaris*, the intermediate host for the stem rust of wheat, was so regionally prolific in the U.S. that the public was enlisted to find and eradicate the shrub across a wide swath of northern states. This public eradication program was spectacularly successful; thousands of landowners and even school children enthusiastically destroyed *B. vulgaris* across tens of thousands of hectares. Success was due in large measure to public recognition of a direct, detrimental consequence to wheat production posed by European barberry, made all the more apparent by U.S. entry into World War I (Hutton 1927).

A rhetorical question that follows from reviewing the control of plant invaders deals with what level and continuity of public response will meet environmental changes of whatever character, and at whatever scale, driven by future increase in the levels of greenhouse gases. This matter is important because no amount of scientific investigation and resulting prediction can be translated into public action unless society is convinced that the undeniable consequences will be detrimental, widely-shared and permanent. If the reaction to most plant invasions is any indication, society will not readily respond measure-for-measure to the level urged by the scientific community.

ACKNOWLEDGEMENTS

I thank the organizers of the NATO Advanced Research Workshop for inviting me to participate and in particular Brian Huntley for his overall supervision of the meeting. In addition I thank W. Amarel, F.A. Bazzaz, R.A. Black, G.S. Hartshorn, and M. Jasienski for their helpful discussions during the course of my writing this manuscript.

REFERENCES

Baker HG (1978) Invasion and replacement in Californian and neotropical grasslands. in Wilson JR (ed) Plant relations in pastures, 368-384. CSIRO East Melbourne, Australia

Baker HG, Stebbins GL (eds) (1965) The genetics of colonizing species. Academic Press, New York

Barrau J (1953) Present-day problems in the utilisation of pasture land in New Caledonia. Eighth Pacific Science Congress 4B:573-576

Baruch Z, Ludlow MM, Davis R (1985) Photosynthetic responses of native and introduced C_4 grasses from Venezuelan savannas. Oecologia 67:388-393

Blydenstein J (1967) Tropical savanna vegetation of the Llanos of Colombia. Ecology 48:1-15

Beerling DJ (1994) Predicting the response of the introduced species *Fallopia japonica* and *Impatiens grandifera* to global climatic change. in de Waal LC, Child LE, Wade PM, Brock JH (eds) Ecology and management of invasive riverside plants, 135-139. Wiley, Chichester

Charman (1990) The current status of *Spartina anglica* in Britain. in Gray AJ, Benham PEM (eds) *Spartina anglica* - a research review. Institute of Terrestrial Ecology Research Publication No. 2

Corlett RT (1992) The ecological transformation of Singapore, 1819-1990. Journal of Biogeography 19:411-420.

Cross JR (1981) The establishment of *Rhododendron ponticum* in the Killarney oakwoods, S.W. Ireland. Journal of Ecology 69:807-824

D'Antonio CM, Vitousek PM (1992) Biological invasions by exotic grasses, the grass/fire cycle, and global change. Annual Review of Ecology and Systematics 23:63-87

Darwin C (1898) Journal of researches into the natural history and geology of the countries visited during the voyage of HMS Beagle round the world, under the command of Capt. Fitz Roy, R.N. Appleton, New York

di Castri F (1989) History of biological invasions with special emphasis on the Old World. *in* Drake JA, Mooney HA, di Castri F, Groves RH, Kruger FJ, Rejmanek M, Williamson M (eds) Biological invasions a global perspective, 1-30. Wiley, Chichester

di Castri F, Hansen AJ, Debussche M (eds) (1990) Biological invasions in Europe and the Mediterranean Basin. Kluwer, Dordrecht

Drake JA, Mooney HA, di Castri F, Groves RH, Kruger FJ, Rejmanek M, Williamson M (eds) (1989) Biological invasions a global perspective. Wiley, Chichester

Eiten G, Goodland R (1979) Ecology and management of semi-arid ecosystems in Brazil. *in* Walker BH (ed) Management of semi-arid ecosystems, 277-300. Elsevier, Amsterdam

Emanuel WR, Shugart HH, Stevenson MP (1995) Climatic change and the broad-scale distribution of terrestrial ecosystem complexes. Climatic Change 7:29-43

Ewel JJ (1986) Invasibility: lessons from south Florida. *in* Drake JA, Mooney HA (eds) Ecology of biological invasions of North America and Hawaii, 214-230. Springer, New York

Ewel JJ, Ojima DS, Karl DA, DeBusk WF (1982) *Schinus* in successional eco-systems of Everglades National Park. South Florida Research Center Report T-676. National Park Service

Fosberg FR (1957) The naturalized flora of Micronesia and World War II. Eighth Pacific Science Congress 4:229-234

Halvorson, WL (1992) Alien plants at Channel Islands National Park. *in* Stone CP, Smith CW, Tunison JT (eds) Alien plant invasions in native ecosystems of Hawai'i, 64-96. University of Hawaii Press, Honolulu, HI

Harper JL (1965) Establishment, aggression, and cohabitation in weedy species. *in* Baker HG, Stebbins GL (eds) The genetics of colonizing species, 245-265. Academic Press, New York

Healy AJ (1974) Introduced vegetation. *in* Williams GR (ed) The natural history of New Zealand, 170-189. Bailey Bros. & Swinfen, Folkestone

Huenneke LF, Vitousek PM (1990) Seedling and clonal recruitment of the invasive tree *Psidium cattleianum*: implications for management of native Hawaiian Forests. Biological Conservation 53:199-211

Huntley B, Birks HJB (1983) An atlas of past and present pollen maps for Europe: 0-13000 years ago. Cambridge University Press, Cambridge

Hutton LD (1927) Barberry eradication reducing stem rust losses in wide areas. The agricultural yearbook, U.S. Dept. of Agriculture, Washington, DC, pp 114-118

Janzen DH (1988) Tropical dry forests: The most endangered major tropical ecosystem. *in* Wilson EO (ed) Biodiversity, 131-137. National Academy Press, Washington, DC

Kloot PM (1987) The naturalised flora of South Australia 2. its development through time. Journal of the Adelaide Botanical Garden 10:91-98

Leopold A (1966) A Sand County almanac. Oxford University Press, New York

Lodge DM (1993) Species invasions and deletions: community effects and responses to climate and habitat change. *in* Kareiva PM, Kingsolver JG, Huey RB (eds) Biotic interactions and global change, 367-387. Sinauer, Sunderland, MA

Lorence DH, Sussman RW (1986) Exotic species invasion into Mauritius wet forest remnants. Journal of Tropical Ecology 2:147-162

Macdonald IAW (1984) Is the fynbos biome especially susceptible to invasion by alien plants? a re-analysis of available data. South African Journal of Science 80:369-377

Mack RN (1981) Invasion of Bromus tectorum L. into western North America: an ecological chronicle. Agro-Ecosystems 7:145-165

Mack RN (1985) Invading plants: their potential contribution to population biology. in White J (ed) Studies on plant demography: a festschrift for John L. Harper, 127-142. Academic Press, London

Mack RN (1989) Temperate grasslands vulnerable to plant invasions: characteristics and consequences. in Drake JA, Mooney HA, di Castri F, Groves RH, Kruger FJ, Rejmanek M, Williamson M (eds) Biological invasions: a global perspective, 155-179. Wiley, New York

Mack RN (1991) The commercial seed trade: an early disperser of weeds in the United States. Economic Botany 45:257-273

Mack RN (1995) Understanding the processes of weed invasions: the influence of environmental stochasticity. in Stirton CH (ed) Weeds in a changing world, 65-74. British Crop Protection Council Symposium Proceedings 64.

Marsh GP (1865) Man and nature; or physical geography as modified by human action. Scribner, New York

Marsh GP (1874) The Earth as modified by human action. Scribner, Armstrong, New York

National Academy of Sciences (1988) Toward understanding global change. Initial priorities for U.S. contributions to the International Geosphere-Biosphere Program. National Academy Press, Washington, DC

Parsons JJ (1970) The "Africanization" of the New World tropical grasslands. Tubinger Geographische Studien 34:141-153

Pierson EA, Mack RN (1990) The population biology of Bromus tectorum in forests: distinguishing the opportunity for dispersal from environmental restriction. Oecologia 84:519-525

Pohl RW (1983) Hyparrhenia rufa (Jaragua). in Janzen DH (ed) Costa Rican Natural History, 256-257. University of Chicago Press, Chicago

Richards JF (1990) Land transformations. in Turner BL, Clark WC, Kates RW, Richards JF, Mathews JT, Meyer WB (eds) The earth as transformed by human action, 163-178. Cambridge University Press, Cambridge

Sasek TW, Strain BR (1990) Implications of atmospheric CO_2 enrichment and climatic change for the geographical distribution of two introduced vines in the USA. Climatic Change 16:31-51

Solomon AM, Prentice IC, Leemans R, Cramer WP (1993) The interaction of climate and land use in future terrestrial carbon storage and release. Water, Air, and Soil Pollution 70:595-614

Stone CP, Smith CW, Tunison JT (eds) (1992) Native ecosystems of Hawai'i. University of Hawaii Press, Honolulu, HI

Turner BL, Clark WC, Kates RW, Richards JF, Mathews JT, Meyer WB (eds) (1990) The earth as transformed by human action. Cambridge University Press, Cambridge

U.S. Congress, Office of Technology Assessment (1993) Harmful non-indigenous species in the United States. OTA-F-565. Washington, DC

Vitousek PM (1990) Biological invasions and ecosystem processes: towards an integration of population biology and ecosystem studies. Oikos 57:7-13

Webb DA (1985) What are the criteria for presuming native status? Watsonia 15:231-236

Wells MJ, Poynton RJ, Balsinhas AA, Musil KJ, Joffe H, van Hoepen E (1986) The history of introduction of invasive alien plants to southern Africa. *in* Macdonald IAW, Kruger FJ, Ferrar AA (eds) The ecology and management of biological invasions in southern Africa, 21-35. Oxford University Press, Cape Town

Invading into an ecologically non-uniform area

Rob Hengeveld and Frank Van den Bosch[1]

Institute of Forestry and Nature Research

P.O. Box 23

6700 AA Wageningen

The Netherlands

INTRODUCTION

In two earlier papers (Van den Bosch et al. 1990, 1992; see also Hengeveld 1994 and Hengeveld & Van den Bosch 1991 for a worked-out calculation) we arrived at the equation describing a species' rate of spatial population expansion, C, under uniform conditions

$$C = \frac{\sigma}{\mu} \sqrt{2\ln(R_0)}$$

(1)

in which μ represents the mean age of child bearing, σ the standard deviation of the marginal density of the sites where the individuals settle relative to their birth place, and R_o, the net reproduction rate of the females in the expanding population.

Figure 1 shows the dependence of the rate of population expansion C on both the net rate of reproduction, R_o, and the dispersal parameter σ (formula 5). In the present paper, we extend this model for spatially non-uniform conditions.

---

[1] Department of Mathematics, Agricultural University, Dreyenlaan 4,
6703 HA Wageningen, The Netherlands

NATO ASI Series, Vol. I 47
Past and Future Rapid Environmental Changes:
The Spatial and Evolutionary Responses of
Terrestrial Biota
Edited by Brian Huntley et al.
© Springer-Verlag Berlin Heidelberg 1997

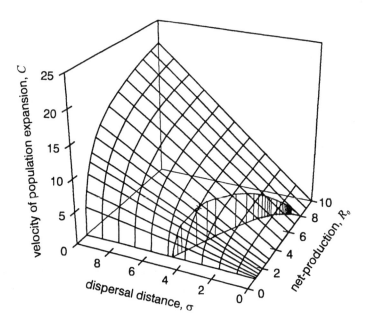

Figure 1: Invasion rate C plotted as a function of both net reproduction rate R_o and dispersal parameter σ. The increasing function results for invasions happening under uniform conditions. C shows a maximum for non-uniform conditions with, in this case, $R_o = 8$ and $\psi/\omega = 0.01$.

EXTENDING THE INVASION MODEL

Assume that the animal or plant moves at random. The location of individuals that have not yet settled around their birth place at time τ after their birth follows an expanding Gaussian distribution with a variance that increases linearly with time (Okubo 1980).

$$\frac{1}{2\pi\omega\tau}\exp\left(-\frac{x_1^2 + x_2^2}{2\omega\tau}\right) \tag{2}$$

Here, x_1 and x_2 are spatial coordinates, and ω is the rate of increase of the variance. A certain fraction of the total area to be invaded is assumed to be favourable to the species concerned. Thus, we introduce the parameter δ that expresses this fraction. However, as the remainder of the area is unfavourable, individuals run a risk of dying before they actually arrive at a suitable site and subsequently reproduce in it. Therefore, by making the landscape ecologically heterogeneous, we also have to allow for a certain mortality rate during dispersal, ψ. The rate of settlement is the product $\psi\delta$ of the rate of increase of the variance, ω, measuring the area traversed per time unit τ, and δ. The time it takes to settle in the area is then exponentially distributed as:

$$(\omega\delta + \psi)\exp(-(\omega\delta + \psi)\tau) \tag{3}$$

For later use, we introduce the ratio ψ/ω which defines the risk of dispersal in non-uniform areas, small values implying that the area is safe for the species and high ones that invading it is hazardous and incurs a great risk. Areas can be safe either because the rate of increase of the variance of dispersion of individuals around the origin (ω) is large – they can spread easily, or because the mortality rate ψ during the invasion is small, or because of a combination of both rates.

The final distribution of settled individuals results from the multiplication of equations (2) and (3) and integrating with respect to τ:

$$D(x_1, x_2) = \frac{\omega\delta + \psi}{2\pi\omega} \int_0^\infty \frac{1}{\tau}\exp\left(-\tau - \frac{x_1^2 + x_2^2}{2\tau\omega/(\omega\delta + \psi)}\right)d\tau \tag{4}$$

It can be shown (Van den Bosch et al. 1988) that:

$$\sigma = \frac{1}{2}\frac{1}{\sqrt{\delta + \psi/\omega}} \tag{5}$$

Having thus defined the parameters and equations necessary for invasions or range expansions into ecologically heterogeneous areas, we can now extend the model formulated by Van den Bosch et al. (1992).

We define $\hat{R}_o$ as the maximal R_o when net reproduction is achieved after settlement, directly after the individual is born. This means that no mortality occurs between birth and settlement with reproduction. In reality, though, only a fraction of the young arrives alive at a suitable breeding place. This fraction equals $\dfrac{\omega\delta}{\omega\delta+\psi}$. Therefore, the realised net reproduction rate is:

$$R_0 = \frac{\delta}{\delta+\psi/\omega}\,\hat{R}_0 \qquad (6)$$

The plane perpendicular to the R_0 -σ plane in Figure 1 shows the relationship between R_0, σ and C as a function of the fraction of habitable sites, δ. To obtain this relationship, equation (6) is used, with $\hat{R}_0 = 8$ and $\psi/\omega = 0\cdot01$. Substituting (5) and (6) in (1), we find

$$C = \frac{1}{2\mu}\,\frac{1}{\sqrt{(\delta+\psi/\omega)}}\sqrt{2\ln\!\left(\frac{\delta}{\delta+\psi/\omega}\,\hat{R}_0\right)} \qquad (7)$$

Figures 2a and b show the dependence of C on the fraction of suitable habitat, δ, for various values of R_0 and ψ/ω. In both cases, the velocity of population expansion – or

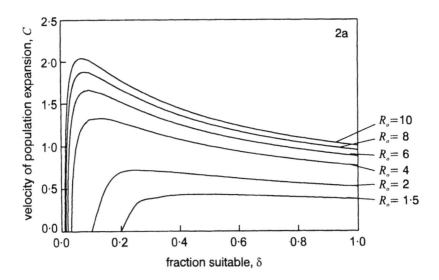

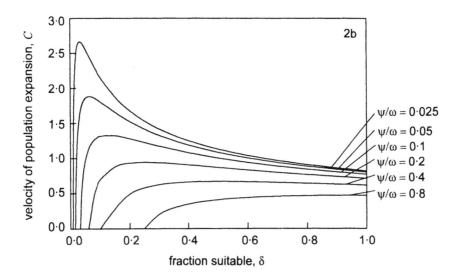

Figure 2: Invasion rate C as a function of the fraction suitable of habitat in non-uniform conditions for various values of R_o The maximum shifts to the left (less suitable habitat) for higher reproduction rates (Figure 2a – facing page). For higher dispersal risk ψ/ω, the maximum shifts to the right (more suitable habitat (Figure 2b – above)).

the species' invasion rate – shows a maximum. Obviously, the rate of invasion is highest for high values of the net reproduction rate R_o in Figure 2a, as well as for the lowest dispersal risk, ψ/ω, in Figure 2b. (The values of δ for which C is maximal shift in both cases from high to low values of R_o and from low to high values of ψ/ω (Figure 3).)

Figure 3 shows the location of these maxima for different values of the dispersal risk ψ/ω, the net reproduction rate R_0 and the fraction of suitable area δ.

Next, we are interested in the parameter combinations for which the expansion velocity is maximal. The location of this maximal invasion rate C is calculated from equation (7) to be

$$\hat{R}_0 = \frac{\delta + \psi/\omega}{\delta} \exp\left(\frac{\psi/\omega}{\delta}\right) \tag{8}$$

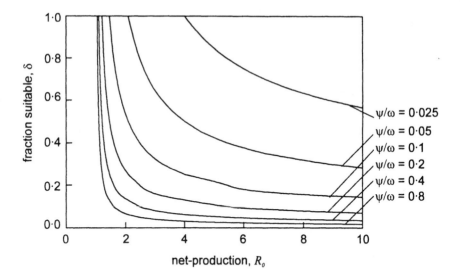

Figure 3: Maximal invasion rates C as a function of the fraction of suitable habitat, δ, net reproduction rate, R_o, and dispersal risk, ψ/ω.

Figure 4 shows the location of these maxima for different values of the dispersal risk ψ/ω, the net reproduction rate R_o, and the fraction of suitable area δ.

It should be clear, however, that the existence of this maximum pertains to the rate of spread of a population into a new area; variation in this rate is determined by the degree of homogeneity of the ecological conditions. It is therefore connected with a process occurring in space. It does not occur in the intrinsic rate of population increase, r, which is non-spatial. For uniform conditions Van den Bosch *et al.* (1990) defined this parameter:

$$r = \frac{\ln(R_0)}{\mu} \tag{9}$$

whereas for the patchily distributed favourable conditions it becomes

$$r = \frac{1}{\mu} \ln\left(\frac{\delta}{\delta + \psi/\omega} \hat{R}_0\right) \tag{10}$$

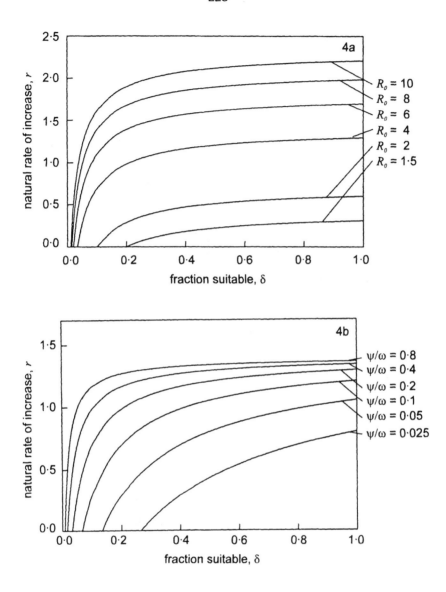

Figure 4: Rate of population growth, r, as a function of suitable habitat δ, net reproduction rate, R_0 (Figure 4a), and dispersal risk, ψ/ω (Figure 4b). In contrast to the curves in Figure 2 concerning spatial spread, this non-spatial measure of population growth does not show maxima.

Figures 4a and b show the relationships between r and the same parameter combinations as in Figures 2a and b. These relationships do not have any maximum value. Thus, the population size increases unhampered by the spatial characteristics of the ecological conditions; these characteristics affect only its rate of spread.

DISCUSSION

Figures 1 and 2 show that the differentiation of a landscape into favourable and unfavourable patches does not merely slow the rate of invasion down to a certain extent, but that the invasion rate shows a maximum. This maximum value depends on several parameters of both the environment and the species' demography. It is generated when, on the one hand, the fraction of habitable area, δ, is low, implying a high mortality, whereas on the other hand, a high value of δ implies that the propagules spread over short distances only. Intermediate values of δ result, as a consequence, in the highest invasion rates.

High values of ψ/ω, expressing the species' dispersal risk, result in the highest invasion rates when the fraction of habitable area δ is high. On the other hand, the invasion rate C can also be maximal in areas with little habitable area – a small value of δ – when the net rate of reproduction R_0 is high.

Thus, different values of the various parameters mentioned can either counteract each other, or they enhance each other, resulting not in a greater or smaller invasion rate under spatially heterogeneous conditions, but also in the non-linearity in C.

The consequence for a rapid expansion into a new area is that the species should combine a small net reproduction rate R_0 with a large fraction of suitable biotope, δ. Alternatively, δ should also be large for those species suffering a high mortality during dispersal.

Schroepfer and Engstfeld (1983) noticed that the muskrat, *Ondatra zibethicus*, invaded into Germany at variable rates. They felt that this could be due to two factors, the patchiness and linearity of their habitat, and the year-to-year variability of the weather and soil conditions. In wet areas and during wet years, the invasion would progress more slowly than in dry areas and during dry years. Their explanation was that this species allocates more of its time to reproduction under wet conditions, but to movement during dry ones. This would mean a general rise or decline of the invasion rate according to the wetness or the dryness of their environment. The results of the present model indicate that, looking at spatially heterogeneous conditions alone, matters may have been more complicated than this. Over time, the pattern of relationships between the parameter values may have complicated the processes even more. Without adequate information, it will be difficult, if at all possible, to reconstruct the expansion of this species at that time in Germany.

The same difficulties will be met in interpreting past processes of immigration into northern Eurasia and North America after the last glacial. Conversely, it will be difficult to predict the rate at which species might be able to follow the changing conditions under future climates.

CONCLUSIONS

The introduction of spatial heterogeneity in a species' living conditions does not necessarily enhance or retard its invasion into a new area, but results in a non-linear rate relative to those parameters. This means that for heterogeneous areas we need more information, either for reconstructing, or for predicting the rate of invasion or range expansion.

REFERENCES

Hengeveld R (1994) Small-step invasion research. TREE 9:339-342

Hengeveld R, Van den Bosch F (1991) The expansion velocity of the Collared Dove Streptopelia decaocto population in Europe. Ardea 79:67-72

Okubo A (1980) Diffusion and ecological problems: mathematical models. Springer, Berlin

Schroepfer R, Engstfeld C (1983) Die Ausbreitung des Bisams (Ondatra zibethicus Linné, 1766, Rodentia, Arvicolidae) in der Bundesrepublik Deutschland. Z angew Zool 70:13-37

Van den Bosch F, Hengeveld R, Metz JAJ (1992) Analysing the velocity of animal range expansion. J Biogeogr 19:135-150

Van den Bosch F, Metz JAJ, Diekmann O (1990) The velocity of population expansion. J Math Biol 28:529-565

Van den Bosch F, Zadoks JC, Metz JAJ (1988) Focus expansion in plant disease. II. Realistic parameter-sparse models. Phytopathol 78:59-64

Migratory birds and climate change

Peter R. Evans
Department of Biological Sciences
University of Durham
Science Laboratories
South Road
Durham DH1 3LE
U.K.

Of all terrestrial organisms, birds might be expected to respond most rapidly and appropriately to the rapid changes in global climate anticipated during the next century. As warm-blooded endotherms, they are unlikely to be affected directly by small increases in temperature or carbon dioxide concentrations, but they will need to respond to shifts in the spatial distributions of their foods, nesting habitats, parasites and predators - or to modify their feeding and nesting behaviours in relation to the changed availability of resources within their present ranges.

Most bird species are highly mobile over short distances and many migrants are capable of prolonged flights of several days' duration. (Migration is used here to signify the twice-yearly regular seasonal movements of individuals of certain species). The movements of migrants between breeding and non-breeding areas usually take place by a sequence of flight stages (varying in length amongst species) interrupted by short but crucial periods of refuelling. Loss or change in the location or seasonal availability of suitable refuelling areas could restrict the ability of some species to adapt to the effects of climate change. Furthermore, any increase in the width of ecological barriers (areas where refuelling is impossible) may also demand flight stages which exceed physiological limits. Additionally, changes in atmospheric circulation patterns associated with climate change could either hinder or help flight performance during migration; they might even allow species to colonize new areas by crossing hitherto effective ecological barriers. This, in turn, could lead to changes in the relative abundances of species in the assemblages ('communities') associated at present with particular habitats in particular continents, even if these bird

NATO ASI Series, Vol. I 47
Past and Future Rapid Environmental Changes:
The Spatial and Evolutionary Responses of
Terrestrial Biota
Edited by Brian Huntley et al.
© Springer-Verlag Berlin Heidelberg 1997

assemblages move their geographical ranges in association with the spatial shifts of key components of the vegetation and/or their foods, whether plant or animal.

This paper seeks to highlight some of the present ecological and physiological constraints on migrant land birds and focuses chiefly on their relevance to aspects of climate change other than global warming. It will not address the responses of 'resident' species, nor the shifts in their breeding ranges that have been documented this century, particularly in relation to the climatic amelioration in Europe in the 1920's and 1930's. These have been reviewed by Kalela (1949) and Merikallio (1951) for central and northern Europe, particularly Finland, Gudmundsson (1951) for Iceland and Williamson (1975) for the British Isles. Williamson was writing at a time when ornithologists in Britain were asking whether the return to colder winters and cooler summers in the 1950's and 1960's was heralding another long climatic recession. He made the important suggestion that extension of breeding range has followed climatic warming more quickly than contraction of range has followed subsequent cooling. Very recently, Huntley (1995) has outlined the implications for several bird species of the major shifts in vegetational zones that are expected in the 21st century.

Berthold (1991) considered the possible effects of climate warming on songbird species breeding in central Europe. He argued that resident breeding species are likely to become more productive and abundant, unless acid rain or other pollutants negate this, and that short-distance migrants are likely to become residents. (Productivity may not, of course, improve if warming is confined chiefly to the winter months.) Berthold also argued that long-distance migrants that have spent the non-breeding season in the tropics would be out-competed when they returned to Europe by the more abundant residents and would decline. This may be true for central Europe but assumes that migrant species are unable to extend their migrations northwards, or that the costs of doing so would outweigh the benefits gained thereby.

MIGRATION STRATEGIES

Migration presumably has evolved and persists because individuals that perform such behaviour increase their potential lifetime reproductive success. The benefit of migration to higher latitudes to breed may be higher productivity resulting (in some species) from larger clutch sizes and from longer hours of daylight in which food can be sought for, or by, chicks of songbirds and shorebirds, respectively. Resident predators also tend to be less numerous and diverse at high latitudes (Larson 1960; Pienkowski 1984). Migration to lower latitudes, which have longer days during the non-breeding season, improves an individual's chances of survival between nesting seasons. Such areas may be perfectly adequate for survival throughout the year; one-, two- or three-year-old individuals of species showing delayed maturity e.g. White Stork *Ciconia ciconia*, often stay throughout the normal breeding season on their 'wintering' areas; some have even bred there (Moreau 1966).

The costs of migration include:

1. The energetic costs related to flight;

2. The risks associated with navigational errors and encountering adverse weather during flight or on arrival at the destination;

3. Increased risks of predation associated with increases in feeding rate achieved at the expense of vigilance (Metcalfe & Furness 1984), or in duration of feeding (Zwarts *et al.* 1990), needed to permit gains in mass before migration, and at refuelling sites (heavier birds may also be more at risk of predation (Witter & Cuthill 1993));

4. Risks associated with the need to visit sites, on a seasonal basis, where the locations of food, competitors and predators may change from year to year and certainly will not be known to juveniles on their first migrations (Pienkowski & Evans 1984, 1985).

In birds, the energetic costs vary with flight speed, following a U-shaped power curve (Pennycuick 1969). Flight to achieve maximum distance for a given fuel load requires a higher speed than that associated with the minimum rate of energy expenditure. The main fuel is fat, which is stored in almost anhydrous form.

Two contrasting migration strategies – energy minimization and time minimization – have been identified by Alerstam and Lindström (1990), though both may be modified by the degree of predation risk during the migratory journey. Energy minimizers should fly at such speeds that they maximise the distance flown for each gram of fuel used. Ideally, they should migrate by a series of short stages, each requiring only a small quantity of fuel. In this way, the fuel is used chiefly to power the forward flight of the almost lean bird rather than to cover the cost of transportation of the fat load. In contrast, time-minimizers should seek to minimize the overall journey time from departure to destination, including the periods of refuelling and the time needed to climb from ground level to the cruising altitude at the start of each flight stage. Since birds may take a few days to adjust to the feeding conditions provided by a new refuelling site before fat deposition can begin, Alerstam and Lindström (1990) argued that time-minimizers should reduce the number of such sites they use and should select only those where they can deposit fat quickly (i.e. those with plentiful resources of appropriate foods). Hence they should use a few long flight stages.

Gudmundsson et al. (1991) demonstrated graphically that time-minimizers should sometimes lay down more fat ('overload') at a refuelling site than they require to reach the next stopover site, or should overfly suitable refuelling sites. An analytical model of these phenomena, developed by Weber et al. (1994), identified some of the conditions in which they should occur, e.g. if sites decrease in quality in the direction of migration. They also considered year-to-year variability in the quality of a site and the implications of this for overloading and overflying.

The degree to which individual species of migrants can be classified as wholly energy- or time-minimizers is not clear at present; nor are all the circumstances under which one or other behaviour would be expected. Indeed, the migratory journey is only one of the elements that affects the lifetime reproductive success and hence fitness of a migrant. Others include the timings of arrival on breeding and non-breeding areas in relation to seasonally varying food resources. More general models of migration, concerned with maximizing overall fitness, are being developed (Weber et al., in press).

CLIMATE CHANGE, FLIGHT STAGES AND REFUELLING

Refuelling areas may be highly localized, as in the case of some wetlands, or cover large areas, as in the case of woodlands with their associated resources of seeds, fruits and insects. The effects of climate change on the location of extensive habitats and their quality as refuelling areas are unlikely to demand major changes in the tactics of energy-minimizing migrants. However, loss or change in the location of scarce but important refuelling habitats could require major reorganization of migratory routes, particularly of time-minimizing migrants that depend upon few but highly resource-rich sites. Changes in rainfall patterns may thus be of great significance.

It has been known for some years that Purple Heron *Ardea purpurea* breeding populations in Europe fluctuate in parallel with the degree of rainfall in the Niger inundation zone in West Africa (Cavé 1983). Recently Peach *et al.* (1991) have shown that the survival of British breeding Sedge Warblers *Acrocephalus schoenobaenus* is also correlated with winter rainfall in that part of Africa. In the 1970's, breeding populations in Britain of several songbird species declined markedly in association with prolonged droughts in the Sahel zone, where they were thought to 'winter' (Winstanley *et al.* 1974). What is not known in all these examples is whether mortality increased on the non-breeding areas in Africa, or on the trans-Saharan migration because birds departed with insufficient fuel. In the case of the Sedge Warblers, recoveries of ringed birds suggest that the Niger inundation zone is used as a refuelling site rather than (or as well as) a 'wintering' area.

Timing of migration may also need to change, but this may not always be possible. Migrants should reach breeding areas in the arctic before, or at least by, the optimal date for egg-laying. If climatic warming advances this optimal date at high latitudes, but leaves conditions in the tropics largely unchanged, some species may not be able to alter the timing of departure from the tropics in an appropriate way. For example, a recent study by Zwarts (1990) has established that Whimbrels *Numenius phaeopus*, sub-arctic breeders which 'winter' on the Banc d'Arguin, Mauritania, cannot begin to store fuel until late March/early April, because only then does their chief prey, the Fiddler Crab *Uca tangeri*, become active and available for a sufficient number of hours in each 24 to permit the Whimbrels' food intake rate to rise to cover fat deposition as well as maintenance metabolism. The timing of availability of prey at refuelling sites may also be important (Lester & Myers 1991). The mass spawning of King Crabs *Limulus* at Delaware Bay in the eastern U.S.A. occurs now in early May, at precisely the time when many species of migrant shorebirds are passing through in spring, *en route* for the arctic from non-breeding areas in South America. These shorebirds are able to gorge themselves on crab eggs and achieve very high rates of fat deposition, as required by time-minimizing migrants. The implications of climate change could be serious if they led to desynchronization in the relative timings of breeding and of availability of important foods on the migration routes of time-minimizing birds. Warming of sub-arctic and arctic areas may relax some of the time constraints presently operating on the length of bird breeding seasons, particularly those of herbivores such as geese. It is less clear what changes might occur in the timing and synchrony of emergence of adult insects, which form the main food of shorebird and songbird chicks at high latitudes. If these emerge earlier in a warmer arctic but the shorebirds cannot leave their 'wintering' areas earlier they would arrive too late to capitalize on their food supplies.

Many long-distance migrants, even if not time-minimizers, have their migration schedules controlled largely by endogenous circannual rhythms (Gwinner 1990). The extents to which such time programmes are modifiable by external and internal factors are incompletely known at present. If a migrant fails to accumulate the usual amount of fat for the next stage of its journey by the normal date of departure, does it delay departure or leave with lower than normal fuel reserves? Presumably this depends in part on the tightness of its breeding schedule. Knot *Calidris canutus* leave a refuelling site in north Norway, bound for Greenland, during a 3-4 day period at the end of May. Birds caught from a final group to depart had body masses averaging 20 g (10%) less than those leaving on the previous day (Evans 1992), which suggests some delay but not a major rescheduling. The implications of departure by Knot at body masses below normal is not known, since these birds normally arrive on their breeding areas with considerable fat reserves unused. This may be a strategy to ensure survival in adverse weather e.g. late snow cover on the breeding sites. Hence, if snow melt occurs earlier in response to global warming, the need for a survival 'insurance' of fat will decrease.

Although departure from the final staging post on migration may be tightly time-programmed, those from earlier refuelling sites may be less severely constrained. Dunlin *Calidris alpina* stayed longer at a staging area in Morocco in a spring when food abundance was low than in the following spring when food was more plentiful (Piersma 1987). It is not known what fat loads were accumulated in the two springs. The possibility of departure with below-optimal loads is particularly important in relation to the crossing of ecological barriers.

CROSSING ECOLOGICAL BARRIERS - LIMITATIONS TO FLIGHT RANGES

Time-minimizing migrants would be expected to fly at speeds greater than the speed which permits maximum range for a given fuel load. Such a risky strategy is unlikely to be used for flights across extensive ecological barriers unless wind-assistance can be relied upon. For a given species, the maximum distance a bird can travel non-stop is determined by the proportion that fat forms of total body mass at take-off. Large species have lean body masses that are closer to their maximum possible take-off masses than small species have; hence, even though maximum-range speed increases with mass, they cannot store such large fat loads, proportionately, and so cannot usually cross such wide ecological barriers as can small species, unless they use energy-sparing soaring flight instead of flapping flight. But thermal soaring is a slow mode of travel and rarely can be used over water. Smaller species that flap can also gain assistance from following winds.

In recent years, the debate has been re-opened as to whether maximum range is limited only by fuel or alternatively by dehydration. If migrants fly higher than the altitude at which air temperature falls below about 10°C, they may be able to dissipate most of the heat produced by flapping flight without recourse to extensive evaporation of body water. At higher air temperatures, with low humidities, water loss could exceed metabolic water production (from oxidation of fat). Indeed, recent models (see review by Klaasen 1995) suggest that dehydration could limit flight duration and hence flight range of migrant species crossing ecological barriers. Biebach (1991) found that, in autumn, migrants crossing the Sahara set out with insufficient fat to complete the journey without the assistance of following winds which, during the day, are found only at relatively low altitudes, where the air is warm and humidity low. Some small birds were found to land at dawn, spend the day in shade and fly only by night, thereby avoiding excessive dehydration. Exhausted migrants nearing the end of the desert crossing had used up their fat but not water.

Wind directions often change seasonally. Moreau (1972) suggested that the reason why many songbirds migrate from Europe to West Africa through Iberia in autumn, but return by a more easterly route in spring, is to avoid headwinds over the Sahara, especially in spring. More recently Piersma and Jukema (1990) have established the importance of wind assistance to the northward migration of Bar-tailed Godwits *Limosa lapponica* from Mauritania to a refuelling area in the Dutch Wadden Sea in spring. This 4000 km non-stop journey requires the migrants to climb to high altitudes to avoid headwinds in the first part of their flight.

The implications of climate change for these long flights over water or desert are several. Many species may be carrying close to their maximum fuel loads at take-off at present. Any increase in the width of deserts could be critical; disappearance of the well-vegetated Maghreb zone across North Africa could prevent refuelling in cases of exhaustion of fat reserves. Changes in circulation patterns of major air masses could alter prevailing wind directions and alter altitude profiles of winds, so that they are more or less favourable to the directions of movement followed by migrants at present. Changes in humidity profiles with altitude could alter the relative importance of fuel exhaustion and dehydration in limiting flight range.

Special attention needs to be drawn to the vulnerability of species using wetlands, either to prepare for migration or as refuelling sites. Without increased precision in the forecasts of changes in rainfall patterns with general climate change, it is impossible to assess how far present networks of inland wetland refuelling areas for migrants will be affected. It is clear, however, that coastal habitats, especially mud- and sandflats and salt marshes, are likely to become severely reduced in area through sea-level rise. This will become a particular problem in countries where centres of human population have been allowed to develop close to the coasts, since these are usually protected by extensive built sea-defences which prevent the movement of the saltmarsh and intertidal zones inland in response to sea-level rise. Thus for shorebirds, many of which breed at high latitudes and whose behaviour is consistent with them being time-minimizing migrants (Alerstam & Lindström 1990 – Figure 9) requiring high quality staging areas, refuelling sites are likely to become scarcer. Such habitats have already been lost extensively to human activities (Evans 1991).

COLONIZATIONS

New centres of breeding of migratory species, as a result of suspension of normal migration patterns by some individuals, have been documented in both Africa (Moreau 1966) and the Americas (Leck 1980). These have not involved changes in overall geographical range. In contrast, colonizations of new areas by range expansion of normally resident species have occurred both naturally, e.g. the westwards spread of the Collared Dove *Streptopelia decaocto* in Europe, and by human assistance, e.g. liberation of European songbirds in Australia in the early 19th century (Hudson 1985). Such colonizations can sometimes lead to conflicts with elements of pre-existing avifaunas, e.g. competition for nest-holes in trees between introduced Starlings *Sturnus vulgaris* and native woodpecker species in the U.S.A.. Thus colonizations of new areas by changes in migration patterns, whether gradual or abrupt, related to climate change, may also be expected to alter at least the relative abundances, if not also the species composition, of the avifauna in the receiving areas. Examples of range expansions related to changes in weather in the short term and over several decades include the following.

The phenomenon of 'over-shooting' of normal breeding ranges in northern Europe by migrants returning from Africa in the spring has been documented regularly. This has normally been associated with unusually warm weather during the typical migration period for a species, often associated with an anticyclone over central Europe bringing southerly winds. If such meteorological conditions occur more frequently in future springs, changes of range could be rapid.

Williamson (1975) reviewed evidence that the position of the storm tracks across the north Atlantic had shifted by between 5° and 10° of latitude at various times during the 20th century, and that the intensity of the airflows between the Azores anticyclone and the Atlantic depressions had also varied. Building on this, Elkins (1979) explained the increasing frequency with which American songbird migrants have reached Europe in autumn as a result of wave depressions moving along a more northerly positioned tropical/polar frontal system, intercepting birds flying over the ocean direct from the eastern states of the U.S.A. to South America and sweeping them eastwards across the Atlantic. (The 'traditional' explanation of migrants caught up in hurricanes is dismissed because the birds would have to remain airborne for too long.) Some of these migrants have survived the European winters; if future winter temperatures are likely to be higher, presumably more will do so and breeding populations may become established. This would parallel the case of Fieldfare *Turdus pilaris*, swept off course by a vigorous depression in the north Atlantic in January 1937, when migrating from Scandinavia to Britain in cold weather, and carried to southwest Greenland where the climatic amelioration of the previous decades allowed them to survive and establish a breeding population (Salomonsen 1951).

At present, it is likely that the very restricted variety of land-birds found breeding in western Greenland and Ellesmere Island is a consequence of the lack of refuelling sites within the maximum flight range of all but relatively small species (Alerstam *et al.* 1986). Increased wind assistance or opening of new staging posts in a warmer arctic could change this.

CONCLUSION

Migration patterns may be affected directly and significantly by changes in atmospheric circulation and rainfall patterns, as well as indirectly by global changes in summer and winter temperatures. Species depending upon wetlands as refuelling sites are likely to be particularly at risk. New patterns, new breeding ranges and colonization from presently distant bird faunas could have implications for the abilities of other taxa to adapt to climate change, because birds are important dispersers of plant seeds, freshwater molluscs, ticks and other ectoparasites, as well as vectors of pathogenic organisms.

REFERENCES

Alerstam T, Hjort C, Högstedt G, Jönsson PE, Karlsson J, Larsson B (1986) Spring migration of birds across the Greenland Inlandice. Medd. om Grønland, Bioscience 21:1-38

Alerstam T, Lindström A (1990) Optimal bird migration: the relative importance of time, energy and safety. *in* Gwinner E (ed) Bird migration: the physiology and ecophysiology, 331-351. Springer, New York

Berthold P (1991) Patterns of avian migration in light of current global "greenhouse" effects: a central European perspective. Proc Int Ornith Cong 20:780-786

Biebach H (1991) Is water or energy crucial for trans-Saharan migrants? Proc Int Ornith Cong 20:773-779

Cavé AJ (1983) Purple Heron survival and drought in tropical west Africa. Ardea 71:217-224.

Elkins N (1979) Nearctic landbirds in Britain and Ireland - a meteorological analysis. British Birds 72:417-433

Evans PR (1991) Introductory remarks: habitat loss-effects on shorebird populations. Proc Int Ornith Cong 20:2197-2198

Evans PR (1992) The use of Balsfjord, north Norway, as a staging post by Knot during spring migration. *in* Piersma T, Davidson NC (eds) The migration of Knots, 126-128. Wader Study Group Bull 64 Supplement

Gudmundsson F (1951) The effects of the recent climatic changes on the bird life of Iceland. Proc Int Ornith Cong 10:502-514

Gudmundsson GA, Lindström A, Alerstam T (1991) Optimal fat loads and long-distance flights by migrating knots, sanderlings and turnstones. Ibis 133:140-152

Gwinner E (1990) Bird migration: the physiology and ecophysiology. Springer, New York

Hudson RW (1985) Range changes. in Campbell B, Lack E (eds) A Dictionary of Birds. Poyser Calton

Huntley B (1995) Plant species' response to climate change: implications for the conservation of European birds. Ibis 117:S127-138

Kalela O (1949) Changes in geographic ranges in the avifauna of northern and central Europe in relation to recent changes in climate. Bird Banding 201:77-103

Klaasen M (1995) Water and energy limitation of flight range. Auk 112:

Larson S (1960) On the influence of the Arctic Fox *Alopex lagopus* on the distribution of Arctic birds. Oikos 11:276-305

Leck CF (1980) Establishment of new population centres with changes in migration patterns. J Field Ornithol 51:168-173

Lester RT, Myers JP (1991) Double jeopardy for migrating wildlife. in Wyman RL (ed) Global Climate Change and Life on Earth. Routledge Chapman and Hall, New York

Merikallio E (1951) Der Einfluss der letzten Wärmeperiode (1930-49) auf die Vogelfauna Nordfinnlands. Proc Int Ornith Cong 10:484-493

Metcalfe NB, Furness RW (1984) Changing priorities: the effect of pre-migratory fattening on the trade-off between foraging and vigilance. Behav Ecol Sociobiol 15:203-206

Moreau RE (1966) The Bird Faunas of Africa and its Islands. Academic Press, London

Moreau RE (1972) The Palaearctic-African Bird Migration Systems. Academic Press, London

Peach W, Baillie SR, Underhill LG (1991) Survival of British Sedge Warblers *Acrocephalus schoenobaenus* in relation to west African rainfall. Ibis 133:300-305

Pennycuick C (1969) The mechanics of bird migration. Ibis 111:525-556

Pienkowski MW (1984) Breeding biology and population dynamics of Ringed Plovers *Charadrius hiaticula* in Britain and Greenland. J Zool Lond 202:83-114

Pienkowski MW, Evans PR, (1984) Migratory behaviour in the western Palaearctic. in Burger J, Olla B (eds) Behaviour of Marine Animals Vol. 6. Shorebirds. Plenum Press, New York

Pienkowski MW, Evans PR (1985) The role of migration in the population dynamics of birds. in Sibly RM, Smith RH (eds) Behavioural Ecology. Symp Brit Ecol Soc 25. Blackwells, Oxford

Piersma T (1987) Hink, stap of sprong? Reisbeperkingen van arctische steltlopers. Limosa 60:185-191

Piersma T, Jukema J (1990) Budgeting the flight of a long-distance migrant. Ardea 78:315-337

Salomonsen F (1951) The immigration and breeding of the Fieldfare in Greenland. Proc Int Ornith Cong 10:515-526

Weber TP, Houston AI, Ens BJ (1994) Optimal departure fat loads and stopover site use in avian migration; an analytical model. Proc Roy Soc Lond B 258:29-34

Weber TP, Ens BJ, Houston AI (in press) Optimal avian migration: a dynamic model of reserve gain and site use. Evolutionary Ecol

Williamson K (1975) Birds and climatic change. Bird Study 22:143-164

Winstanley D, Spencer R, Williamson K (1974) Where have all the Whitethroats gone? Bird Study 21:1-14

Witter MS, Cuthill IC (1993) The ecological cost of avian fat storage. Phil Trans Roy Soc B 340:73-92

Zwarts L (1990) Increased prey availability drives premigration hyperphagia in Whimbrels and allows them to leave the Banc d'Arguin, Mauritania, in time. Ardea 78:279-300

Zwarts L, Blomert A-M, Hupkes R (1990) Increase of feeding time in waders preparing for spring migration from the Banc d'Arguin, Mauritania. Ardea 78:237-256

Tree demography and migration: What stand level measurements can tell about the response of forests to climate change

Dale S. Solomon, William B. Leak and David Y. Hollinger
USDA Forest Service
Northeastern Forest Experiment Station
P.O. Box 640
Durham, NH 03824
U.S.A.

INTRODUCTION

Renewed interest in plant migration has developed due to the potential impacts of global climate change. Predictions from general circulation models and bog/lake pollen analyses suggest major shifts in species ranges, and even extinctions, over the next 50 to 200 yr in response to temperature shifts of up to 4·5°C (e.g., Davis & Zabinski 1992; Overpeck *et al.* 1991; Pastor & Post 1988; Peters 1990). Using computer simulation, Overpeck *et al.* (1991) predicted shifts in plant ranges of as much as 500 to 1000 km within periods as short as 200 yr. Others (e.g. Davis & Zabinski 1992) suggest the possibility of near extinction of species such as sugar maple *Acer saccharum* Marsh. and American beech *Fagus grandifolia* Ehrh. because establishment in newly suitable sites would fail to keep pace with a changing climate. In developing simulation models, relationships between species' occurrence and climatic variables are developed or inferred using modern or historical data (Denton & Barnes 1987; Gajewski 1987; Prentice *et al.* 1991; Spear *et al.* 1994). Future ranges are then predicted assuming a strong relationship between species and climate, which can be compared to estimated migration rates by species as evidenced in the pollen record.

Trends in important climate variables, such as temperature or precipitation, may vary widely across a region complicating the broad scenarios based on present species

NATO ASI Series, Vol. I 47
Past and Future Rapid Environmental Changes:
The Spatial and Evolutionary Responses of
Terrestrial Biota
Edited by Brian Huntley et al.
© Springer-Verlag Berlin Heidelberg 1997

distributions. A spatial interpolation of the regression-derived slopes (corrected for auto-correlation) of the historical temperature records of the northeastern United States (Figure 1) provides an example of the complexity of climate trends. Looking at data arbitrarily truncated to the last 100 yr of record suggests that mean annual and summer temperatures have generally increased, especially along the coast (calculated from Vose *et al.* 1992). The most recent 50 yr, however, are dominated by a slight cooling trend in the interior, but a warming along the coast. The spatial variability in the record should caution strongly against the extrapolation of a single climate station record to represent a region. Figure 1 also demonstrates another dilemma when interpreting species' distribution changes based on climate trends: selecting the appropriate time interval and climatic measure. Clearly, for recent changes the latest trends (e.g. Figure 1b & 1d) are more relevant than longer-term trends. We also argue that growing season (summer) temperatures are more likely relevant to the physiological processes affecting germination and growth than mean annual temperatures. However, minimum winter temperatures are also known to affect tree growth and survival (Federer *et al.* 1989).

Although some field studies have suggested actual changes in the elevational distribution of red spruce *Picea rubens* Sarg. and other species due to temperature and climate change (Grabherr *et al.* 1994; Hamburg & Cogbill 1988), other studies show little or no detectable response (Leak & Smith in press; Solomon & Leak 1994) that could be attributed to changes in climatic factors. This paper attempts to address this inconsistency by an examination of the various approaches used to measure, detect or predict species migration: succession, resurveys, remeasured plots and age/distance trends.

Figure 1. Interpolation of temperature trends of New england, U.S.A. area climate stations. Station locations shown by circles. a – Annual trend 1887 – 1987 (n = 42 climate stations). b – Annual trend 1937 – 1987 (n = 97). c – Summer (June, July, August) trend 1887 – 1987 (n = 56). d – Summer trend 1937 – 1987 n = 100).

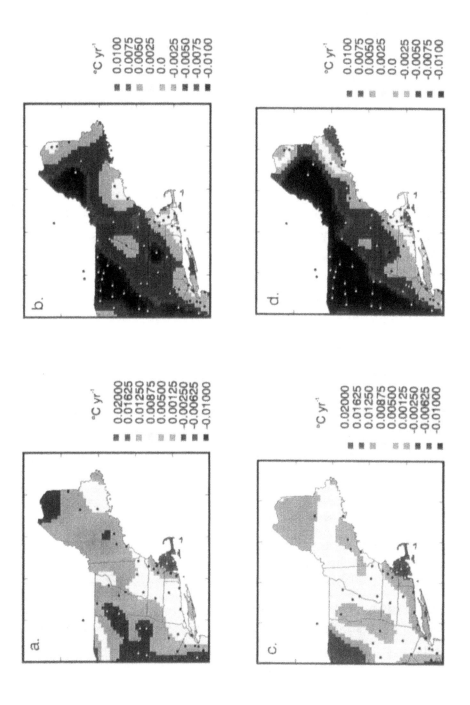

SUCCESSION

The term migration means the movement of a species into a region where it has not recently been able to grow because of climatic or edaphic conditions. Succession, on the other hand, refers to the appearance or changing dominance of a species within a region where the species naturally occurs. Following the terminology of Colinvaux (1986), succession may be either autogenic (succession driven by factors within the community) or allogenic (driven by external factors). To detect the influence of climate change, we need indicators of allogenic succession.

In the northeastern United States, successional change commonly follows Egler's (1954) initial floristic model (Bormann & Likens 1979) where the successional process consists of changes in species dominance over time rather than replacement of one unique community by another. Under autogenic succession, we would anticipate that the original climax forest would replace itself. Under allogenic succession driven by presently anticipated climate change, we would expect that successional change would trend towards increased proportions of species suited to warmer climates.

In addition to remeasurements or resurveys over time, anticipated successional direction may be inferred from one-time surveys of species composition by diameter or age class. In northeastern forests, where exogenous disturbance from fire or windthrow is minimal, future species composition is almost completely determined by the advanced regeneration of shade-tolerant or semi-tolerant species (Leak & Wilson 1958). Furthermore, all tree species in this region have survivorship curves that approximate the negative exponential, or one of its near relatives, such as a rotated sigmoid or power function (e.g. Hett 1971). It follows that the species mix in lower diameter or age classes will be indicative of the future species mix of the overstorey. A similar argument holds for even-aged stands: the commonality in survivorship curves implies that the species mix in young stands will reflect the species mix in later-successional stands. For example, 60-yr records from Connecticut stands (Stephens & Ward 1992) shows that red maple *Acer rubrum* L. comprised 14·7 percent of the basal area in 20- to 40-yr old stands in 1927, and 15·3 percent of the basal area in 1987. The comparable figures for yellow birch *Betula alleghaniensis* Britton were 20·5 and 24·1, respectively.

Examination of species percentages (based on numbers of stems per species per diameter size class vs. total stems per size class) from the 1982 – 83 Forest Surveys of Maine and New Hampshire, U.S.A., showed marked differences in species composition among stem size classes ranging from seedlings to mature trees (Table 1) (Frieswyk & Malley 1985; Powell & Dickson 1984). In both states, balsam fir *Abies balsamea* shows the strongest tendency toward successional increase, based on the high percentage of seedlings and saplings relative to the percentage of balsam fir in one or more of the larger size classes. Based on elevational distribution (Figure 2), balsam fir is one of the more boreal, high-elevational species in the region. On the other hand, white pine *Pinus strobus* and hemlock *Tsuga canadensis* (both lowland species generally occurring below the elevations in Figure 2), show consistently lower percentages of small stems as compared to large

Table 1: Successional trends of forest species in New Hampshire (NH) and Maine (ME), U.S.A. (not all species listed): percent of stem numbers in 1982 – 83 by size class (from Frieswyk & Malley 1985 and Powell & Dickson 1984).

Species	Seedlings – 7·4 cm		7·5 – 22·6 cm		22·7 – 73·7+ cm	
	NH	ME	NH	ME	NH	ME
Balsam fir *Abies balsamea*	13·7	26·5	16·1	26·7	5·6	12·2
Red spruce *Picea rubens*	5·0	5·8	8·0	13·9	6·1	18·0
White pine *Pinus strobus*	3·2	0·7	7·3	2·6	18·2	6·8
Hemlock *Tsuga canadensis*	3·0	1·6	5·9	3·7	8·3	6·5
Red maple *Acer rubrum*	12·8	10·9	19·3	10·3	15·0	9·5
Sugar maple *Acer saccharum*	11·0	9·6	4·9	3·3	7·9	6·0
Yellow birch *Betula alleghaniensis*	3·7	5·2	5·2	2·3	6·3	5·5
Beech *Fagus grandifolia*	6·7	4·6	4·2	4·1	5·6	5·2
Red oak *Quercus rubra*	4·2	0·8	3·9	1·2	6·7	1·6

stems. Other species reflect somewhat stable or variable trends. Although many other factors must be considered, such as disturbance patterns referred to elsewhere in this paper, the successional trends depicted in Table 1 provide support for a recent cooling trend in New Hampshire and Maine.

RESURVEYS

Under this topic, we refer to remeasurements where the initial plots cannot be relocated. This type of information might include resurveys of general areas where earlier survey data are available (e.g. Grabherr *et al.* 1994) or surveys representing a sequence over time in comparable (but spatially distinct) locations (e.g. Hamburg & Cogbill 1988). Resurveys have at least two special problems. First, species/area considerations make it necessary to use the same sampling protocol at each inventory

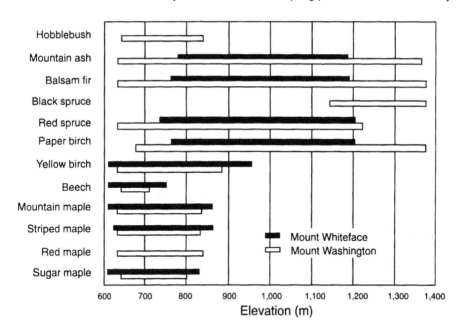

Figure 2. Elevational ranges of woody species on Mount Whiteface and Mount Washington, based on occurrence of stems over 1 year old. Transects began at 610 and 632 m above sea level and ended at 1,201 and 1,373 m respectively (from Leak & Graber 1974a).

so that the appearance or loss of species does not simply reflect a change in methodology. Second, in areas such as New England, with high variability in environmental conditions on small spatial scales, imprecise identification of the original survey sites may complicate interpretations. In regions with large landscapes that are uniform in soil/climatic conditions, imprecise location of original plots is less problematical.

REMEASURED PLOTS

Remeasured permanent plots, while costly to install and laborious to maintain, provide an ideal way to monitor species migration, and to provide the data needed to test migrational models. One complicating factor is that existing vegetation, on a regional scale, may be responding to factors other than climate change. In New England, for example, one dominant factor is natural successional change following prior disturbances from agricultural clearing and logging (Solomon & Leak 1994). This factor could easily mask any tendencies toward climatically-driven changes in vegetation. A 24-year record on U.S. Department of Agriculture (USDA), Forest Service, Forest Inventory and Analysis (FIA) plots in Maine (approximately five hundred 0·08 ha plots), indicated that white pine and balsam fir were declining in both average latitude and elevation (Figure 3, Table 2). However, analysis of land-use patterns in the area suggested that these changes were due to the invasion of abandoned agricultural land by these species rather than a reflection of climate change.

Another approach for detecting migration with spatially-dispersed remeasured plots is to examine the inventory data within the context of the geographical pattern of recent climate change. For example, the more temperate white pine would be expected to be moving into plots in areas which have experienced a recent summer warming trend, whereas the more boreal balsam fir should be behaving in an opposite manner. We use the temperature trend contours from Figure 1d to divide plots into those experiencing warming or cooling trends (Figure 3). Because of the low frequency of new occurrences, data from the regions were amalgamated into two groups split by the $-0·0025°C \ yr^{-1}$ contour. The null hypothesis is that the frequency of appearance would not be related to the temperature trend of the plots. In the case

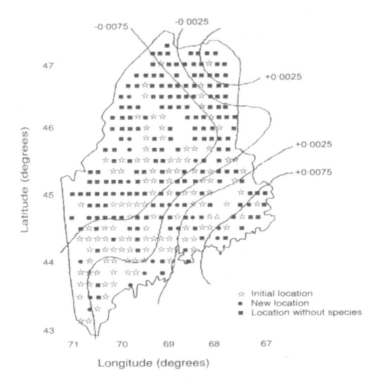

Figure 3: Initial locations in Maine with living white pine, new locations after an average 24-year period and locations without white pine at either point in time (from Solomon & Leak 1994). Contours show temperature trends from Figure 1d.

of white pine, chi-square values for new occurrences were significant at $p < 0.05 (\chi^2_{(1)} = 4.77)$ with the pine appearing (migrating) more frequently than expected into plots experiencing no temperature trend or a recent warming. However, there was no significant difference in the frequency of new occurrences of balsam fir in plots from which it had previously been absent that were warming or cooling $(\chi^2_{(1)} = 1.90)$. Although recent temperature changes may have influenced the movement of white pine, these results support our contention that other significant factors, such as response to disturbance, may affect species movement.

Local sets of long-term plot data exist in unmanaged or lightly managed conditions that allow for some assessment of both successional and migrational tendencies (Table 3). A 60–yr record (441 0·1ha plots) from the Bartlett Forest in New

Hampshire provided information on species changes related to management, disturbance, land type and elevation. On coniferous, unmanaged land types, eastern hemlock increased from 13% to 25% of the basal area in the 200–350 m elevational class, from 3% to 8% in the 500–650 m class, and from 1% to 3% in the 650–820 m class. The results indicate that hemlock shows only a slight tendency to

Table 2: Average elevation (m) and latitude and longitude (degrees) of Maine plots at three remeasurement periods (spanning 24-year average) from the 1950's and 60's to 1980's (from Solomon & Leak 1994).

Measure-ment period	Red spruce	Hemlock	Balsam fir	White pine	Sugar maple	Beech	Red oak
			Elevation				
1	242(7)[a]	136(8)	238(7)	150(9)	265(10)	243(11)	121(16)
2	240(8)	135(8)	241(7)	142(9)	273(10)	250(11)	117(16)
3	237(7)	136(7)	231(7)**	134(8)**	272(10)	245(11)	119(14)
			Latitude				
1	45·66(·04)	44·98(·06)	45·69(·04)	44·82(·07)	5·64(·07)	45·41(·06)	44·15(·10)
2	45·68(·04)	44·96(·06)	45·71(·04)	44·73(·07)	45·62(·07)	45·42(·07)	44·12(·09)
3	45·64(·04)	44·95(·06)	45·65(·04)**	44·72(·06)*	45·58(·06)	45·40(·06)	44·10(·07)
			Longitude				
1	69·05(·04)	69·18(·08)	69·08(·04)	69·50(·08)	69·30(·07)	69·25(·07)	70·00(·14)
2	69·01(·04)	69·20(·08)	69·09(·04)	69·53(·08)	69·36(·07)	69·27(·07)	70·01(·14)
3	69·02(·04)	69·23(·07)	69·07(·04)	69·44(·07)	69·39(·06)**	69·26(·07)	70·05(·12)

[a]Standard error in parenthesis; **significant at 0·01 level; *significant at 0·05 level.

increase its elevational range, a result developed from an independent study of age/distance/elevational trends at Bartlett, NH (Solomon & Leak 1994). Summer and annual temperatures at Bartlett have dropped somewhat in recent decades (Figure 1). Effects of a temperature change, if any, on expansion or contraction of the elevational range of eastern hemlock appear minimal or non-existent at present. Red spruce, on the other hand, increased from 34% to 43% of the basal area in the 650 – 820 m class (Table 3), and increased from 22% to 26% in the 500 - 650 m class.

Apparently red spruce populations are not declining but maintaining themselves well at these elevations despite warnings about growth decline or winter injury due to acid deposition (e.g. Hornbeck *et al.* 1988). In summary, permanent plots are a long-term solution requiring careful selection and analysis to confirm migrational tendencies or serve as a database for model testing.

Table 3: Percent of basal area by species (not all species listed), elevation class (E), measurement year (Y) and deciduous or coniferous land type (T) for unmanaged stands on Bartlett Experimental Forest[1], with significance tests[3].

Species[2]	Deciduous 200 to 350 m			Coniferous 200 to 350 m			500 to 650 m			650 to 820 m		
	1931	1940	1992	1931	1940	1992	1931	1940	1992	1931	1940	1992
BE	18·4	19·6	23·0	12·1	12·1	10·9	17·9	19·7	19·0	9·1	9·3	11·1
YB	14·5	11·5	6·9	12·9	12·1	6·6	12·9	14·0	14·3	8·4	9·5	12·5
SM	6·2	5·9	6·6	2·9	2·8	2·6	11·7	11·0	12·7	9·2	8·6	7·9
RM	19·3	22·4	25·7	21·9	24·6	29·2	9·3	8·3	8·2	8·2	7·4	6·4
PB	14·3	16·4	8·7	11·7	12·1	5·9	15·8	13·0	5·5	21·6	21·8	6·6
WA	4·7	6·1	6·3	3·3	3·3	4·0	0·7	0·9	0·4	0·2	0·2	0·3
ASP	10·8	5·6	2·9	8·2	5·0	1·3	0·2	0·0	0·0	0·0	0·0	0·0
EH	6·9	7·9	14·9	13·3	13·5	24·8	3·0	3·4	7·8	1·1	1·2	3·0
RS	2·5	2·6	2·9	5·3	5·5	6·6	22·1	23·0	25·9	34·5	34·9	43·4
BF	0·3	0·6	0·3	2·7	2·6	1·8	0·8	0·8	0·5	5·8	5·4	6·2

[1] Leak and Smith (in press)

[2] BE = beech, YB = yellow birch, SM = sugar maple, RM = red maple, PB = paper birch (*Betula papyrifera* Marsh.), WA = white ash (*Fraxinus americana* L.), ASP = aspen (*Populus tremuloides* Michx.), EH = eastern hemlock, RS = red spruce, BF = balsam fir.

[3] Significant at 0·05 level (main effects and interactions): BE (T,Y), YB (Y, EY), SM (E), RM (E, EY), PB (Y), WA (E, EY), ASP (EY), EH (E,Y,EY), RS (T,E).

AGE/DISTANCE TRENDS

This approach assumes that migrating species exhibit a sequential relationship between age and distance or elevation; in other words, young plants will be out in front of old plants (Leak & Graber 1974a; Solomon *et al.* 1990; Solomon & Leak 1994) (Figure 4). The advancing front theory is based on the premise that regenerating species appear to move gradually away from a concentration of seed-producing trees, a concept that aligns with what we know about seed fall distribution and sprouting/suckering behaviour (Leak & Graber 1974b). It is most successful in carefully selected areas where there is a steep climatic gradient (e.g. on a mountain slope), the species have elevational limits as illustrated in Figure 2, there are no significant barriers to plant movement and the stands are unmanaged. The age–location pattern at such sites permits estimation of the rate of migration and how migration rates vary with site, stands and climatic factors. At other sites, species movement may be better modeled by discontinuous jumps (e.g. Davis 1987).

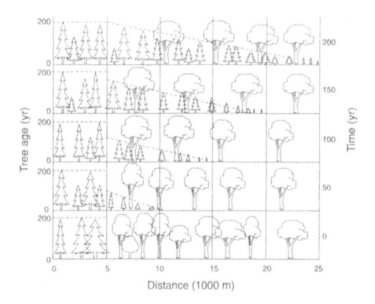

Distance (1000 m)

Figure 4: Schematic of a developing advancing front for a species moving at a constant rate of 1000 m per decade. The dashed line represents fits of distance over maximum age (from Solomon *et al.* 1990).

The advancing front requires fairly intensive sampling of actual tree ages or predicted tree ages based on age/diameter relationships over a distance/elevational transect. Species migration is calculated as:

$$Migration\ rate\ =\ \frac{\Delta distance}{\Delta time}\ =\ \frac{(d_{i+1} - d_i)}{(a_i - a_{i+1})}\ =\ f(site_i, climate_i) \qquad (1)$$

where:

i = measurement number $0, ..., n$,

d_i = distance from parent stand,

a_i = maximum tree age at d_i,

$site_i$ = site characteristics from d_i to d_{i+1},

$climate_i$ = climatic factors during a_i to a_{i+1}.

Percent of species composition at any point in the advancing front is:

$$Species\ percent_{i+1} = f(site_i, climate_i, species\ percent_i, a_i) \qquad (2)$$

where:

a_i = an indication of species percentage as a function of age.

Migration rate for any given time would be estimated as a function of site and climate, based on either hypothesised values or predictions from climate models. Applications of this approach for several species, and both advancing and retreating fronts, can simulate stand-level dynamics in terms of species mix, age structure and location along corridors or from isolated pockets in any direction. Data from Haystack Mountain, Bartlett, NH, indicated that age trends of hemlock resembled a slowly-moving advancing front (Figure 5); the front is variable due to the episodic nature of hemlock regeneration and development. Hemlock readily grows on the parent materials shown as gravel/loose rock and dry pan; thus, these materials were not an obstacle to hemlock movement. The decline in hemlock age is not due to site differences because hemlock readily regenerates and grows on both sites, and longevity is seldom hampered by site conditions; i.e. old trees are commonly found on less-than-optimum sites. Other species in the Bartlett study, such as red spruce, showed a stationary (non-moving) front in equilibrium with site and climate. This was reflected by maximum ages, minimum ages and seedlings occurring at the same elevation points, usually at the limits of a site change (Figure 5).

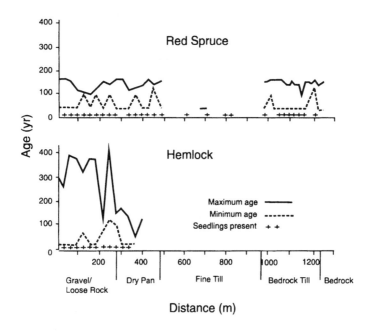

Figure 5. Example of an advancing front from the Bartlett Experimental Forest, Bartlett, NH, U.S.A. (Solomon & Leak 1994). Maximum and minimum ages over horizontal distance for hemlock and red spruce on Haystack Mountain.

DISCUSSION AND CONCLUSIONS

Each of the methods discussed above has certain advantages and disadvantages. Two methods require only a single measurement, successional trends inferred from surveys of species composition by size class and age/distance/elevation trends. The former approach is amenable to broad regional surveys, but interpretation can be difficult due to past disturbance. The latter requires careful site location and is more like a carefully controlled field experiment. Past climatic records can readily be correlated with the trend from age/distance/elevation studies.

The other two methods require extended time periods for resurveying and remeasuring previously established plots. Resurveys of previously sampled areas, when the original plots cannot be located, are imprecise at best; the main attraction of this approach is that it utilises the very long-term information recorded in some of the early literature. A system of remeasured plots, such as those maintained by the USDA Forest Inventory and Analysis Unit, appear necessary to truly monitor changes in species occurrence and dominance by location and elevation. The plot network reported herein (e.g. *ca.* 500 plots in 6.9×10^6 ha of forest in Maine) would appear sufficient to document some significant changes in elevation and location over time. However, it appears necessary to eliminate plots in historically disturbed landscapes to provide clear interpretations.

Although several of the methods described (e.g. permanent plots, advancing fronts) may indicate change in the composition of a community or even migration, such measures do not establish the cause for observed changes. To further bolster the case for a climatic cause requires the evaluation of community changes in many geographically dispersed plots, as well as a clear understanding of the climatic parameters that regulate species composition and of the climatic trends at those plots. As an example, if mean summer temperature is an important regulator of species occurrence, the geographical pattern of change in New England over the last 50 years should broadly parallel the temperature trends in Figure 1d. Species typical of warmer climates might be expected to be newly found along coastal areas but be disappearing from sites further in the interior. Even so, it is difficult (perhaps impossible) to unequivocally implicate climatic change as a causative agent for community change when so many other stressors occur with a geographically complex pattern. For example, in New England there are strong regional north–south and east–west gradients in precipitation pH, nitrogen deposition and ozone concentration (Ollinger *et al.* 1995). Future indications of climatic change from stand–level measurements will come from data analysis conducted across climatic zones to assess species movement and separate out confounding effects across regional gradients.

REFERENCES

Bormann FH, Likens GE (1979) Pattern and Process in a Forested Ecosystem. Springer, New York

Colinvaux P (1986) Ecology. Wiley, New York

Davis MB (1987) Invasions of forest communities during the Holocene: beech and hemlock in the Great Lakes region. *in* Gran AJ, Crawley MJ, Edward PJ (eds) Colonization, succession and stability, 373-393. Blackwell Scientific Publications, Oxford

Davis MB, Zabinski C (1992) Changes in geographical range from greenhouse warming effects on biodiversity in forests. *in* Peters RL, Lovejoy TE (eds) Proceedings, World Wildlife Fund conference on consequences of global warming for biological diversity, 297-308. Yale University Press, New Haven

Denton SR, Barnes BV (1987) Tree species distributions related to climatic patterns in Michigan. Canadian Journal of Forest Research 17:613-629

Egler FE (1954) Vegetation science concepts. I. Initial floristic composition--a factor in old-field vegetation development. Vegetatio 4:412 -417

Federer CA, Tritton LM, Hornbeck JW, Smith RB (1989) Physiologically based dendroclimate models for effects of weather on red spruce basal-area growth. Agricultural and Forest Meteorology 46:159-172

Frieswyk TS, Malley AM (1985) Forest Statistics for New Hampshire – 1973 and 1983. *in* Broomall PA (ed) Northeastern Forest Experiment Station Resource Bulletin NE-88, 100. U.S. Department of Agriculture, Forest Service.

Gajewski K (1987) Climatic impacts on the vegetation of eastern North America during the past 2000 years. Vegetatio 68:179-190

Grabherr G, Gottfried M, Pauli H (1994) Climate effects on mountain plants. Nature 369:448

Hamburg SP, Cogbill CV (1988) Historical decline of red spruce populations and climatic warming. Nature 331:428-431

Hett JM (1971) A dynamic analysis of age in sugar maple seedlings. Ecology 52:1071-1074

Hornbeck JW, Smith RB, Federer CA (1988) Growth trends in 10 species of trees in New England. Canadian Journal of Forest Research 18:1337-1340

Leak WB, Graber RE (1974a) Forest vegetation related to elevation in the White Mountains of New Hampshire. Broomall, PA. U.S. Department of Agriculture, Forest Service, Northeastern Forest Experiment Station. Research Paper NE-299. 7p.

Leak WB, Graber RE (1974b) A method for detecting migration of forest vegetation. Ecology 55:1425-1427

Leak WB, Smith ML (in press) Sixty years of management and natural disturbance in New England forested landscape. Forest Ecology and Management

Leak WB, Wilson RW Jr (1958) Regeneration after Cutting of Old-Growth Northern Hardwoods in New Hampshire. *in* Northeastern Forest Experiment Station, Station Paper No. 103. U.S. Department of Agriculture, Forest Service

Ollinger SV, Aber JD, Federer CA, Lovett GM, Ellis JM (1995) Modeling physical and chemical climate of the northeastern United States for a geographic information system. *in* Radnor, PA Northeastern Forest Experiment Station. General Technical Report NE-191. 30. U.S. Department of Agriculture, Forest Service

Overpeck JT, Bartlein PJ, Webb T III (1991) Potential magnitude of future vegetation change in eastern North America: comparisons with the past. Science 254:692-695

Pastor J, Post WM (1988) Response of northern forests to CO_2 induced climate change. Nature 334:55-58

Peters RL (1990) Effects of global warming on forests. Forest Ecology and Management 35:13-33

Powell DS, Dickson DR (1984) Forest statistics for Maine: 1971 and 1982. *in* Broomall PA Northeastern Forest Experiment Station. Resource Bulletin NE-81, 194. U.S. Department of Agriculture, Forest Service,.

Prentice IC, Bartlein PJ, Webb T III (1991) Vegetation and climate change in eastern North America since the last glacial maximum. Ecology 72:2038-2056

Solomon DS, Leak WB (1994) Migration of tree species in New England based on elevational and regional analysis. *in* Northeastern Forest Experiment Station. Research Paper NE-688. U.S. Department of Agriculture, Forest Service

Solomon DS, Leak WB, Hosmer HA (1990) Detecting and Modeling the Migration of Tree Species in Response to Environmental Change. *in* Burkhart HE, Bonner GM, Lowe JJ, (eds) Research in Forest Inventory Monitoring Growth and Yield Proceedings IUFRO S4.01 and S4.02. Virginia Polytechnical Institute and State University, Blacksburg, VA. Publication FWS-3-90:230-239.

Spear RW, Davis MB, Shane LCK (1994) Late Quaternary history of low- and mid-elevation vegetation in the White Mountains of New Hampshire. Ecological Monographs 64:85-109

Stephens GR, Ward JS (1992) Sixty Years of Natural Change in Unmanaged Mixed Hardwood Forests. Connecticut Agricultural Experiment Station, New Haven. Bulletin 902

Vose RS, Schmoyer RL, Steurer PM, Peterson TC, Heim R, Karl TR, Eischeid JK (1992) The global historical climatology network: Long-term monthly temperature, precipitation, sea level pressure, and station pressure data. Energy, Environment, and Resources Center, University of Tennessee, Knoxville, TN. Environmental Sciences Division, Department of Energy. Publication No. 3912.

Structural changes in the forest–tundra ecotone: A dynamic process

Annika Hofgaard
Norwegian Institute for Nature Research
Tungasletta 2
N-7005 Trondheim
Norway

INTRODUCTION

In recent decades there has been increasing discussion of the possible effects of anthropogenic changes in climate. The biological consequences of anticipated changes have been discussed mostly in terms of ecosystem reaction to an increasingly warmer climate (Emanuel *et al.* 1985; Bonan *et al.* 1990; Botkin & Nisbet 1992; Chapin *et al.* 1995), involving a northward movement of boreal forest into areas currently covered by tundra (cf. Prentice *et al.* 1991). This generalized view can and needs to be discussed in depth in order to reach a more reliable prediction. This paper focuses on structural changes in time and space in the forest-tundra ecotone (i.e. the transition zone from closed forest to tundra, crossing the treeline up to the krummholz limit). The dynamic nature of the structural changes is discussed in relation to a hypothetical climate–vegetation equilibrium.

The boundary between the closed boreal forest and the tundra marks a clear separation in species composition as well as in life-forms (Sveinbjörnsson 1992). Global climate change may alter ranges of the species and life-forms, as well as restructuring the ecosystem (Campbell & McAndrews 1993). This means that the altitude or latitude of various limits may change in the future. Limits have moved back and forth in the past, depending on prevailing climatic conditions, and the locations of these limits at the present moment may not necessarily be in harmony with present climate. Forest stands and/or individual trees may persist under adverse climatic conditions without sexual reproduction through longevity and layering of branches (Sveinbjörnsson 1992; Payette 1993; Payette & Lavoie 1994; Kullman

NATO ASI Series, Vol. I 47
Past and Future Rapid Environmental Changes:
The Spatial and Evolutionary Responses of
Terrestrial Biota
Edited by Brian Huntley et al.
© Springer-Verlag Berlin Heidelberg 1997

1995a). The longevity of trees and their potential for vegetative reproduction produces community inertia, bridging the ecosystem over into a future more favourable period, and providing an extended time for adjustment to new environmental conditions. Despite this ecosystem inertia, treeline ecotonal communities are sensitive indicators of environmental change because phenotypic plasticity and regeneration provide a rapid response to changed conditions (Kullman 1986, 1988; Hofgaard et al. 1991; Hofgaard 1993a, b; Payette 1993; Lavoie & Payette 1994). It is often argued that the forest–tundra ecotone (and also other transition zones between plant communities) represent a delicate balance between opposing forces of nature and ought therefore to be modified even by slight changes in the environment – including climate change (Emanuel et al 1985; Bonan et al 1990; Atkinson 1992).

Disturbance is a central factor in vegetation dynamics (Oliver & Larsen 1990) and in translating climatic change into vegetational response (Overpeck et al. 1990). Consequently, knowledge of the disturbance régime is essential for understanding and modelling system responses. A commonly used definition of disturbance is "any relatively discrete event in time that disrupts ecosystem, community, or population structure and changes resources, substrate availability, or the physical environment" (Pickett & White 1985). When considering changes in the forest–tundra ecotone in a long-term perspective, e.g. centuries to millennia, this definition of disturbance is not satisfactory or complete, as "disturbance is a highly scale-dependent concept, and consideration must be given to the spatial and temporal extent of any disturbance as well as to its intensity or severity" (Engelmark et al. 1993; see also discussion in van der Maarel 1993). Natural disturbances occur on widely different spatial and temporal scales and their apparent importance depends on the time scale at which we are observing the ecosystem (Tausch et al. 1993). Many different kinds of disturbances may interact over time.

Forest–tundra ecotonal communities are shaped through episodic disturbances (e.g. fire and insect outbreaks) and longer term, chronic disturbances (e.g. climatic stress and grazing) occurring at all spatial scales from landscape or region to individual forest stands or single trees. Additionally, the frequency of disturbance fluctuates through time due to climatic variability and change. Consequently, I use a definition of disturbance that is not restricted to discrete events, but also includes chronic disturbance (cf. Engelmark *et al.* 1993).

Studies of effects of natural disturbance régimes require areas where human impact has been continuously low or absent. Areas within the forest–tundra ecotone in e.g. Canada and Scandinavia provide this opportunity. Additionally, observations that are detailed in space and extend over a long period of time are available from these areas (Kullman 1990; Payette 1993; Hofgaard 1993c; Lavoie & Payette 1994). These observations indicate substantial change on both long and short time scales. A detailed historical perspective is a prerequisite for understanding the dynamic inter-relationships between vegetation and environmental change, especially in cases where disturbance has been fluctuating in kind, intensity and spatial extent (van der Maarel 1993).

LONG- AND SHORT-TERM STRUCTURAL CHANGES

Since the mid-Holocene, the forest–tundra ecotonal vegetation in the northern hemisphere displays generally decreasing stand density, retreating forest limits and tree limits, and changes in species composition. These changes are a consequence of climate change. The tree populations that are farthest north and/or at highest elevations may be seen as survivors of a previously denser distribution due to more favourable conditions in the past. The inherent inertia of the communities has allowed these remnants to persist. The withdrawal and/or density reduction during the late Holocene is associated with changed fire régimes and subsequent regeneration failure (Payette & Morneau 1993; Payette 1993), and changed seasonality towards cooler summer conditions and greater snowfall (Kullman 1990, 1995b). In northern Scandinavia this change in seasonality caused the previous pine (*Pinus sylvestris*) forest to retract to lower elevations, while the subalpine birch (*Betula pubescens*) forest belt emerged. This shift in species dominance resulted in an ecosystem with a wide spectrum of regenerative options (Kullman 1990), which favour its long-term survival in a changing environment.

In a short-term perspective (years to decades) structural changes result from a series of mechanisms, governed directly or indirectly by climate variability and small-scale disturbances. The resulting stand density fluctuations do not normally cause a shift in species composition, but rather a change in species performance, involving shifts in seed production, regeneration pattern, survival, tree vitality and physiognomy. Mechanisms that induce such changes include large year-to-year variations in mean summer temperature (Alexandersson & Eriksson 1989), late thawing of the soil, or the development of episodic permafrost (Kullman 1991), heavy snow load (Hofgaard *et al.* 1991), winter frost injury (Kullman 1988; Sveinbjörnsson 1992; Lavoie & Payette 1992) and insect outbreaks (Tenow & Nilssen1990; Veblen *et al.* 1991).

CLIMATE–VEGETATION EQUILIBRIUM LINE – A CONCEPTUAL MODEL

Vegetation changes continuously to adjust to its abiotic environment, especially climate. The resulting limits to stand density in time and space can be imagined as a potential *'climate-vegetation equilibrium line'* (Figure 1). The position of the line is set by the prevailing climatic régime translated into vegetational response by disturbance, acting at both stand and landscape levels (see also Bradshaw & Hannon 1992 – they discuss dynamics of the nemoral-boreal transition in similar terms). All single disturbances (vertical arrows in Figure1) will move the relative position of an ecosystem relative to the equilibrium line. The position of the equilibrium line, however, changes with time in response to millennial-scale trends in climate. In this representation I have chosen a slope of the line that fits the course of the Holocene, in other words with deteriorating long-term environmental conditions for tree growth (most species), resulting in decreased stand density over time at all elevations.

From a starting point close to the 'equilibrium line', absence of disturbances may cause ecosystems to deviate away from the equilibrium condition. This occurs through processes that allow stands to persist under harsh environmental conditions (see above). As the distance between the ecosystem condition and the equilibriium condition increases with time, the ecosystem becomes a relict of previous conditions. Consequently, the effect of a subsequent disturbance will be larger than for equilibrium vegetation. Under such conditions the ecosystem will not recover to its previous condition, but instead a condition in equilibrium with present day climate. The inability to regain the previous stand density should not be misinterpreted as resulting from a recent environmental change. It results instead from long-term and continous environmental change.

Episodic disturbances such as insect outbreaks and fires can cause forests and open woodlands to disappear either temporarily or more permanently, depending on whether the forest before destruction was in equilibrium with the current climate or a relict of a previous climatic régime. It can be argued that recovery will occur, given sufficient time without disturbances (Magnuson 1990), but recovery is conditional on the previous condition of the system relative to the equilibrium line.

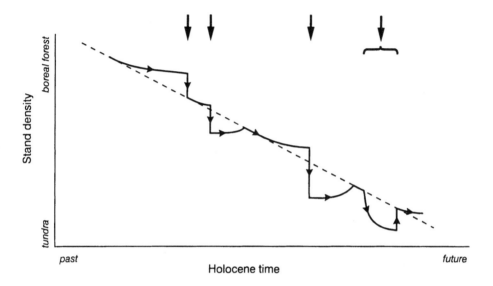

Figure 1. Stand density change in forest–tundra ecotone over long and short time scales. The dotted straight line shows theoretical stand density in equilibrium with the long-term climate trend. This equilibrium line slopes downward with time, because climate has undergone a steady deterioration at these latitudes throughout the mid- and late- Holocene. Actual stand density (solid line) departs from the theoretical equilibrium line due to inertia or response to disturbance. Disturbances are shown by the vertical arrows. Single arrows indicate episodic disturbances, and arrow and bracket chronic disturbance.

Chronic disturbances (arrow and bracket in Figure1), such as grazing and climatic stress (e.g. the Little Ice Age), may force and keep communities at a low stand density level during substantial periods of time until the suppression force eventually ceases. When released from such suppressed conditions the recovery process may appear rapidly through revitalization processes and regeneration. In the case of grazing this apparent progressive process in the forest–tundra ecotone can easily be misinterpreted as a result primarily of climate change, rather than of changes in land-use.

Discrete disturbance events of low intensity do not necessarily change the system in a way that brings it back into 'equilibrium' with climate. They can cause small fluctuations in the ecosystem structure. Again, this depends on how we define the time scale.

The average values of environmental factors govern the regional position of the vegetation-climate equilibrium line. The frequency of climatic extremes moulds the communities in a more or less unique way for every area. These climatic extremes, as well as sporadically occurring disturbances of various kinds, magnitude and persistence, are not predictable. Still we have to take them into account. The effect of single events may linger for many decades, centuries or millennia, all depending on kind, magnitude and duration (cf. van der Maarel 1993). In addition, the effects depend on the preceding position of the ecosystem in relation to the equilibrium with the long-term climatic trend.

CONCLUDING REMARKS

Patterns in one space and time compartment can not be applied to other scales in space and time, nor to other points in space or time, as every point is special (Sprugel 1991). It can be questioned whether equilibrium conditions between climate and vegetation can ever occur. Evidence for equilibrium sometimes is found, as an artifact of the temporal and spatial scale of the observations (De Angelis & Waterhouse 1987). The principles and concepts that ecologists use for interpretation of changes are largely based on the expectation of equilibrium conditions. If these are artificial constructs, due to our limited perspective in time, then the results of our descriptions, interpretations and models for the future may be misleading.

Consequently, there is a need to discuss the effects of climate change both in a long-term perspective (e.g. the Holocene), and with a shorter perspective, which includes interactions between greenhouse warming and natural long-term climatic trends. Additionally, when considering the forest-tundra ecotone, it is essential to distinguish anthropogenic from climatic determination of the present tree-line. A rewarding way to achieve this and to decide whether a specific stand density decline or increase is anomalous or expected, is to use comprehensive stand history analyses and detailed long-term monitoring in permanent plots (Kullman 1986, 1988; Hofgaard *et al.* 1991; Lavoie & Payette 1992, 1994; Hofgaard 1993c).

Structural changes within the forest–tundra ecotone are naturally and continuously occurring as responses to natural climatic changes. Such changes are also occurring in response to anthropogenic impacts on the environment. Present-day communities have been shaped by former disturbance régimes, and today's disturbance régimes will shape the communities of tomorrow. Certain disturbances of apparently small effect and extent may have wide-ranging consequences at other spatial scales. Thus, detailed and deeper knowledge of actual processes and their interactions is needed to increase the reliability of predictions of future changes.

ACKNOWLEDGEMENTS

This paper was partly prepared during my postdoctoral fellowship (NSERC) at the University of Québec in Montréal, Canada. I have greatly appreciated discussions with M. Davis, L. Kullman, O. Engelmark, Y. Bergeron and H. Hytteborn and their useful comments on earlier drafts of this paper. Additionally, I am grateful for support from the Department of Physical Geography, Umeå University, Sweden.

REFERENCES

Alexandersson H, Eriksson B (1989) Climate fluctuations in Sweden 1860-1987. SMHI Reports Meteorology and Climatology 58:1-54

Atkinson D (1992) Interactions between climate and terrestrial ecosystems. Trends Ecol Evol 7:363-365

Bonan GB, Shugart HH, Urban DL (1990) The sensitivity of some high-latitude boreal forests to climatic parameters. Climatic Change 16:9-29

Botkin DB, Nisbet RA (1992) Forest response to climatic change: effects of parameter estimation and choice of weather patterns on the reliability of projections. Climatic Change 20:87-111

Bradshaw R, Hannon G (1992) Climatic change, human influence and disturbance régime in the control of vegetation dynamics within Fiby Forest, Sweden. J of Ecology 80:625-632

Campbell ID, McAndrews JH (1993) Forest disequilibrium caused by rapid Little Ice Age cooling. Nature 366:336-338

Chapin III FS, Shaver GR, Giblin AE, Nadlehoffer KJ Laundre JA (1995) Response of arctic tundra to experimental and observed changes in climate. Ecology 76:694-711

De Angelis DL, Waterhouse JC (1987) Equilibrium and nonequilibrium concepts in ecological models. Ecological Monographs 57:1-21

Emanuel WR, Shugart HH, Stevenson MP (1985) Climatic change and the broad-scale distribution of terrestrial ecosystem complexes. Climatic Change 7:29-43

Engelmark O, Bradshaw RHW, Bergeron Y (1993) Disturbance dynamics in boreal forests: Introduction. J Veg Sci 4:729-732

Hofgaard A (1993a) Structure and regeneration patterns in a virgin *Picea abies* forest in northern Sweden. J Veg Sci 4:601-608

Hofgaard A (1993b) Seed rain quantity and quality, 1984-1992, in a high altitude old-growth spruce forest, northern Sweden. New Phytol 125:635-640

Hofgaard A (1993c) 50 years of change in a Swedish boreal old-growth *Picea abies* forest. J Veg Sci 4:773-782

Hofgaard A, Kullman L. Alexandersson H (1991) Response of old-growth montane *Picea abies* (L.) Karst. forest to climatic variability in northern Sweden. New Phytol 119:585-594

Kullman L (1986) Recent tree-limit history of *Picea abies* in the southern Swedish Scandes. Can J For Res 16:761-771

Kullman L (1988) Subalpine *Picea abies* decline in the Swedish Scandes. Mountain Research and Development 8:33-42

Kullman L (1990) Dynamics of altitudinal tree-limits in Sweden: a review. Norsk Geografisk Tidsskrift 44:103-116

Kullman L (1991) Ground frost restriction of subarctic *Picea abies* forest in northern Sweden. A dendroecological analysis. Geografiska Annaler 73A:167-178

Kullman L (1995a) New and firm evidence for Mid-Holocene appearance of *Picea abies* in the Scandes Mountains, Sweden. J of Ecology 83:439-447

Kullman L (1995b) Holocene tree-limit and climate history from the Scandes Mountains, Sweden. Ecology 76:2490-2502

Lavoie C Payette S (1992) Black spruce growth forms as a record of a changing winter environment at treeline, Québec, Canada. Arctic and Alpine Research 24:40-49

Lavoie C, Payette S (1994) Recent fluctuations of the lichen-spruce forest limit in subarctic Québec. J of Ecology 82:725-734

Magnuson JJ (1990) Long-term ecological research and the invisible present. BioScience 40:495-501

Oliver CD, Larsen BC (1990) Forest Stand Dynamics. McGraw-Hill, Inc., New York

Overpeck JT, Rind D, Goldberg R (1990) Climate-induced changes in forest disturbance and vegetation. Nature 343:51-53

Payette S (1993) The range limits of boreal tree species in Quebec-Labrador: an ecological and palaeoecological interpretation. Review of Palaeobotany and Palyonology 79:7-30

Payette S, Lavoie C (1994) The arctic tree line as a record of past and recent climatic changes. Environ Rev 2:78-90

Payette S, Morneau C (1993) Holocene relict woodlands at the eastern Canadian Treeline. Quaternary Research 39:84-89

Pickett STA, White PS (1985) The ecology of natural disturbance and patch dynamics. Academic Press, London

Prentice IC, Sykes MT, Cramer W (1991) The possible dynamic response of northern forests to global warming. Global Ecology and Biogeography Letters 1:129-135

Sprugel DG (1991) Disturbance, equilibrium, and environmental variability: What is 'natural' vegetation in a changing environment? Biological Conservation 58:1-18

Sveinbjörnsson B (1992) Arctic tree line in a changing climate. *in* Chapin III FS *et al.* (eds). Arctic ecosystems in a changing climate. An ecophysiological perspective, 239-256. Academic Press.

Tausch RJ, Wigand PE, Burkhardt JW (1993) Viewpoint: Plant community thresholds, multiple steady states, and multiple successional pathways: legacy of the Quaternary? J Range Manage 46:439-447

Tenow O, Nilssen A (1990) Egg cold hardiness and topoclimatic limitations to outbreaks of *Epirrita autumnata* in northern Fennoscandia. J of Appl Ecol 27:723-734

van der Maarel E (1993) Some remarks on disturbance and its relations to diversity and stability. J Veg Sci 4:733-736

Veblen TT, Hadley KS, Reid MS, Rebertus AJ (1991) The response of subalpine forests to spruce beetle outbreak in Colorado. Ecology 72:213-231

Modelling the structural response of vegetation to climate change

Herman H. Shugart, Guofan Shao, William R. Emanuel and Thomas M. Smith
Department of Environmental Sciences
University of Virginia
Charlottesville, Virginia 22903
U.S.A.

INTRODUCTION

Prediction of the response in vegetation either locally or globally to a change in the climate presents a rich array of scientific challenges. One must understand the dynamics of ecological systems at different time and space scales (Delcourt *et al.* 1983). One must understand how to handle the mutual causality between the structure of ecosystems and their functioning (Shugart in press). One must extrapolate from the physiological responses of the few species that we know fairly well to a vast array of species that we hardly know at all.

Probably the central challenge in understanding the manner in which vegetation might respond to novel climatic conditions (particularly in concert with changes in the concentration of atmospheric CO_2) is to reverse the fascination with reductionism that drives much of modern science and to engage a meaningful synthesis of the knowledge that we have accumulated. Principal tools for such a synthesis are mathematical models. Models can be used to bring together, in the form of a synthetic hypothesis, what we know (and think we know) about how vegetation functions.

THE IMPORTANCE OF ECOSYSTEM STRUCTURE IN UNDERSTANDING ECOSYSTEM CHANGE

As ecology has moved from a largely descriptive science to a science with greater emphasis on dynamics, the importance of understanding the structural features of

NATO ASI Series, Vol. I 47
Past and Future Rapid Environmental Changes:
The Spatial and Evolutionary Responses of
Terrestrial Biota
Edited by Brian Huntley et al.
© Springer-Verlag Berlin Heidelberg 1997

ecosystems has taken something of a lesser position to the importance of understanding the processes or the functions of ecosystems. It is sometimes easy to forget that 'structure and function' (as a classic biologist might use the term) or 'pattern and process' (a phraseology more frequently used by ecologists), are portrayals of two mutually causal agents. Biology has a central tenet – the concept that the form or shape of entities is both modified by and a creator of function. Processes cause patterns to occur; patterns alter the magnitude and the direction of processes.

For example, Körner (1993) reviewed over 1000 published papers to determine the response of plant systems at several different levels (single plant, cultivated plants, natural vegetation) to conditions of elevated CO_2. He found that the higher the levels of organisation one considered in judging the response of plant systems (leaf photosynthesis, plant growth, ecosystem yield) and the longer the period of time of observation (from hours to years), the greater was the magnitude of the attenuation of the positive effects of elevated levels of CO_2. The causes of these responses are potentially many, and include everything from a tendency for plants to outgrow their pots in longer term greenhouse studies, and thus slow their growth, to a 'down regulation' of photosynthesis in high CO_2 conditions.

The responses that Körner (1993) documents are to be expected in treating a fundamental response of complex systems at progressively greater levels of interactions (Delcourt et al. 1983; Rosswall et al. 1988). The structure and variation of structure in natural systems is one of the principal sources of these single-effect-altering interactions. One would expect that the structure of ecosystems should have the effect of altering the changes in rates or processes in terrestrial ecosystems. Indeed, there appears to be a hierarchical spectrum of structural effects that can alter the response that attends a change in a fundamental process (Table 1). These structurally controlled modifications of system response are nontrivial in many cases – in some of the example cases shown in Table 1, they can reduce the potential response of the system by 70% or more. Other changes in system structure can similarly amplify potential responses.

Table 1. Structure of systems as a mediating factor in the system response to changes in fundamental processes.

Level of Response	Structural Change	Functional Implication
Leaf Tissue	Change in stomatal index (number cm^{-2}) of plants grown in different CO_2-environments. Effect can be seen in plant material collected before the industrial revolution and can be induced under laboratory conditions (Woodward 1987).	Alteration of the stomatal conduction response of the plant. Implication that plant responses historically may be different from present responses (due to change in stomatal index induced by ambient atmospheric conditions).
Individual Plant	Rates of leaf photosynthesis can increase on the order of 50% for C3 plants in response to a doubling of CO_2 but rates of whole plant growth are often less than 20% of those for control conditions (Körner 1993).	Structural considerations including photosynthate allocation and internal interactions (e.g. with nutrients such as nitrogen) can moderate the carbon fixation at the leaf level.
Plant Stand	The increase in stand biomass is less by a factor of about 0·30 (Shugart & Emanuel 1985) than the increase in growth of the individual plants comprising the stand.	Stand interactions (competition, shading, etc.) reduce the stand level biomass increase (or yield) in response to increased growth rates of individual plants.
Landscape	Mosaic properties of landscapes alter the stand biomass response to changed conditions (Borman & Likens 1979).	Landscapes can be thought of as mosaics in different stages of recovering from natural disturbances. Changes in plant and stand processes are mediated by the local state of disturbance recovery.
Region	The terrestrial surface can alternate between being a source to a sink of carbon in the transient response to environmental change – even in cases in which the long-term response to change is similar to the initial condition (Smith & Shugart 1993).	Shifts in vegetation in response to change are delayed by large scale processes involving dispersal, recovery and other inertial effects.

This table presents examples of what are a wide range of structural responses that can either attenuate or amplify the process response (from Shugart, in press).

A class of models that appears to have some capability to simulate ecosystem structure is individual-based, mixed-species, mixed-aged simulation models called gap models (Shugart & West 1980). These models project changes in the landscape by simulating the birth, growth and death of individual plants in an interactive plant community. Developed initially for forests and derived from earlier forestry models (Munro 1974), gap models and their descendants have now been applied to a variety of ecosystems including grasslands and savannas. They have also been applied to boreal, temperate and tropical forest systems around the world. Gap models have been recently reviewed with regard to their role in predicting vegetation response to climate change (Shugart *et al.* 1992b; Smith *et al.* 1992). The models simulate (and have been tested with regard to predicting) structural features of ecosystems.

GAP MODELS

As is the case with many of the earlier individual-based models used in forestry, gap models simulate the establishment, diameter growth and mortality of each tree in a given area. Calculations are on a weekly to annual time step. At least initially, gap models were developed for plots of a fixed size. Many of the models focus on a size unit (*ca.* 0·1 ha) approximately that of a forest canopy gap (Shugart & West 1980). Gap models feature relatively simple protocols for estimating the model parameters (Botkin *et al.* 1972; Shugart 1984). For many of the more common temperate and boreal forest trees, there is a considerable body of information on the performance of individual trees (growth rates, establishment requirements, height – diameter relations) that can be used directly in estimating the parameters of such models. The models have simple rules for interactions among individuals (e.g. shading, competition for limiting resources, etc.) and equally simple rules for birth, death and growth of individuals. The simplicity of the functional relations in the models has positive and negative consequences. The positive aspects are largely involved in the ease of estimating model parameters for a large number of species; the negative aspects with a desire for more physiologically or empirically 'correct' functions (see Pacala *et al.* 1993).

The more recent gap models in many cases have functional relationships that are different from those used in the earlier gap models (Bugmann *et al.* in press). Gap models differ in their inclusion of processes which may be important in the dynamics of particular sites being simulated (e.g. hurricane disturbance, flooding, formation of permafrost, etc.), but share a common set of characteristics. These latter characteristics involve an emphasis on the demography and natural history of plant species, relatively general rules for physiological trade-offs among species, and emphasis on the understanding of plant processes at the plant level. Each individual plant is simulated as an independent entity with respect to the processes of establishment, growth and mortality. This feature is common to most individual-tree based forest models and provides sufficient information to allow computation of species- and size-specific demographic effects. Gap model structure emphasises two features important to a dynamic description of vegetation pattern: firstly, the response of the individual plant to the prevailing environmental conditions, and secondly, how the individual modifies those environmental conditions. The models are hierarchical in that the higher-level patterns observed (i.e. population, community, and ecosystem) are the integration of plant responses to the environmental constraints defined at the level of the individuals.

MODEL TESTS

One of the values of models in predicting vegetation response to climate change is in testing and evaluating theory. To be credible in this role, models must be tested against independent data regularly. Mankin *et al.* (1977) and Shugart (1984) divide model testing into two basic types of procedures (verification and validation) and see model application as a measure of a model's usefulness. In verifying a model, the model is tested on whether it can be made consistent with some set of observations. In validation procedures, a model is tested on its agreement with a set of observations independent of those observations used to structure the model and estimate its parameters. When testing a model, it is important that it can simulate the pattern of the system under the constraint that all the parameters in the model are realistic. In the rationale of gap models it is appropriate to have a high level of realism in the model parameters. In the construction of most gap models, initial model verifications involve determining the model's ability to reproduce the general features of forest pattern while constrained to using model parameters that are reasonable.

A wide range of tests have been applied to gap models that test the ability of the models to produce appropriate vegetation structure. The models have been tested in their ability to predict forest-level features (total biomass, leaf area, stem density, average tree diameter, etc.) given a priori parameter estimation for species. Results of long (quasi-equilibrium approaching) simulation runs have been compared with patterns of mature forests in a region. Models have been calibrated on stands of a given age, and compared with independent data from stands of different ages. 'Natural experiments' have been used to test the models by inspecting the models' ability to predict forest composition after the introduction of a disease that eliminates a species, to predict forest composition change in response to single or multiple environmental gradients, or to reconstruct composition of vegetation under past climates (palaeo-reconstruction). One can also test the models' ability to predict independent tree diameter increment data or forestry yield tables. More detailed examples of each of these tests are elaborated in Shugart (1984).

Significantly, most of these model tests have involved predicting structural attributes such as composition, diameter distributions, tree densities, and other, similar variables. The models are able to simulate such functional attributes as net production or basal area increment, but they have not been tested to as great a degree with respect to these attributes.

APPLYING MODELS TO UNDERSTAND FOREST RESPONSE TO NOVEL CLIMATIC CONDITIONS

There has been some success in reproducing vegetation composition in response to past climates, but palaeoecological data often are not detailed (typically genus-level reconstructions of the vegetation) and, thus, are not as difficult a challenge as one might wish. Many of our current models have been tested reasonably well at single points in space. The problem of reconstructing changes in the patterns of vegetation structure over large areas remains a problem of substantial difficulty. One difficulty lies in reproducing the expected pattern of vegetation for fundamentally novel situations. The increasing CO_2 level of the atmosphere is a prime example, in that it could result in differences in the water-use efficiencies of plants, among other possible effects. However, changes in the pattern of covariation among other important environmental variables with climate changes represent equally daunting problems.

Chimeras that combine different models (sometimes with different underlying assumptions) in ways that backstop one another's weaknesses have been produced as solutions to these problems. The HYBRID model, which uses a gap model to change the physical structure of the plant canopy (based on the allometry and performance of individual plants) and a canopy model to capture the productivity of the canopy as well as to determine the water use and carbon fixation, is an example of such a development. Friend et al. (1993) initially inspected the HYBRID model by simulating the responses of forests in two cases — a lodgepole pine Pinus contorta forest under climate conditions associated with Missoula, Montana, and a white oak Quercus alba forest with a Knoxville, Tennessee climate. In this application, the responses of the two simulated forests were compared to observations from nearby sites (leaf areas, gross primary production, water yield, etc.). Subsequently, Stevens et al. (in press) applied the HYBRID model to the prediction of the response of an evergreen needle-leaved forest Picea rubens for the British Isles and the Iberian Peninsula to inspect the simultaneous and independent effects of increased atmospheric CO_2 and of climate change.

Gap model parameterisations have been developed for the northern boreal forest (Shugart et al. 1992a), and models have been developed for most of the temperate deciduous forests of the northern hemisphere (see Shugart et al. 1992b for review and reference citations). The models have also been applied on sample grids to simulate possible changes at continental scales to climatic change (Solomon 1986, Smith et al. 1995). These large scale applications all point to the potential of these models to produce results at a scale of interest to those seeking to understand global change, and to provide a capability to better understand the vegetation structure and consequences of structure that effect the response of vegetation to large-scale change in the environment.

REFERENCES

Bormann FH, Likens GE (1979) Pattern and Process in a Forested Ecosystem. Springer-Verlag, New York

Botkin DB, Janak JF, Wallis JR (1972) Some ecological consequences of a computer model of forest growth. J Ecol 60:849-872

Bugmann HKM, Xiaodong Y, Sykes MT, Martin P, Lindner M, Desanker PV, Cumming SG (in press) A comparison of forest gap models: Model structure and behaviour. Climatic Change

Delcourt HR, Delcourt PA, Webb T III (1983) Dynamic plant ecology: The spectrum of vegetation change in time and space. Quaternary Science Reviews 1:153-175

Friend AD, Shugart HH, Running SW (1993) A physiology-based model of forest dynamics. Ecology 74:792-797

Körner C (1993) CO_2 fertilization: The great uncertainty in future vegetation development. in Solomon AM, Shugart HH (eds) Vegetation Dynamics and Global Change, 53-70. Chapman and Hall, New York

Mankin JB, O'Neill RV, Shugart HH, Rust BW (1977) The importance of validation in ecosystem analysis. in Innis GS (ed) New Directions in the Analysis of Ecological Systems. Part I. Simulation Councils of America, 63-71. LaJolla, California

Munro DD (1974) Forest growth models: A prognosis. in Fries J (ed) Growth Models for Tree and Stand Simulation, 7-21. Res. Note 30, Royal College of Forestry, Stockholm

Pacala SW, Canham CD, Silander JA Jr (1993) Forest models defined by field measurements: I. The design of a northeastern forest simulator. Can J For Res 23:1980-1988

Rosswall T, Woodmansee RG, Risser PG (eds) (1988) Scales and Global Change. SCOPE 35. John Wiley, Chichester

Shugart HH (in press) The importance of structure in understanding global change. in Walker BH (ed) Global Change in Terrestrial Ecosystems. Cambridge University Press, Cambridge

Shugart HH, West DC (1980) Forest succession models. BioScience 30:308-313

Shugart HH, Emanuel WR (1985) Carbon dioxide increase: The implications at the ecosystem level. Plant, Cell and Environment 8:381-386

Shugart HH (1984) A Theory of Forest Dynamics: The Ecological Implications of Forest Succession Models. Springer-Verlag, New York

Shugart HH, Leemans R, Bonan R (eds) (1992a) A Systems Analysis of the Global Boreal Forest. Cambridge University Press, Cambridge

Shugart HH, Smith TM, Post WM (1992b) The application of individual-based simulation models for assessing the effects of global change. Annual Reviews of Ecology and Systematics 23:15-38

Smith TM, Shugart HH (1993) The transient response of terrestrial carbon storage to a perturbed climate. Nature 361:523-526

Smith TM, Shugart HH, Bonan GB, Smith JB (1992) Modeling the potential response of vegetation to global climate change. Advances in Ecological Research 22:93-116

Smith TM, Halpin PN, Shugart HH, Secrett CM (1995) Global Forests. in Strzepek KM, Smith JM (eds) If Climate Changes: International Impacts of Climate Change, 146-179. Cambridge University Press, Cambridge

Solomon AM (1986) Transient response of forests to CO_2-induced climate change: Simulation experiments in eastern North America. Oecologia 68:567-579

Stevens AK, Friend AD, Mobbs DC (in press) Ecosystem responses to climate and atmospheric CO_2 change: A physiological approach. Climatic Change

Woodward FI (1987) Stomatal numbers are sensitive to increases in CO_2 from pre-industrial levels. Nature 327:617-618

Section 4

Evolutionary responses to past changes

Species' habitats in relation to climate, evolution, migration and conservation

Elisabeth S. Vrba
Department of Geology and Geophysics
Yale University, P. O. Box 6666
New Haven, Connecticut 06511
U. S. A.

The theory of how organisms and species relate to their habitats is deeply relevant not only to the study of long-term evolution but also to future conservation efforts. A species' habitat consists of resources which comprise physical variables such as temperature, light, substrate and inorganic ions, and biotic variables such as prey, including plants for herbivores, host organisms and mates. Although species may be flexible in the range of resources under which they can live, no species can live in all environments. The limits of species' habitat specificities with respect to variables such as temperature, rainfall, substrate, food and vegetation cover, can in principle be estimated and quantified for living species.

I have elsewhere explored and tested several habitat-related hypotheses and predictions that together form a compatible 'habitat theory' (Vrba 1992, 1995a, b). Relevant to these hypotheses are three background sets of observations.

1. Species have commonly had long durations, surviving through successive strong global climatic oscillations. For instance, Plio-Pleistocene climatic and vegetational changes of large scale recurred (contributions in Vrba et al., 1995) with periodicity about 1/20th or less of the duration of the average mammalian species (about 2 Ma; longer for many other taxa, Stanley 1979). It is well documented that species of diverse kinds survived many climatic cycles by the passive responses of geographic shifting and vicariance (fragmentation into separated or allopatric populations) of their distributions, while maintaining habitat fidelity (review in Vrba 1992).

2. There is strong evidence that habitat adaptations commonly have been constant features of entire clades for millions and even billions of years (e.g. Vrba 1995a, b; rigorous evidence of this kind comes from mapping the limits of species' habitat

NATO ASI Series, Vol. I 47
Past and Future Rapid Environmental Changes:
The Spatial and Evolutionary Responses of
Terrestrial Biota
Edited by Brian Huntley et al.
© Springer-Verlag Berlin Heidelberg 1997

specificities with respect to temperature, food, etc. onto cladograms with radiometrically dated branching nodes). Together 1. and 2. imply that species and their habitat adaptations are recalcitrant to change and that only unusual combinations of circumstances can provoke the evolution of new species and altered habitat specificities.

3. There is widespread consensus among evolutionists that allopatric speciation has strongly predominated in the history of life. That is, vicariance into allopatric or separated populations usually precedes speciation. Allopatry can result from dispersal over pre-existing barriers, but most often it results from *in situ* fragmentation of populations by physical changes. Thus, if allopatric speciation has been predominant in the history of life, then initiation of speciation by tectonic and climatic changes must have been the predominant initiating cause. However, allopatry on its own is clearly not sufficient, and something additional must be required to bring about speciation. I have suggested (Vrba 1995a) that small populations must be isolated in altered new environments *for a sufficiently long time* for the required evolutionary change.

In this brief communication I can only list several of the hypotheses and predictions of habitat theory with some comments on implications for conservation. I expand on one of the hypotheses, the one concerning availability of migration routes during climatic cycles.

SOME PREDICTIONS OF HABITAT THEORY

1. Changes in geographic distributions by vicariance and/or latitudinal and altitudinal shifting have been much more prevalent responses to climatic changes than speciation and extinction.
   There are disturbing implications for conservation on a future earth that is changing progressively towards a 'greenhouse': Most living species have suffered human-induced shrinkage of their geographic distributions, and many of the corridors through which their distributions could spread or move to new areas in response to past climatic changes are now obliterated. That is, the incidence of survival by distribution changes is expected to decrease on a greenhouse earth at the expense of a large increase in the incidence of extinction.

2. The 'eco-shuffle' prediction. To some extent the ecological associations of taxa are predicted to be different at different extremes of the climatic cycles. That is, if species are moved about as climatic change moves their habitat components, then their communities in ecosystems are expected to be 'shuffled' in terms of species content.

One reason is that each species has a particular tolerance range for each habitat component that may or may not be identical to that of other species, and a combination of such ranges for all its requirements that is unique. Due to the complex interactions of factors like topography with climatic regimes, the associations of different environmental variables underwent recombination during past times, and some recombination among taxa in communities necessarily followed suit. In agreement with these expectations, there is some evidence that extant species with Pleistocene records did occur in past taxonomic associations different from their modern ones (review in Vrba 1992). For this reason, biome islands that have resulted from vicariance are unlikely to be true community refugia in the strict sense of providing refuge for an association of taxa that is identical to that of the more widespread parent community. An implication for the future translocation of endangered taxa to chosen conservation areas on a greenhouse earth is this: Such decisions need to be based on careful research on habitat specificities, and on projections of which species are expected to occur together in a given area under the altered environmental circumstances. The species membership of current communities is unlikely to be a reliable guide.

3. The 'resource-use hypothesis' predicts that clades of species whose resources persisted as environmental extremes came and went during their histories, had low rates of vicariance, speciation and extinction. Such eurybiomic lineages include generalists, that can flexibly use different resources in different environments, and specialists whose specialist patches persisted in alternative environments. In contrast, clades of species whose resources tended to disappear during one of the recurrent environmental extremes during their histories, had high rates of vicariance, speciation and extinction. This stenobiomic category includes clades of specialists on one or more kinds of resources that are confined to a narrow biome range.

There is a great deal of support for this (see analysis and review in Vrba 1987).

An obvious implication for conservation on a greenhouse earth is that specialists on the cooler biomes on earth will be particularly at risk. Failure to understand the subtle habitat-specificities of endangered species has resulted in some costly failures of conservation efforts. For instance, the Tatra Mountain Ibex, *Capra ibex ibex*, became extinct in Czechoslovakia but was successfully reintroduced from Austria. Subsequently, *C. i. aegagus* from Turkey and *C. i. nubiana* from Sinai were added to the Czech herd. The resulting hybrids expressed the native seasonal habitat specialization of their middle-eastern parents and rutted in early autumn, rather than in winter as the Czech ibex did. As a result, their young were born in February, the coldest month of the year, which led to the extinction of the entire Czech population (Greig 1979).

4. Latitudinal biases. During periods of strong latitudinal thermal contrasts and little change in amplitude of the Milankovitch cycles, biomes closest to the equator should in general have higher speciation and extinction rates than biomes at adjacent, higher latitudes. At the first major intensification of the cold cyclic extremes, after a longer period of roughly constant amplitude, a massive preponderance of extinctions over speciations occurs at low latitudes. Conversely, at such intensification of the warm cyclic extreme, more extinctions than speciations occur at high latitudes.

The implication of the last part of this prediction for a greenhouse earth is that the high latitude areas are at particularly high risk of massive extinctions, especially if they are geographical 'cul-de-sac areas'.

5. Topographic biases. Of two areas of similar large size, both subject over the same time to Milankovitch cyclic extremes, that remain habitable for organisms, the area that is more diverse in topography will have higher incidences of selection pressures and vicariance per species. (Topographic diversity refers to the number of patches of high and low elevation, and of different substrates, per area. It generally implies climatic diversity, and on a land surface vegetational diversity). The prediction is that higher rates of speciation, extinction and distribution change will occur in the topographically more diverse area than in the one that is less so. The sub-prediction of particular interest is that, *ceteris paribus*, topographically diverse areas are more likely to be factories of species diversity because there is constant fragmentation of species' habitats and distributions.

The implications for a greenhouse earth is that the high altitude areas are at particularly high risk of massive extinctions.

6. Major shifts in the periodicity pattern of the astronomical cycles, given that prolonged vicariance is important for turnover (speciation, extinction and migration), are predicted to promote bursts of speciation and extinction (Vrba 1992). Changes in dominance from a cyclic pattern of shorter period to one of longer period, together with progressive severity of the cyclic extremes, are especially significant for prolongation of vicariance across many lineages and therefore for increase in speciation and extinction rates.

   deMenocal and Bloemendal (1995) documented examples of such major global shifts: A shift occurred prior to 2·8 Ma from dominant climatic influence at 23 - 19 ka periodicity to dominance at 41 ka variance thereafter, with a further shift to dominance at 100 ka variance after 0·9 Ma.

7. Larger trends in the mean of the climatic cycles, given that prolonged vicariance is important for turnover, are predicted to elicit strong turnover in a particular sequential pattern. During a cooling trend with decreasing maxima of the climatic cycles, the warmer-adapted species are predicted to enter vicariance, become extinct and speciate before the cooler-adapted ones do. The reverse sequence is predicted during a warming trend.

   The pattern in African antelopes (Bovidae) during the cooling trend after 2·9 Ma supports this. Even if a cooling or warming trend has a gradual mean, the model that long-term vicariance is needed for speciation predicts a punctuated response in the responding biotic variables via threshold effects.

8. The 'Turnover Pulse Hypothesis'. If physical changes initiate most speciations, extinctions and migrations (which together comprise turnover), then turnover involving different kinds of organisms should be concentrated non-randomly in time and in predictable association with the climatic record for that area.

   This hypothesis is more complex than is suggested by the statement above, including the following elements (Vrba 1985, 1995a): Global climatic and tectonic changes are causal agents of turnover; pulses occur at different scales of time, geography and numbers of lineages involved; biotic interactions are important in influencing the nature of turnover, although the initiation is reserved for physical change; and lineages respond differently depending on intrinsic properties. Similar phenomena at different scales are relevant to this hypothesis. Scientists are accustomed to recognize hierarchical nesting of similar phenomena, or similarity among phenomena at different scales, especially recently with the growing exploration of fractal structures and processes – from those involved in the sutures of ammonites, through tree growth, to coastlines (Mandelbrot *et al.* 1979; Gleick 1987). There is an hierarchical *geographical* layering relevant to turnover: 'habitat plates' shifting rapidly over the continental plates that drift more sedately beneath them. There is also hierarchical nesting of *climatic change* which contains several tiers of higher-frequency cycles that are nested within lower-frequency cycles. The variation among taxa in breadth of habitat tolerance dictates that the more ecologically-specialised should respond with turnover to smaller climatic changes and more often than the more generalized taxa. In fact, both the hierarchically-nested pattern of the climate signal itself, and the variation in breadth of habitat tolerance among taxa, suggest that *biotic responses to climate should present an hierarchically-nested pattern at different scales as well.*

9. Migration biases. On an earth with polar ice, trans-continental migration will overall favour cooler-adapted terrestrial species, while migration between marine basins will favour warmer-adapted marine species. For the case of latitudinal migration, over longer time a larger number of cooler-adapted terrestrial species will migrate to land at lower latitudes than will warmer-adapted species to higher latitudes, while a larger number of warmer-adapted marine species will migrate to marine basins at higher latitudes than *vice versa*. With respect to the smaller scales of seasonal migration, land taxa will be more disrupted by the sea-level rise expected during global warming than will marine taxa.

Conservationists need to understand the details of migrations both at the larger geographic scales of habitat changes implied by global warming (by models based on past and present evidence), and at the smaller seasonal scales. An example of a conservation effort that failed due to lack of such information is the isolated elephant *Loxodonta africana* population in Knysna at the tip of Africa that was by 1990 on the verge of extinction (Dudley 1990). Analysts concluded in retrospect that a strong contributory factor was that the reserve allotted to these animals separated them from part of their resource spectrum by cutting off their previous seasonal migration routes. In this case the agency that interrupted the migration corridor was not climate but a highway. Nevertheless, the principles remain the same. I shall next expand on the predictions for migration at larger geographic and temporal scales, discussed previously as the 'Traffic Light Model' (Vrba 1995a).

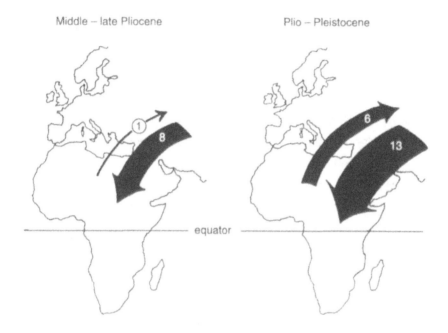

Figure 1: Imbalance in migration of antelopes (Bovidae) between Eurasia and Africa during the Miocene (Thomas 1984) and Plio-Pleistocene (Vrba 1995b). Numbers on the arrows are numbers of species that migrated between the continents.

THE 'TRAFFIC LIGHT MODEL' OF MIGRATION

Darwin (1859, p. 379) was intrigued by the "remarkable fact ... that many more ... forms have apparently migrated from the north to the south, than in the reversed direction." He speculated that this was largely due "to the northern forms having ... been advanced through natural selection and competition to a higher stage of perfection or dominating power, than the southern forms. And thus, ... the northern forms were enabled to beat the less powerful southern forms." Precisely the same explanation was suggested more recently for the 'Great American Interchange' of mammals after the Isthmus of Panama emerged (Marshall *et al.* 1982); but a reanalysis suggests that an explanation based on habitat theory, that does not draw on competitive advantages of northern forms, may fit the data better (Vrba 1992). Another example is shown in Figure 1: The Miocene imbalance between antelope (Bovidae) migrants from Eurasia to Africa and migrants in the opposite direction was 8 to 1, and during the Plio–Pleistocene it was 13 to 6. In the following proposal of a model to explain imbalances among inter-landmass migrants and among migrants between marine basins, I shall discuss the terrestrial case first and in more detail, and exemplify it by migration between Eurasia and Africa (Figure 2; the basic principles apply also to marine imbalances). Land migration can only occur when both a landbridge is present and a suitable habitat extends across it from the ancestral habitat. The model uses evidence (Denton 1995) that, in each case of cooling or warming on an earth with polar ice, the ice-volume response (and with it eustatic sea-level change) is delayed relative to land cooling or warming and biome response. For the African–Eurasian case, I here accept that during global warming the landbridge was significantly reduced from its maximal glacial extent, and at times flooded. During glacial maxima the land connection was much more extensive than it is today, especially southwards.

During global cooling, changes to new, seasonally-cooler habitats on either side of the persistent water barrier precede polar ice buildup, sea-level lowering and landbridge enlargement. Thus, the first arrivals from North to South are limited by landbridge enlargement, not by habitat in the immigrant area. The 'traffic light turns green' for migrants only once the habitat and landbridge conditions are both met, and then stays green for a long time while cooler habitat conditions and lowered sea-level coincide throughout most of a glacial period (Figure 2). On a cooling earth, migrants that traverse towards the equator, here Eurasia to Africa, are strongly favoured because the vector of moving habitats carries along more higher latitude migrants to the landmass at a lower latitude than *vice versa*. Fewer lower-latitude species, that were confined to African high altitudes during the previous warm phase, would also gain habitat passage to Asia as their habitats link up through lowlands with similar ones moving in from the North.

During global warming, polar melting lags behind land warming and biome change: The vegetation and other habitat changes should spread to higher latitudes and altitudes relatively fast (e.g. Dupont & Leroy 1995). The short delay in melting of ice caps allows a brief interval during which landbridges remain open *and* are already covered by suitable habitat for some species. Thus, the first arrival in Eurasia of African immigrants should record the spread to higher latitudes of mesic habitats before landbridge reduction. The abrupt ending of such a migration episode would signify the time when ice melting has progressed sufficiently to sever the landbridge or at least those parts of it that were formerly *habitable* and thus traversable by migrants. This lag period, between land warming and sea-level rise, is the only time during the entire warmer phase of the cycle when both conditions for passage of warm-adapted species are met simultaneously (Figure 2).

This brief interval favours migrants *away from* the equator, the direction of moving habitats. But such immigrants to Eurasia are likely to be few because this 'traffic light' turns red again quickly as sea-level rise reduces landbridges.

Given a suitable habitat, and a means to traverse to a new area, mammals can move very quickly in geological terms (Kurten 1957). Thus, the earliest arrivals of immigration episodes can date landbridge emergence during cooling and land biome response during warming. In sum, there has been a strong bias for an equatorward migration-vector for terrestrial species, dictated by the linked climatic effects of global temperature and sea level. An added factor during the Cenozoic is that there have been more frequent and averagely-longer global cooling events than warming excursions. On an earth with most landmasses situated at high northern latitudes, through long time a preponderance of North to South movements results as Darwin (1859) noted. However, this has little to do with the northern forms having "been advanced ... to a higher stage of perfection or dominating power, than the southern forms" (Darwin, 1859, p. 379), but rather with the relative lengths of cooler and warmer time during which thoroughfares of suitable habitat were available.

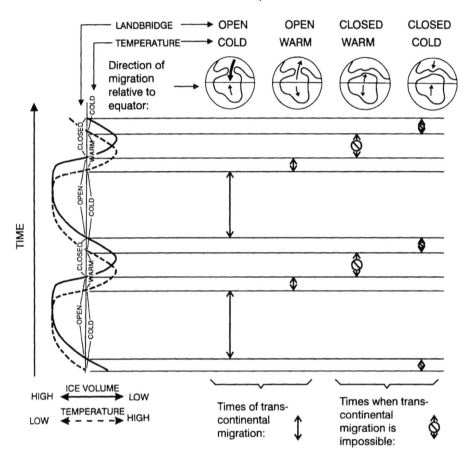

Figure 2. The 'traffic light' model of trans-land migrations. When the earth is cold, the landbridge is open (the 'traffic light' is green) for a long time because cool habitats extend to the equator *and* sea-level is low. In contrast, when the earth is warm, the landbridge is reduced or gone for most of the time. Only during the short lag between onset of global warming and ice-volume response can migration towards higher latitudes of warmer-adapted forms occur. Even if interglacials and glacials are equivalent in length, the interplay of eustasy and temperature would ensure that more species move towards lower latitudes than *vice versa*.

One can similarly explore land migration between East and West, and marine migration between North and South and between East and West. Essentially, land migration overall favours cooler-adapted species, and the warmer-adapted forms generally have more restricted opportunities for land passage, namely only during the first onset of global warming before polar ice masses melt sufficiently to obliterate landbridges. For instance, land passage across the Bering landbridge by temperate forms is expected to have been restricted to the earlier parts of interglacials. In contrast, marine migration overall favours warmer-adapted species, and the cooler-adapted forms have briefer opportunities for passage between marine basins, during the first onsets of global cooling before polar ice masses build up sufficiently to sever sea connections.

CONCLUSION

Research on how organisms and species relate to their habitats is absolutely basic to understanding both the causes of evolutionary turnover (speciation, extinction and migration) and conservation biology. Current estimates suggest that of the 10 to 31 million species on the planet, and considering only projected human population growth – without starting to consider the effects of greenhouse warming – we must anticipate losing between 3 to 16 million species. Among vertebrates alone some 2000 species require the most drastic conservation measures, while the situation is far worse for the vastly more numerous invertebrates, and for plants. While the above evolutionary predictions symmetrically concern both shrinkage *and* spreading of geographic distributions, and both extinction *and* speciation, the analogous predictions for conservation on a greenhouse earth heavily emphasize distribution shrinkage and extinction. Whether the hypotheses and predictions that I have here suggested turn out to be largely correct or are replaced by alternative ones, all such hypotheses necessarily have as a cornerstone the interactions between species and their habitats. There is a general and urgent need that research institutions everywhere should increase their involvement in empirical and theoretical research on the habitat specificities of organisms and species.

REFERENCES

Darwin C (1859) On The Origin of Species by Means of Natural Selection, or the Preservation of favoured Races in the Struggle for Life. John Murray London deMenocal P, Bloemendal J (1995) Plio-Pleistocene subtropical African climate variability and the paleoenvironment of hominid evolution: a combined data-model approach. *in* Vrba ES, Denton GH, Partridge TC, Burckle LH (eds) Paleoclimate and Evolution, with Emphasis on Human Origins, 262-288. Yale University Press, New Haven, Connecticut

Denton GH (1995) The problem of Antarctic Pliocene climate and ice-sheet evolution. *in* Vrba ES, Denton GH, Partridge TC, Burckle LH (eds) Paleoclimate and Evolution, with Emphasis on Human Origins, 213-229. Yale University Press, New Haven, Connecticut

Dudley JP (1990) A chronicle of extinction: the Knysna elephants *Loxodonta africana africana*. Abstract Proceedings of IV Intnl Cong Syst and Evol Biol, University of Maryland, College Park, July 1990

Dupont LM, Leroy SA (1995) Steps towards drier climatic conditions in North-Western Africa during the Upper Pliocene. *in* Vrba ES, Denton GH, Partridge TC, Burckle LH (eds) Paleoclimate and Evolution, with Emphasis on Human Origins, 289-298. Yale University Press, New Haven, Connecticut

Gleick J (1987) Chaos. Making a New Science. Viking, New York

Greig JC (1979) Principles of genetic conservation in relation to wildlife management in Southern-Africa. S Afr Wildl Res 9:57-78

Kurten B (1957) Mammal migrations. Cenozoic stratigraphy, and the age of Peking man and the australopithecines. J Paleontol 31:215-227

Mandelbrot B, Laff A, Hubbard D (1979) Fractals and the rebirth of iteration theory. *in* Mandelbrot B (ed) The Beauty of Fractals, 151-160. Academic Press, London

Marshall LG, Webb SD, Sepkoski JJ, Raup DM (1982) Mammalian evolution and the Great American Interchange. Science 215:1351-1357

Stanley SM (1979) Macroevolution: pattern and process. WH Freeman and Co, San Francisco

Thomas H (1984) Bovidae (Artiodactyla:Mammalia) du Miocene du Sous-Continent Indien, de la Peninsule Arabique et det l'Afrique, biostratigraphie, biogeographie et ecologie. Palaeogeography, Palaeoclimatology, Palaeoecology 45:251-299

Vrba ES (1985) Environment and evolution:alternative causes of the temporal distribution of evolutionary events. S Afr J Sci 81:229-236

Vrba ES (1987) Ecology in relation to speciation rates: some case histories of Miocene–recent mammal clades. Evolutionary Ecology 1:283-300

Vrba ES (1992) Mammals as a key to evolutionary theory. J Mamm 73:1-28

Vrba ES (1993) Turnover-pulses, the Red Queen, and related topics. Am J Sci 293-A:418-452

Vrba ES (1995a) On the connections between paleoclimate and evolution. *in* Vrba ES, Denton GH, Partridge TC, Burckle LH (eds) Paleoclimate and Evolution, with Emphasis on Human Origins, 24-45. Yale University Press, New Haven, Connecticut

Vrba ES (1995b) The fossil record of African antelopes (Mammalia, Bovidae) in relation to human evolution and paleoclimate. *in* Vrba ES, Denton GH, Partridge TC, Burckle LH (eds) Paleoclimate and Evolution, with Emphasis on Human Origins, 385-424. Yale University Press, New Haven, Connecticut

Vrba ES, Denton GH, Partridge TC, Burckle LH (eds) (1995) Paleoclimate and Evolution, with Emphasis on Human Origins, Yale University Press, New Haven, Connecticut

The evolutionary response of vertebrates to Quaternary environmental change

Adrian M. Lister
Department of Biology
University College London
London WC1E 6BT
U.K.

INTRODUCTION

It has often been stated that the amount of evolutionary change occurring in the Quaternary was greater than in previous periods, as a result of the profound environmental changes which occurred. In very general terms, the mammalian evidence bears out this suggestion. Comparison of faunal lists for different stages of the British Quaternary, for example (taken from Stuart 1982), indicates not a single mammalian species common to the earliest Pleistocene (*ca.* 1·7–1·5 Ma) and late Pleistocene (*ca.* 120 ka BP to present, including both 'cold' and 'warm' faunas). These changes are a conflation of extinctions, immigrations and evolutionary transitions, but the implication of substantial and rapid faunal turnover is clear.

Moreover, the Quaternary saw the origin of many species whose adaptations can be seen to be related to environments which arose or became much more widespread within that period. The most celebrated of Quaternary mammals, such as the woolly mammoth *Mammuthus primigenius* and woolly rhinoceros *Coelodonta antiquitatis*, are among many which appeared in conjunction with the vast and peculiarly Quaternary ecosystem of the tundra-steppe, which covered much of the northern continents (Guthrie 1990). Their woolly coats and strongly grazing-adapted feeding organs were clearly adaptations to this environment. Among the living fauna, the polar bear *Ursus maritimus* is a famous example of a species which arose *de novo* in the Pleistocene, apparently from a coastal population of brown bear *U. arctos* (Kurtén 1968).

NATO ASI Series, Vol. I 47
Past and Future Rapid Environmental Changes:
The Spatial and Evolutionary Responses of
Terrestrial Biota
Edited by Brian Huntley et al.
© Springer-Verlag Berlin Heidelberg 1997

The patterns of faunal turnover seen in the Quaternary are the result of complex interactions among a range of causal factors and responses. Table 1 lists some of the main categories of response by vertebrates to environmental change. Two of these are being treated by other contributors to this volume, distributional shifts by Graham (1996) and extinction by Sher (1996). Others which are not strictly 'evolutionary' are behavioural and ecophenotypic responses. All of these factors interact, so that an understanding of the evolutionary responses of Quaternary vertebrates to environmental change would be incomplete without considering them.

Table 1: Responses of vertebrates to environmental change

- Behavioural accommodation
- Distributional shifts
- Ecophenotypic modification (non-genetic)
- Evolution (genetic)
  anagenesis (morphological or physiological change within an unbranching lineage)
  cladogenesis (division of one species into two)
- Extinction

BEHAVIOURAL EFFECTS

Behavioural flexibility is especially important among 'higher' vertebrates such as birds and mammals. If the environment changes, these animals may be able to survive by behavioural means, for example in sheltering from inclement weather or manipulating unusual foods. By this means they may 'avoid the need' to migrate or to evolve new adaptations. Nonetheless, behavioural accommodation can be an important precursor of evolutionary change itself. By allowing the species to remain in the habitat, animals will be subjected to natural selection to develop morphological or physiological adaptations to the new conditions, the so-called Baldwin Effect (Bateson 1988). An amusing example of behavioural flexibility came in a report from the Pacific island of Tokelau, where domestic pigs swim in the sea to catch fish (Anonymous 1983). Given enough time, one might speculate that this could lead to the evolution of suids with amphibious adaptations. Groups such as seals and whales presumably originated in this way.

Processes founded in behaviour may be impossible to observe directly in the fossil record, but I believe that an appreciation of their potential role is important for a realistic view of the responses of Quaternary vertebrates to past and present environmental changes.

ECOPHENOTYPIC EFFECTS

When different samples within a given lineage of Quaternary mammals are compared, statistically significant morphological differences are very commonly observed (Lister 1992, 1993a; Chaline & Werdelin 1993; Martin & Barnosky 1993). Not all such differences, however, are properly termed evolutionary, operationally defined here as phenotypic modification resulting from genetic change. Phenotypic change may also result from a direct effect of the environment on growth of the organism. Such changes are generally termed 'ecophenotypic'. Ecophenotypic effects, far from being trivial, have an important relation to adaptation and evolution.

Ecophenotypic modifications are most likely to be seen in simple features such as body size, and are very fast; changes can take effect in one or a few generations. The changes are also readily reversible. It is likely that many of the short-term size fluctuations observed in Quaternary mammals have a significant ecophenotypic component. Red deer *Cervus elaphus* in Britain, for example, decreased in body size through the Holocene (AM Lister, unpublished data), probably as a result of anthropogenic forest clearance, but living Scottish animals of small size, moved to the lush pastures of New Zealand, increased in mean body weight by a factor of two to three in the space of 15–20 years (Huxley 1931), coming to equal in size their early Holocene precursors.

Experiments show how body proportioning may also be affected directly by environmental effects. By transplanting eggs between regions, James (1983) showed that body proportions in red-winged blackbirds were influenced by environmental as well as genetic factors. In the laboratory, Weaver and Ingram (1969) found that pigs raised at 5°C had markedly stockier limbs than their siblings raised at 35°C. Variable limb stockiness is a common feature of Quaternary mammal species (e.g. in red and giant deer *Megaloceros giganteus*: Lister 1994), and although this may in many cases have a significant genetic basis, an ecophenotypic component cannot be ruled out. Changes in more complex characters are usually genetically-controlled, but even these can sometimes be influenced by ecophenotypic change if they are developmentally linked to, say, body size; an example is provided by the reduction in complexity of deer antlers in developmentally-stunted individuals (Lister 1995, 1996).

An understudied but very interesting aspect of direct developmental (ecophenotypic) effects is their possible adaptive aspect (Via *et al.* 1995). For example, reduction of body size on poor quality nutrition is advantageous because it reduces the volume of food the animal requires to survive (Lister 1995). Similarly, shortening of limbs in response to cold climate reduces the surface area available for heat loss. There are two possible links between these phenomena and genetically entrenched evolutionary changes. First, ecophenotypic effects may be the precursors of evolutionary change by the process described by Waddington (1953) as 'genetic assimilation', i.e. selection for animals genetically predisposed to respond ecophenotypically in the adaptive morphological direction. This process has been invoked in the explanation of dwarfing in island mammal populations, where initial ecophenotypic effects due to low nutritional plane will subsequently have become genetically entrenched and enhanced (Lister 1995).

Second, organisms may have become genetically adapted by natural selection to respond ecophenotypically in an adaptive way to environmental change. This phenomonon has been termed 'facultative adaptation' (Underwood 1954). The red deer, for example, is a species of broad environmental tolerances which, at the present day, is found in a range of vegetational habitats, and in the Quaternary persisted in northern and central Europe through both forested interglacials and open, grassy phases (Lister 1984). This may in part have been due to migrational alternation of differently-adapted populations. In addition, however, analysis of modern red deer shows that when raised on graze they develop a larger rumen space with unevenly distributed papillae, typical of grazing mammals such as the cow. Raised on browse, they show a smaller but more evenly papillated rumen (Hofmann 1983). By contrast the moose *Alces alces* lacks this facility and has a permanently 'browse-adapted' rumen. In consequence its range, both now and in the Quaternary, is seen to be more restricted to areas of forest than that of the red deer (Lister 1984). The ability to accommodate environmental change by facultative adaptation could, depending on the circumstances, either pre-empt the need for further evolutionary responses, or conversely, encourage evolution by enabling the animal to survive in a changed habitat where other selective forces will impinge on it. The facultative capacity in species such as the red deer may itself have evolved as a response to the constantly fluctuating environments of the Quaternary.

SIZE FLUCTUATION: A TEMPERATURE OR VEGETATIONAL EFFECT?

The most commonly observed difference between samples of a Quaternary mammalian lineage is in body size. As discussed above, this may have either an ecophenotypic or genetic basis. On an adaptive level, the explanation for such changes has been a subject of debate.

Many mammalian species, for example, show a marked reduction in body size at the Pleistocene/Holocene boundary (Kurtén 1968; Davis 1981). Davis (1981) shows that for a number of species in Israel (foxes, wolves and boars) this was due to a combination of environmental change and anthropogenic influence. The environmental factor in body-size reduction has generally been regarded as a response to temperature increase, according to the theory that body size responds inversely to ambient temperature as an adaptation for heat regulation by altering the surface area to volume ratio. This model, inspired by 'Bergmann's Rule' of alleged increased size with latitude within modern species, has been widely applied to body size differences in Quaternary mammals. Kurtén (1968) found oscillation in body size of several mammalian species, large forms occurring in cold stages, smaller ones in interglacials, and attributed the changes to 'Bergmannian' effects.

Others, however, have contested this interpretation, suggesting instead that the duration and abundance of the plant growing season is the primary determinant of herbivore body size, which in turn selects for corresponding changes in carnivore size (McNab 1971; Geist 1987; Guthrie 1990). Under this interpretation, the Pleistocene tundra–steppe is regarded as a richer habitat for many large herbivores (and hence carnivores) than the interglacial forests, and the size reduction found by Davis (1981) in the Middle East could result from the vegetational response to increased aridity. The dwarfing of mammals on islands can be seen in part as an extreme case of the same process, adaptive to a finite food resource (Lister 1995).

Guthrie (1990) has further suggested that the extinction of some mammals at the Pleistocene/Holocene boundary, and the size reduction in others, are parts of the same phenomenon, brought about by habitat deterioration (see Lister 1992 for further discussion).

RATES OF EVOLUTION: THE MODERN PERSPECTIVE

In considering the origin of significant, genetically-based evolutionary modifications, the rate at which organisms can adapt to environmental chage is clearly of crucial importance.

The capacity of animal and plant populations to respond very rapidly to environmental selective pressure is forcefully illustrated, as Darwin realised, in the origin of domestic strains. The diversity of morphology among domestic dogs, largely achieved within a few hundred years, is a graphic example of this. Numerous experiments demonstrate the possible rapidity of response; for example, Falconer (1973) achieved substantial genetically-based changes in body size of laboratory mice under strong artificial selection over only a few generations. In nature, significant morphological differences have been observed in a single generation. On the Galapagos island of Daphne Major, for example, Grant (1986) observed a significant increase in mean bill size of a finch species *Geospiza fortis* from one year to the next as a result of strong natural selection imposed by the differential survival of large-seeded plants in response to a drought.

Genetics also demonstrates that, in principle at least, relatively large phenotypic changes could occur by a simple, and potentially rapid, genetic switch. Stanley (1979) gives the example of achondroplasia, a condition resulting in dwarfing and short limbs among domestic animals, which occasionally arises as a single spontaneous mutation, and speculates that a similar mechanism might have produced short-limbed rhinoceros species known in the fossil record.

The geneticist's perspective leads to an expectation for rapid evolutionary transitions in relation to the geological timescale, although low selective pressures and genes of small individual effect could also be modelled, producing slower evolutionary change. In fact, a plausible genetic mechanism can be posited for almost any rate or pattern of evolution observed by a palaeontologist. Only by directly observing fossil sequences, therefore, can we observe what actually happened on Quaternary timescales.

RATES OF EVOLUTION: QUATERNARY EXAMPLES

Detailed studies of selected lineages of Quaternary mammals have illustrated a wide range of patterns of evolutionary change in terms of rate, the implication of speciation, and so on. Quaternary evolutionary studies, because of the time-scale in which they operate, offer the great promise of forging a link between the short-term processes observed by population geneticists, and the coarser evolutionary patterns of the more distant palaeontological past.

The fossil record of the moose, *Alces* spp., provides an instructive example of a relatively coarsely-resolved sequence. Lister (1993a,b) described a sequence of four samples spanning the entire Quaternary (*ca.* 1·7 Ma) which show directional reduction in antler length in each of the three steps. With half a million years or more between sampling points, however, the drawing of a straight, 'gradualistic' line between these four points is not justified. Eldredge and Gould (1972) were correct to point out that the assumption of straight-line evolution between a few known points has been a persistent feature of past palaeontological practice, where in fact the actual transitions may have been much more rapid in the intervening intervals. For the moose, we do not have evidence one way or the other. However, it is important not to lose sight of the fact that, whether or not it included speciation events or episodes of rapid phenotypic change, even a sequence of four samples is telling us something important about the pattern of evolution which is not deducible from population genetics. The moose took nearly two million years to reduce its antler beam length by two-thirds, and did so in a fashion which in an important sense is gradualistic, since it apparently progressed through stages of intermediate morphology, and not in a single jump. Nor should we be misled by calculations showing that such trends cannot be distinguished from a random walk (Lande 1976). This statistical observation does not mean that the transitions, with developmental, adaptive and biomechanical implications for the animal, were anything other than deterministic, driven primarily by natural selection.

A more finely-sampled sequence is that of the mammoth, *Mammuthus* spp., where fourteen dated samples have been analysed through the European Quaternary (Lister 1993a,b, in press). There is a strong trend for heightening of the tooth crown and increase in the number of molar enamel plates. Superimposed on this trend are minor reversals, due no doubt to population movements and evolutionary fluctuations. Fortey (1985) suggested that an increasing number of chronological points in a gradualistic sequence makes the postulation of unseen punctuations increasingly unnecessary. This is valid as a null hypothesis, but speciation events may nonetheless have intervened and we need to look out for them. In the case of the mammoths, the generally gradualistic sequence may have included two speciation events (cladogenesis), evidenced in Europe by temporal overlap between ancestral and descendent morphologies (Lister, in press). Even so, the morphological picture is one of relatively slow change through a series of intermediates, and it is uncertain whether morphological shifts were associated with the speciation events as punctuated equilibrium theory (Eldredge & Gould 1972) implies. Martin (1993) argued for a lack of correlation between speciation and morphological change in North American Quaternary rodents.

On an adaptive level, the morphological trends in the mammoth teeth are almost certainly connected with an observed progressive shift from temperate forest to cold tundra-steppe through the Pleistocene. It is difficult on present evidence to see if there are detailed correlations between episodes of morphological transition and specific environmental events.

An even more finely-resolved example of an evolutionary trend is provided by the water vole *Arvicola terrestris* through the European Middle and Late Pleistocene (van Kolfschoten 1990; Lister 1992). This lineage, sampled from more than 20 horizons across the period *ca.* 300 to 10 ka BP, displays a reduction in enamel differentiation index (the ratio of enamel thickness between the trailing and leading edges of the molar loops), probably linked to an increasingly grassy diet. There is an approximately gradualistic trend in index-reduction in central and north European voles from the Holsteinian (*ca.* 300 ka BP) to late-Weichselian (*ca.* 10 ka BP); this was interrupted by a reversal in the late Saalian which may have been due to an evolutionary reversal or a migration event (see below).

In general, reviewing Quaternary mammal evolution, I do not perceive clear evidence to support the theory that small mammals evolve faster than large ones. In terms of cladogenesis, it is clear that small mammals have produced greater numbers of species, for rather obvious demographic and ecological reasons. But in terms of rates of morphological change, there is no very clear difference between the dental sequences of the water vole and the mammoth, for example. I believe this is to be expected because neither of these lineages was evolving at anything near the maximal rate implied by laboratory selection experiments. Only closer to this limit might factors such as the faster generation time of small mammals have an effect. In the field, voles, mammoths and moose took of the order of $10^5 - 10^6$ years to complete significant adaptive morphological changes, and the slowness of these net rates must have been imposed by other factors such as environmental conditions, developmental constraints, and co-adaptation of the animal's total phenotype.

DISTRIBUTIONAL SHIFTS AND EVOLUTION

Distributional shifts can in some respects be perceived as a kind of 'alternative' to evolution in that, by tracking a preferred environment and thus maintaining themselves in an approximately constant habitat, individuals of a species, in teleological terms, 'avoid the need' for evolutionary modification. However, this perspective underplays the intimate links which exist between geography and evolution. Changes in distribution will often entail (i) widespread selective deaths, which may lead to adaptive evolution, and (ii) subdivision of populations, which may result in differentiation and even speciation.

An example of a distributional shift is provided by the European fallow deer *Dama dama* whose distribution in the Quaternary 'cold stages' was restricted to the Mediterranean region, but which in the interglacials expanded north as far as Britain and southern Scandinavia (Lister 1984). As the northern environment deteriorated at the end of each interglacial, the fallow deer's distribution shrank to the south. If we assume that the population density in southern Europe during the 'cold stages' was of the same order of magnitude as that throughout Europe during the interglacials, this range shrinking cannot be regarded simply as 'migration', but must have involved many selective deaths in the northern and central regions of Europe, implying high selection pressure if the change was rapid. In this situation, there is potential for individuals to survive within the 'old' range if they have appropriate morphological, behavioural or physiological traits – evolution will have occurred.

The opposite situation, where a species is *expanding* its range in response to environmental change, may also lead to evolutionary change. It has been suggested, for insects, that Quaternary shifts in environment resulted principally in niche-tracking, and that evolutionary change was thereby obviated. While this may be the case for insects if precisely the same microhabitat recurred in widely distant areas, it is unlikely to be the case for mammals which perceive the environment in a more 'fine-grained' way. For example, the environment encountered by the musk ox *Ovibos moschatus* when its range expanded into the tundra–steppe of mid-latitude Europe during the Quaternary cold stages was certainly different from that which it occupies in the Arctic tundra today in terms of climate, daylight regime and plant and animal communities (Guthrie 1990). Such changes of habitat alter the selective regime and may result in adaptive evolution. Geist (1971) has argued that the expansion of large mammals, such as bison and deer, into extensive, vacant new territories in the Quaternary was particularly potent in producing an r-selective regime resulting in evolutionary changes such as increases in body size and size of display organs.

A likely example of the interplay of distributional and evolutionary change is provided by the water vole, *Arvicola terrestris*, described above (van Kolfschoten 1990; Lister 1992). The approximately gradualistic trend of reduction in enamel differentiation index in central and north European voles through the Holsteinian and Saalian was interrupted by the appearance of more 'primitive' (high-index) animals in the Eemian, which then resumed the trend toward reduced index until the late Weichselian (Lister 1992, Fig. 8). The explanation appears to be that the late Saalian ice advance led to local extinction of *A. terrestris* in northern and central Europe, followed in the Eemian by re-immigration from the south and east, where even today water voles have a higher (more 'primitive') index (Röttger 1987). Once in the northern realm, a process of adaptive change re-commenced.

It is important to note that the apparently very 'rapid' late Saalian reversal of the morphological trend in European water voles was not a case of rapid evolution, but of immigration. Many such apparently rapid or instantaneous morphological jumps seen in the fossil record of a given region may be the result of migration events of already-differentiated populations. The genuine evolutionary changes which earlier produced that differentiation could have occurred more slowly. In practice, it can be difficult to separate the effects of migration and evolution, even in modern examples. Corbet (1975) described a population of bank voles *Clethrionomys glareolus* near Loch Tay, Scotland, which showed a significant change in dental morphology in the space of twenty years, but it is uncertain to what extent this was due to *in situ* evolution, population replacement, or gene flow from another population.

SPECIATION

Where Quaternary environmental changes resulted in the subdivision of species ranges, allopatric speciation sometimes resulted. This could occur by dispersal leading to the founding of a new, allopatric population, or by geographical subdivision of an existing population. There are various genetic and ecological factors likely to promote rapid evolution in small, isolated populations (e.g. Stanley 1979). The classic example of the Quaternary origin of the polar bear (Kurtén 1968) has been mentioned in the Introduction. This species, *Ursus maritimus,* is unknown before the Late Pleistocene, and almost certainly arose from a population of brown bear *U. arctos.* The two species are very closely related genetically, and indeed will produce viable offspring if forced to interbreed. Although there is only limited fossil evidence of the transition, Kurtén (1968) suggested that a northern coastal population of brown bears took to feeding on seals, and progressively developed the webbed feet, white coat and other adaptations which are characteristic of the polar bear.

Some of the clearest examples of mammalian speciation in the Quaternary are provided by the evolution of endemic species from populations isolated on islands. In offshore islands formed by sea level rise, isolation of the mammalian population can be seen as a direct result of Quaternary environmental change. Lister (1996) contrasts the effects of short-term (10^3–10^4 yr) versus long-term (10^5–10^6 yr) isolation on these populations. Examples of the former are the dwarf red deer *Cervus elaphus* which arose on Jersey in the last interglacial (Lister 1995); and the Holocene dwarf mammoths *Mammuthus primigenius* of Wrangel Island (Vartanyan *et al.* 1993). These populations became dwarfed in a few thousand years at most, but the evidence suggests that this time-span was not sufficient for full speciation relative to the mainland ancestor . Pressure for size reduction appears to have dominated, with minor morphological changes which were direct developmental or allometric effects of the dwarfing. On re-connection of the island (e.g. Jersey in the last cold stage), the nascent form was lost. On permanently isolated islands, on the other hand, where mammalian populations had much longer to evolve in isolation, adaptive morphological changes are seen and there is evidence of speciation. Examples are found on islands in the Mediterranean, where dwarf hippos, elephants and deer, and giant rodents and insectivores, evolved over several hundred thousand years of the Quaternary (Sondaar 1977). Adaptive changes in limb proportions and dentition, and unique antler constructions among the deer, are seen in these cases (Lister 1996).

In the Holocene, some mammalian populations show changes due to isolation, forming an analogy to the short-term evolutionary experiments seen in the fossil record of Jersey and Wrangel islands. Stanley (1979) cites Cameron's work on the mammal fauna of Newfoundland, where 10 of the 14 species have evolved into distinct subspecies since the island's isolation *ca.* 12 ka BP, and points out that the actual duration of change could have been much shorter than this. Similarly, Lister (1995) noted that subspecies of several European mammals had developed on Britain and its offshore islands during the Holocene, but that none had reached full speciation. In a few cases there is evidence of good biological species arising among vertebrates within a similar timespan, as in the endemic cichlid fishes of Lake Nabugabo, Uganda, derived from species of adjacent Lake Victoria within the 4 ka that the lakes have been separate (Greenwood 1965; Stanley 1979).

The consideration of speciation on islands leads to the important question of whether analogous 'islands' of terrestrial habitat, occurring on the mainland, were a significant stimulus for mammalian evolution there. Many modern species distributions do provide circumstantial evidence of allopatric speciation in the Quaternary. It is difficult to observe such processes palaeontologically, however, because a new species evolving from an allopatric isolate on the mainland would soon spread its range, and unless we have tightly correlated fossil sampling in the area of origin as well as outside, we would not resolve the speciation event. There is evidence that certain broad areas may have acted as foci for species formation in the Quaternary, however. Sher (1986) has suggested that the landmass of Beringia, spanning modern northeast Siberia, Alaska and adjacent shelf, where Quaternary cooling had its earliest and most persistent effect, and which yet remained largely ice-free, was the area of origin of many species adapted to the tundra–steppe. The new forms subsequently spread east and west into North America and Eurasia. On the other hand, the repeated nature of Quaternary climatic cycles may in some cases have acted against speciation by homogenising nascent species as ranges expanded and contracted (Bennett 1990; Hafner & Sullivan 1995).

VERTEBRATE RESPONSES TO ENVIRONMENTAL CHANGE: PAST, PRESENT AND FUTURE

In the Quaternary, it is clear that behavioural flexibility, genetic evolution, ecophenotypic effects and population movements interacted in complex ways to produce very significant and varied changes within vertebrate species in response to environmental changes, even if the effects of these different processes are difficult to separate in practice in the fossil record.

In moving from past observations to future predictions, it is important first to resolve the apparent paradox between the extremely fast rates of observed historical transitions and laboratory experiments on the one hand, and generally slower Quaternary evolution on the other, not to mention the even slower rates observed over longer geological timescales. Gingerich (1983, 1993) has gone some way to resolving this paradox by analysing evolutionary rates measured over different timescales. He showed that fast rates are maintained for only a short time, and the longer the observed timespan, the lower the measured rate. Long evolutionary sequences will have included short episodes of change as fast as the modern examples, but over time these were averaged out with periods of slower change or stasis. This effect is enhanced by the reversals of trend, seen in Quaternary examples, which act as 'negative rate' and slow the net progress. Moreover, many of the fast transitions observed in short fossil or historical sequences are probably not evolutionary at all, but result from ecophenotypic or migrational effects.

It is in any case doubtful whether the very fast rates of morphological change observed in laboratory experiments, or even through history in the development of domestic breeds, could be duplicated under natural conditions. The stringent conditions of adaptation and survival in the wild would normally render such rapidly altered forms inviable, except for relatively minor modifications. Longer timescales are required for the production of genotypically and phenotypically co-adapted organisms, whose development and structure are adjusted for both 'internal' integrity, and function in relation to the outside world.

Future environmental changes, whether climatically or anthropogenically forced, will have severe effects on vertebrate faunas. On the evidence of Quaternary examples, few vertebrate species will be able to evolve rapidly enough to adapt to the new conditions through natural selection, since the environmental changes will be intense and very rapid (10^3 yr or less) in relation to evolutionary timescales. In the large majority of cases, more likely responses will be behavioural accommodation, distributional shifts and extinction. Unlike long-term evolution, which is almost limitless in its ability to adapt species to virtually any available habitat, behavioural and distributional accommodation can go only so far in protecting organisms from drastic habitat change or loss. Where they fail, extinction will be the result.

REFERENCES

Anonymous (1983) Pigs can swim in Tokelau. The Guardian, London, May 6th

Bateson PPG (1988) The active role of behaviour in evolution. *in* Ho M-W, Fox SW, (eds) Evolutionary processes and metaphors, 191-242. Wiley, Chichester

Bennett KD (1990) Milankovitch cycles and their effects on species in ecological and evolutionary time. Paleobiology 16:11-21

Chaline J, Werdelin L (eds) (1993) Modes and tempos of evolution in the Quaternary. Quaternary International 19:1-116

Corbet GB (1975) Examples of short- and long-term changes of dental pattern in Scottish voles (Rodentia; Microtinae). Proc Zool Soc Lond 143:191-217

Davis SJM (1981) The effects of temperature change and domestication on the body size of Late Pleistocene to Holocene mammals of Israel. Paleobiology 7:101-114

Eldredge N, Gould SJ (1972) Punctuated equilibria: an alternative to phyletic gradualism. *in* Schopf TJM (ed) Models in Paleobiology, 82-115. Freeman, San Francisco

Falconer DS (1973) Replicated selection for body weight in mice. Genet Res 22:291-321

Fortey RA (1985) Gradualism and punctuated equilibria as competing and complementary theories. Spec Pap Palaeontology 33:17-28

Geist V (1971) The relation of social evolution with dispersal in ungulates during the Pleistocene, with special reference on the old world deer and the genus *Bison*. Quat Res 1:283-315

Geist V (1987) Bergmann's rule is invalid. Can J Zool 65:1035-1038

Gingerich PD (1983) Rates of evolution: effects of time and temporal scaling. Science 222:159-161

Gingerich PD (1993) Quantification and comparison of evolutionary rates. Am J Sc 293A:453-478

Graham RW (1996) The spatial response of mammals to Quaternary climate changes. *in* Huntley B, Cramer W, Morgan AV, Prentice HC Allen JRM (eds) Past and future rapid environmental changes: The spatial and evolutionary responses of terrestrial biota, 153-162. Springer-Verlag, Berlin

Grant PR (1986) Ecology and evolution of Darwin's finches. Princeton University Press, Princeton

Greenwood PH (1965) The cichlid fishes of Lake Nabugabo, Uganda. Brit Mus Nat Hist Bull (Zool) 12:315-357

Guthrie RD (1990) Late Pleistocene faunal revolution - a new perspective on the extinction debate. *in* Agenbroad LD, Mead JI, Nelson LW (eds) Megafauna and man: discovery of America's heartland, 42-53. The Mammoth Site of Hot Springs South Dakota, Hot Springs, South Dakota

Hafner DJ, Sullivan RM (1995) Historical and ecological biogeography of Nearctic pikas (Lagomorpha: Ochotonidae). J Mammal 76:302-321

Hofmann RR (1983) Adaptive changes of gastric and intestinal morphology in response to different fibre content in ruminant diets. *in* Wallace G, Bell L (eds) Fibre in human and animal nutrition. The Royal Society of New Zealand, Bulletin 20

Huxley JS (1931) The relative size of antlers in deer. Proc Zool Soc Lond 1931:819-864

James FC (1983) Environmental component of morphological differentiation in birds. Science 221:184-186

Kolfschoten T van (1990) The evolution of the mammal fauna in The Netherlands and the middle Rhine area (western Germany). Meded Rijks Geol Dienst 43:1-69

Kurtén B (1968). Pleistocene mammals of Europe. Weidenfeld and Nicolson, London

Lande R (1976) Natural selection and random genetic drift in phenotypic evolution. Evolution 30:314-334

Lister AM (1984) Evolutionary and ecological origins of British deer. Proc R Soc Edinb 82B:205-229

Lister AM (1992) Mammalian fossils and Quaternary biostratigraphy. Quat Sc Rev 11:329-344

Lister AM (1993a) Patterns of evolution in Quaternary mammal lineages. Linn Soc Symp Ser 14:71-93

Lister AM (1993b) 'Gradualistic' evolution: its interpretation in Quaternary mammal lineages. Quaternary International 19:77-84

Lister AM (1994) The evolution of the giant deer, *Megaloceros giganteus* (Blumenbach). Zool J Linn Soc 112:65-100

Lister AM (1995) Sea levels and the evolution of island endemics: the dwarf red deer of Jersey. Geol Soc Spec Pubs 96:151-172

Lister AM (1996) Dwarfing in island elephants and deer: process in relation to time of isolation. Symp Zool Soc Lond 69:277-292

Lister AM (in press) Evolution and taxonomy of Eurasian mammoths. *in* Shoshani J, Tassy P (eds) The Proboscidea: Trends in Evolution and Paleoecology. Oxford University Press, Oxford

Martin RA (1993) Patterns of variation and speciation in Quaternary rodents. *in* Martin RA, Barnosky AD (eds) Morphological change in Quaternary mammals of North America, 226-280. Cambridge University Press, Cambridge, UK

Martin RA, Barnosky AD (eds) (1993) Morphological change in Quaternary mammals of North America. Cambridge University Press, Cambridge UK.

McNab BK (1971) On the ecological significance of Bergmann's Rule. Ecology 52:845-854

Röttger U (1987) Schmeltzbandbreiten an Molaren von Schermäusen (*Arvicola* Lacépède, 1799). Bonner Zool Beitr 38:95-105

Sher AV (1986) On the history of mammal fauna of Beringida. Quartärpaläontologie 6:185-193

Sher AV (1996) Late Quaternary extinction of large mammals in Northern Eurasia: A new look at the Siberian contribution. *in* Huntley B, Cramer W, Morgan AV, Prentice HC Allen JRM (eds) Past and future rapid environmental changes: The spatial and evolutionary responses of terrestrial biota, 319-340. Springer-Verlag, Berlin

Sondaar PY (1977) Insularity and its effect on mammalian evolution. *in* Hecht MK, Goody PC, Hecht BM (eds) Major patterns in vertebrate evolution, 671-707. Plenum, New York

Stanley SM (1979) Macroevolution. Freeman, San Francisco

Stuart AJ (1982) Pleistocene vertebrates in the British Isles. Longman, London

Underwood G (1954) Categories of adaptation. Evolution 8:365-377

Vartanyan SL, Garutt VE, Sher AV (1993) Holocene dwarf mammoths from Wrangel Island in the Siberian Arctic. Nature 362:337-340

Via S, Gomulkiewicz R, De Jong G, Scheiner SM, Schlichting CD, Van Tienderen PH (1995) Adaptive phenotypic plasticity: consensus and controversy. Trends Ecol Evol 10:212-217

Waddington CH (1953) Genetic assimilation of an acquired character. Evolution 7:118-126

Weaver ME, Ingram DL (1969) Morphological changes in swine associated with environmental temperature. Ecology 50:710-713

The weight of internal and external constraints on *Pupilla muscorum* L. (Gastropoda: Stylommatophora) during the Quaternary in Europe

Denis-Didier Rousseau[1]
Université Montpellier II
Institut des Sciences de l'Evolution
Paléoenvironnements & Palynologie
place Eugène Bataillon, case 61
34095 Montpellier cedex 05
France

INTRODUCTION

Pupilla muscorum L. is a terrestrial gastropod which is very common in European Quaternary deposits. It shows great morphological variability, mainly in shell size. To understand how this variability is expressed, a survey of modern and Quaternary European specimens of *Pupilla muscorum* was undertaken using multivariate analyses to characterize the morphology. This was followed by an ontogenic study of shell sections to determine the changes in the growth timing. The first step relates to intra-population variability which is mainly expressed in shape variation. The shape of the shell lies between two extremes: a fat shell with a slightly protruding aperture and a slim shell with a strongly protruding aperture. The second step describes inter-population variability, both temporal (for fossil populations) and spatial (for modern populations), which is mainly in size. During the Quaternary, the size variation parallels the climatic cycles: large shells during pleniglacial phases, small shells during interglacial ones. At present, N – S and E – W size gradients are present with the adult size varying between 2·5 and 4·7 mm. The final step addresses the occurrence of this growth variability by studying morphological indices, which

---

[1] Also at: Lamont-Doherty Earth Observatory of Columbia University, Palisades, N.Y. 10964, U.S.A.

NATO ASI Series, Vol. I 47
Past and Future Rapid Environmental Changes:
The Spatial and Evolutionary Responses of
Terrestrial Biota
Edited by Brian Huntley et al.
© Springer-Verlag Berlin Heidelberg 1997

characterise size and spire of the shell, against the number of whorls taken as age reference. The variations in these indices show that heterochronic processes (*sensu* McNamara 1986), acceleration and hypermorphosis, induce a supernumerary part-whorl of the spire in the juvenile phases of the growth. Because *P. muscorum* is in evolutionary stasis during the Quaternary, such morphological variations, channelled by external and internal constraints, are a good example of ecophenotypic iterative changes at the intraspecific level.

What does the ontogeny of a fossil organism represent? The study of the palaeontologist focuses on the effect of ontogeny on morphology which results from the interaction between internal (direct influences) and external (indirect influences) that express themselves by channelling phenotypic potentialities (Waddington 1974; Gould 1980; Alberch 1982).

An example of these constraints can be provided by the analysis of morphological variability in the terrestrial molluscs during the Quaternary in Europe. Quaternary deposits are frequently rich in malacofauna which permits precise reconstructions of past environments (Rousseau 1987). However, among the numerous terrestrial mollusc species, the pulmonate *Pupilla muscorum* L. is particularly common, especially in loess sections, and shows great morphological variability, mainly in shell size. The aim of this paper is to characterise both intra- and inter-population morphological variations and to explain this variability by studying changes in timing of the ontogeny of the shell.

MATERIALS AND METHODS

P. muscorum is a small Holarctic species (2·5 to 4·7 mm high) living in open environments. The number of whorls, the size and the shape of the shell show a large variability. The species first appeared in the Early Pliocene (Wenz 1923).

Whatever the climatic conditions, *P. muscorum* is one of the more common species in Pleistocene malacofaunas. Generally, the populations have abundant individuals,

Figure 1: Morphometry of *Pupilla muscorum* used for the biometric study. The number of whorls is counted after the embyonic shell. All these parameters are studied by correspondence analysis.

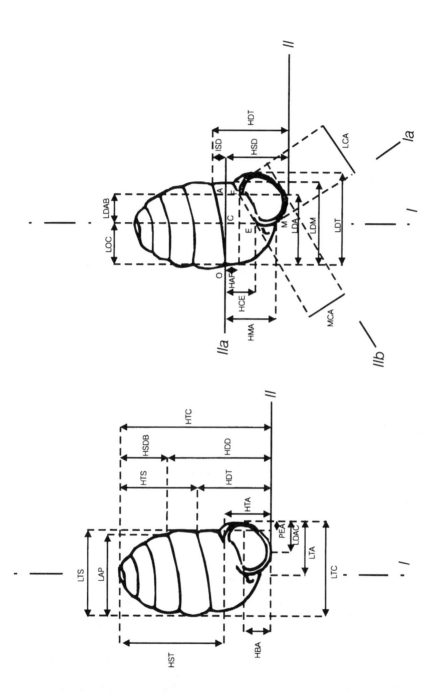

and show variations in the size of the shell. Among the Upper and Middle Pleistocene deposits analyzed in Achenheim (Alsace, France) there are more than 55 stratigraphic levels providing many fossil populations in which *P. muscorum* is abundant. From these populations, 709 individuals (up to 30 individuals per population, randomly selected from the total number) have been measured. Other fossil populations from French sites were added to those from Achenheim in order to improve the analysis of inter-population variations. The fossil variations were then compared to those in recent populations.

Measurements proceeded as indicated on Figure 1. In addition, in order to understand the morphological variations, the ontogeny was anaysed by studying indices characteristic of the spire (spire index SpL = H/L ,Cain 1977), the size (size index SI = H+L) and ß, the angle that each section of whorl spire makes with regard to the coiling axis.

RESULTS

Morphological intra-population variation during the Pleistocene in Europe

The spire fluctuations influence the morphology of the shell. They mainly affect the embryonic and infantile phases, which constitute the more sensitive growth stages of the animal to the environment. Certain shell size variations involve the shell dimensions and/or the whorl number within each population.

Shape variation of the shell is also evident within populations. It results from parallel tendencies between the spire and final whorl dimensions which establish two

---

Figure 2: Size variations in *Pupilla muscorum*. **Recent populations:** 1. Cracow RA10 (HTC = 4·38 mm); 2. Arcy-sur-Cure YC16 (HTC = 3·53 mm); 3. La Rochepot RO17 (HTC = 2·41 mm); 4. Villers-la-Faye LF22 (HTC = 2·39 mm). **Fossil populations:** 5. Achenheim CL01 (HTC = 2·80 mm); 6 Schiltigheim SJ22 (HTC = 3·43 mm); 7, Biache-St Vaast (HTC = 4·10 mm) and for comparison 8. a recent individual of Cracow RE03 (HTC = 4·38 mm). Note the similarity of the size between the Polish (8) and fossil (7) individuals. **Extreme shapes of the morphology in *Pupilla muscorum***: 9. slim shell with a high protusion of the aperture Individual AG29 (HTC = 3·35 mm); 10. fat shell with a low protusion of the aperture Individual CG08 (HTC = 3·61 mm); a = general view, b = detail of last whorl. 11. Difference in ornamentation between protoconch and the post-protoconch whorls.

extreme morphologies: a fat shell and a slimmer shell. The fat shells present an aperture with a low protrusion and distant parietal crests (Figure 2). The slim shells have an aperture with a high protrusion and close parietal crests (Figure 2). On the other hand, analysis of the last whorl demonstrates the independence of the aperture dimensions. The shape of the aperture, rounded or ellipsoidal, does not correspond to a particular morphology of the rest of the shell.

No size–shape relation is observed, in spite of these two characters being both dependent upon the shell architecture. Intra-population variation depends on both size fluctuations and the expression of morphological tendencies.

Morphological inter-population variation during the Pleistocene in Europe

Grouped by populations, the individuals determine overlapping 'subclouds' as indicated by the extreme populations AB and CI (Figure 3). The small morphologies of the former (AB) are plotted close to the 'large ones' of the latter (CI). This distribution shows an inter-population variation which is mainly due to shell size. Each population is plotted on the first factorial plane according to a size gradient represented by the first axis.

Analysis of the individuals from Achenheim reveals great intra- but also inter-population variability occurring during the Middle and Upper Pleistocene. Multivariate analyses of 496 individuals (Rousseau 1985) from other French deposits indicates a range of morphological variation that lies within the previously determined range of variability (Rousseau 1989). These results imply that morphological variations obtained for the Achenheim deposits are due neither to local environmental conditions, nor do they correspond to variations in isolated populations, but have to be generalized to European Quaternary deposits.

Figure 3: Correspondence analysis of Pleistocene samples (populations) using all the parameters. Distribution of the individuals from morphological criteria expressed by the first two axes. The triangles represent the individuals belonging to the populations CI (small shell), the squares those belonging to the population AB (tall shells); stars correspond to the drawn individuals.

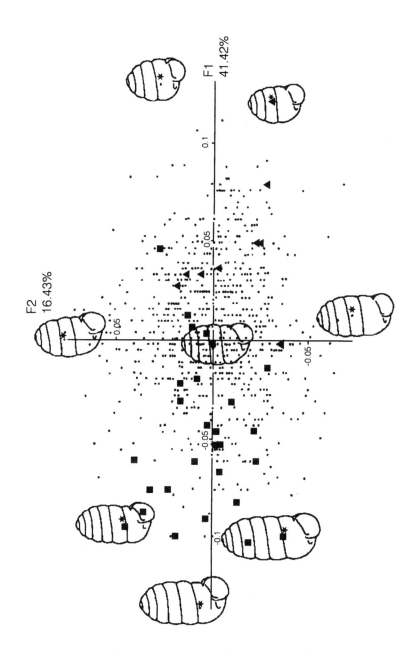

Modern morphological variation in Europe

In modern populations, as with fossil ones, the dimensions that describe the size (heights) of the shell discriminate the different morphologies more effectively. Among the parameters describing the aperture, those expressing the protrusion are dissociated from the other ones. As observed in fossil individuals, we see the contrast between fat shells, with slight protusion of the aperture, and slim shells, with high protrusion (Figure 2).

The local fluctuations lie within a more general biogeographic variation. In western and central Europe, *P. muscorum* is subject to very variable climatic and environmental conditions. Over the large area where a temperate climate prevails, the west is under an oceanic influence, whereas the climate becomes more and more continental towards the east. This continental trend of the climate induces a discrepancy in size variations, the small western shells having a morphology similar to the large eastern individuals.

There are great variations in shell size and this is observed in all Pleistocene localities studied. Several times during the Middle and Upper Pleistocene *P. muscorum* increased in shell size. After these increases there were always corresponding decreases. A parallelism has been demonstrated between climatic glacial–interglacial cycles and increasing–decreasing phases of shell size (Rousseau 1985, 1989).

By their reversibility, size variations do not express any evolutionary trends as classically defined. The rhythm over time is characteristic of iterative variations of increase and decrease of shell size during the Middle and Upper Pleistocene. Shape variations are independent of size fluctuations. Their link with time does not result from particular trends *a priori*. They seem to represent the random expressions of an architectural potential which has two poles:

- a fat shell with a small protrusion of the aperture;

- a slim shell with a large protrusion of the aperture.

If the size of the shell increases, all the other parts of the shell grow proportionally in order to conserve the architectural framework of *P. muscorum*. This conclusion demonstrates again the importance of the first stages of ontogeny and also emphasizes the subtle inter-relationships of the various characters.

The fossil variability expressed in the Pleistocene deposits is, for this period, similar to that recognized in the recent populations. The dimensions of the aperture indicate the same variations in fossil as in recent individuals. An identical interpretation concerning the cephalopodium of individuals must be proposed. According to Gould (1968), two strategies are available for changing in shell size, from one population to another:

- an increase or a decrease of the shell size correlative to a proportional variation of the size of the organism,

- an internal structural strengthening without any size modification of the organism.

Such a correlation between recent and fossil populations implies certain consequences for the evolution of *P. muscorum*. During the Middle and Upper Pleistocene, *P. muscorum* was the subject of morphological variations identical to those recently observed, demonstrating the lack of any evolutionary change. Nevertheless, the species present a stable global morphological equilibrium, as a result of internal and external (i.e. climatic) constraints, which the analysis of the shell ontogeny can help us to characterise.

Ontogeny of Pupilla muscorum

Two main points must be recalled:

1. an individual has its own trajectory corresponding to changes in the growth rate during ontogeny with the acquisition of the pupiform morphology;

2. differences appear in ontogenic trajectories of individuals from the same population.

These two conditions determine the limits in the global timing of growth for *P. muscorum*.

Size has to be disregarded initially in order to characterise the (heterochronic) process(es) (*sensu* Gould 1977; McNamara 1986) responsible for these morphological variations. The shells studied were at the adult stage (see above for adult criteria). The number of whorls was taken as a standard reference for age because they are more representative of the growth pattern, and adult characters are expressed on the shell at the end of the growth. In order to provide reliable and valuable comparisons between the different morphologies, a reference value is necessary. 3·5 whorls is the lower limit of the adult morphological variability and this value was therefore chosen as the standard value for the 'age character'.

If we consider the spire index (SpL) variations with weighted whorl number (Figure 4a), we note that the values, from common or relatively close origins and through the adult stage, are clearly different. This illustrates a morphological acceleration: compared to the standard trajectory; that studied shows an accelerated rate of morphological development during the juvenile stages of the growth. A similar result is obtained for size index (SI) vs. weighted whorl number (Figure 4b). If we consider the Spire (SpL) and the size indexes (SI) variation with unweighted whorl number, we note that an acceleration is linked with a hypermorphosis: extension of the juvenile growth period caused by a delay in maturation. This provides the size variation (as a consequence of the acceleration, Figure 4). Acceleration-hypermorphosis is one of the possibilities for hyperamorphosis *sensu* Dommergues *et al.* (1986). At this stage of the analysis, it is tempting to be satisfied with these results. One may argue the importance of the role of two heterochronic processes. Nevertheless when variations of angle ß are examined with respect to weighted whorl number (Figure 5), clear differences exist at the end of the protoconch with similar magnitude to those occurring in the adult stage. This result implies that with a relatively equal height, the protoconchs have a comparable width. Angle ß is not the same in one morphology as another. Consequently, the shape of the aperture of the protoconch is not the same in the different cases, which has an immediate repercussion on the following whorls. These initial variations of angle ß permit understanding of how the shell shape is channelled, leading to fat or slim morphologies. If the angle is high, according to the pupiform schema, the value of the angle will become progressively reduced and the shell will be slimmer. On the

Figure 4: Reference ontogenetic trajectories established. 1. with weighted whorl number: a) variation in spire index SpL; b) variation in size index SI. 2. with non-weighted whorl number: c) variation in spire index Spl, d) variation in size index SI. (1 = 3·5, 2 = 4, 3 = 4·5, 4 = 5 and 5 = 5·5 whorls.)

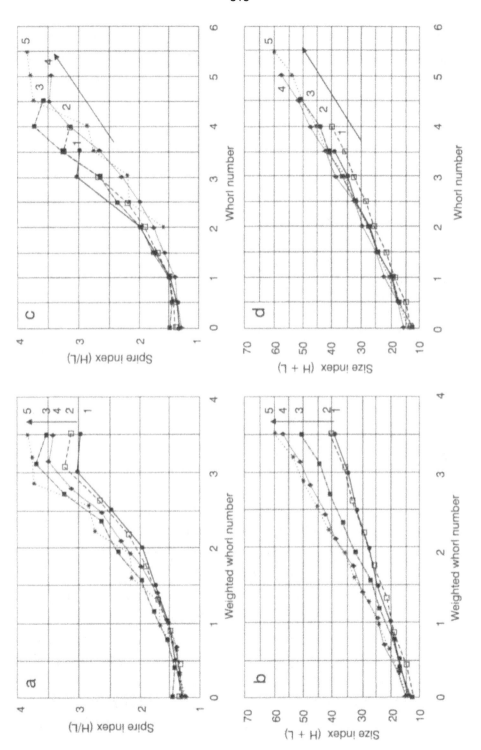

other hand, if the angle is proportionally low, the value of the angle will become progressively larger and consequently, the shape morphology will be fat. There are no heterochronic processes involved to explain the shape variations. All are determined by the protoconch morphology.

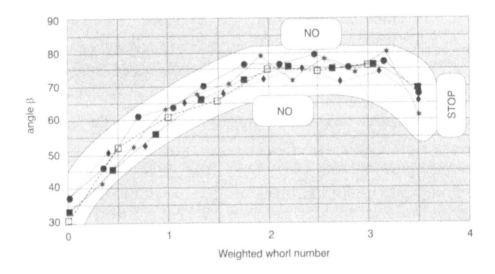

Figure 5: Channelling of the morphology as expressed by angle β in its variations versus the weighted whorl number. (Same conventions as in Figure 4)

DISCUSSION

Size variations between individuals or between populations seem to be the result of the combined interplay (more or less pronounced) of acceleration and hypermorphosis, if a shell with 3·5 whorls of spire is considered as a standard reference. Also, this variation will be more evident in an inter-population than in an intra-population context because of environmental contraints. The shape variation (slim–fat) is directly induced right from the protoconch by the angle β which contrains the coiling of the rest of the shell. Variations of β with regard to the weighted whorl number demonstrate that β also has an impact on the size of the shell and that it characterises the internal constraints that act on the shell of *P. muscorum*. The smallest shells have 3·5 whorl spires, the largest 5·5. Considering the mean values at the end of the protoconch, we note that these two limits correspond to the lowest values for the ß variations. Intermediate morphologies show higher ß values. All of them determine the structural limits imposed by the pupiform scheme of growth. This explains why recent morphological variations of *P. muscorum* are identical to those observed during the glacial–interglacial cycles, although a pleniglacial climate cannot be attributed to present-day central Europe. It appears that in the two types of situation, the organism approaches the two limits of its morphological variation, which it cannot overstep. Based on these results, although it does not appear in the literature, the intra-population variation may also be interpreted as the result of the occurrence of heterochronic processes.

After eliminating the problem of the standard reference of age, absolute age correlations must be attempted by choosing the biological age (represented by the weighted whorl number) rather than the size of the shell. The real duration of post-embryonic growth in *P. muscorum* can be expected to be annual, based on observations made on living individuals in the laboratory (DD Rousseau, unpublished data). How can the modifications of the growth rate be explained? The existence of modifications in the growth cycle have been confirmed at different latitudes and altitudes in pulmonate gastropods (Tompa 1984). In *Arianta arbustorum*, a taller snail (HTC = 10–22 mm and LTC = 14–28 mm) than *P. muscorum*, individuals in central Europe, for example, are sexually mature at 3 or 4 years. Populations from France and western Germany, however, are adult at the first or second year of life (Terhivuo 1978). Uminski (1975) has demonstrated that *Vitrina pellucida* and *Semilimax kotulai* in Poland have two different growth cycles depending on the altitude at which they are found on the same slope. Below 1200 m, the growth cycle is annual, whereas above 1200 m it is bi-annual. These facts confirm the dependence of maturation and growth of pulmonates on temperature (Raven 1975). Logically, a gradient, both N–S and E–W, seems to appear where the age of the organisms at maturity is doubled. The same rule can plausibly be applied to *P. muscorum*. The non-weighted number of whorls is thus the most representative characteristic of the absolute age of the organism rather than its size. In these conditions, if we refer to the previously defined trajectories, it is evident that morphological acceleration during the juvenile stage becomes dominant. However, there is a paradox, because this acceleration actually corresponds to a graphic morphological prolongation of growth in response to environmental stress. This acceleration characterises the modification of the rate of calcium carbonate secretion during the infant stage. The final hypermorphosis is only the apparent consequence of the modification of the growth cycle initiated by temperature.

Intra-population variation is reflected in both size and shape fluctuations of the shell. Numerous intermediate morphologies exist between a fat shell (low protrusion of the aperture) and a slim shell (high protrusion of the aperture). These morphologies are due to differences occurring right from the protoconch in terms of the aperture angle, without any indication of heterochrony. The inter-population variation is mainly characterised by important fluctuations in shell size due to the combined interplay of acceleration and hypermorphosis, or of neoteny and progenesis, in accordance with the morphological type taken as reference. This variation is identical to that demonstrated among recent populations for *P. muscorum* and is related to the environment.

From a temporal point of view, identical variations have been recorded during the Quaternary, in parallel with climatic variations corresponding to successions of cooling and warming. In such conditions, a 'mutation–selection' mechanism, as described in the literature, is not necessarily the explanation of these temporal variations. Simply an alteration of the ontogenic sequence in response to environmental stress can be proposed. During the infant phase, which corresponds to the more stenoecious stages of the animal, important structural variations occur which consequently lead to a larger or smaller size with the presence or absence of a supplementary part of a whorl. We remain in the intraspecific realm, and on the basis of results obtained for modern shells, these morphological oscillations have to be described as ecophenotypic (Rousseau 1985, 1989; Rousseau and Laurin 1984; Solem 1986).

CONCLUSIONS

Based on previous results, we can characterise the constraints acting on *P. muscorum* morphology. In the morphological space of *Pupilla*, *P. muscorum* morphology is channelled by a pupiform framework (fat or slim), which represents the phylogenetic background. Organisation and stimulation of the mantle cells (structural constraints) permits the elaboration of a supernumerary portion of a whorl (functional constraints) in response to climatic stress (environmental constraints). Morphological variation of *P. muscorum* must be understood in a dynamic way as the result of heterochronic processes (acceleration–hypermorphosis or neoteny–progenesis) regulating and channelling the internal and external constraints.

Such morphological stability reinforces the adaptative potential of this species with regard to its environment. Variations in recent *P. muscorum* have been related to environmental parameters. Two complementary data sets confirm this interpretation in fossil populations. In this way the morphological variations in *P. muscorum* during the Middle and Upper Pleistocene do not express any evolutionary trend. The species is in a global morphological and evolutionary stasis. Nevertheless, the phenotypic expression presents important intra- and inter-population variations which, analysed using ontogenic trajectories, define the potential morphological landscape.

ACKNOWLEDGEMENTS

Special thanks to Drs. V. Lozek, F. Nash, J.J. Puisségur and H. Tintant for helpful discussions and comments about this study. Borrowed material came from the Institute of Systematic Zoology in Cracow (Polish specimens) and from Dr. Lozek's collection in Praha (Czech specimens). Thanks to Dr. Gittenberger who allowed me to examine the collections of the Rijskmuseum in Leiden and J.P. Garcia for measurement assistance. Drs. R. Lotti, D. Keen and N. Landmann improved the English text. The author greatly thanks Dr. Huntley for inviting him to the NATO meeting, and Dr. Morgan for the last text improvements. ISEM contribution 95-093. This work received support from the EU Environment program EV5V-CT92-0298

REFERENCES

Alberch P (1982) Developmental contraints in evolutionary processes. *in* Bonner JT (ed) Evolution and development, 313-332. Springer-Verlag, Berlin
Cain AJ (1977) Variation in the spire index of some coiled gastropod shells and its evolutionary significance. Phil Trans Roy Soc London Ser B 956:377-428
Dommergues JL, David B, Marchand D (1986) Les relations ontogenèse-phylogenèse: applications paléontologiques. Geobios 19:335-356
Gould SJ (1968) Ontogeny and the exploration of form: an allometric analysis. J Pal 42 (suppl 5):80-93
Gould SJ (1977) Ontogeny and Phylogeny. Belknap Press Cambridge
Gould SJ (1980) The evolutionary biology of constraint. Daedalus 109:39-52
McNamara KJ (1986) A guide to the nomenclature of heterochrony. J Pal 60:4-13

Raven CP (1975) Development. *in* Fretter V, Peake J (eds) Pulmonates 1 - Functional Anatomy and Physiology, 367-400. Academic Press, London

Rousseau DD (1985) Structure des populations Quaternaires de *Pupilla muscorum* (Gastropode) en Europe du Nord: Relations avec leurs environnements. PhD thesis, Institut des Sciences de la Terre Dijon

Rousseau DD (1987) Paleoclimatology of the Achenheim series (Middle and Upper Pleistocene Alsace France): A malacological analysis. Palaeogeogr Palaeoclimatol Palaeoecol 59:293-314

Rousseau DD (1989) Réponses des malacofaunes terrestres Quaternaires aux contraintes climatiques en Europe septentrionale. Palaeogeogr Palaeoclimatol Palaeoecol 69:113-124

Rousseau DD, Laurin B (1984) Variations de *Pupilla muscorum* L (Gastropoda) dans le Quaternaire d'Achenheim (Alsace): une analyse de l'interaction entre espèce et milieu. Geobios mém sp 8:349-355

Solem A (1986) Pupilloid land snails from the South and Mid-west coasts of Australia. J Malacol Soc Australia 24:143-163

Terhivuo J (1978) Growth reproduction and hibernation of *Arianta arbustorum* (L) (Gastropoda Helicidae) in southern Finland. Ann Zool Fennici 15:8-16

Tompa AS (1984) Land snails (Stylommatophora). *in* Tompa AS, Verdonk NH, Van Den Biggelaar JAM (eds) The Mollusca 7 - Reproduction, 47-140. Academic Press, London

Uminski T (1975) Life cycles of some Vitrinidae from Poland. Ann Zool (Poland): 2317-33

Waddington CH (1974) A catastrophe theory of evolution. Ann New York Acad Sci 231:32-42

Wenz W (1923) Gastropoda Extramarina Tertiara Fossilium catalogus 1 Animalia: *Pupilla* pars 20:952-967 Diener C (ed) W Junk Berlin

Late-Quaternary extinction of large mammals in northern Eurasia:
A new look at the Siberian contribution

Andrei V. Sher

Severtsov Institute of Ecology and Evolution

33 Leninskiy Prospect

117071 Moscow

Russia

Although extinction occurred throughout the Quaternary, various calculations of extinction rates agree that the end-Pleistocene extinction was among the largest by the number of species involved and probably one of the most abrupt in time (Kurtén & Anderson 1980; Martin & Klein 1984). Generally, extinction can be considered as the failure of a species to adapt to changing biotic or abiotic conditions. The transition from the Late Pleistocene to the Holocene was marked by wide-scale vegetational restructuring, which broadly correlated with the dramatic changes in the spatial distribution of some mammalian species and the extinction of others. Since the last century this correlation has been used as the main argument for environmentally caused ('climatic') extinction hypotheses.

The well known fact that the end-Pleistocene extinction hit mostly large species of mammals, such impressive creatures as woolly mammoth *Mammuthus primigenius*, mastodon *Mammut americanum*, giant deer *Megaceros giganteus*, woolly rhinoceros *Coelodonta antiquitatis* and ground sloth *Megalonyx* among them, together with the apparent abruptness of their disappearance, suggested the uniqueness of this event and forced a search for other explanations. Co-occurrence of human artifacts with the fossils of extinct Pleistocene mammals has given rise to the human overkill paradigm[1]. Eleven years ago the state of the 'climate' versus 'overkill' debate was summarized by Marshall (1984) and Grayson (1984). It was shown that both paradigms had essential weak points. The human paradigm had become so

---

[1] Using the term 'paradigm' I follow Marshall (1984) who demonstrated that from the methodological point of view neither the 'climatic' nor 'overkill' concepts can be considered as 'theories' or even 'hypotheses'.

NATO ASI Series, Vol. I 47
Past and Future Rapid Environmental Changes:
The Spatial and Evolutionary Responses of
Terrestrial Biota
Edited by Brian Huntley et al.
© Springer-Verlag Berlin Heidelberg 1997

resilient, notes Grayson, 'that it could withstand virtually any factual onslaught'. The climatic paradigm is still stuck upon the basic question: why were repeated earlier Pleistocene transitions from glacial to interglacial periods by no means so destructive for large mammals ? In this paper I attempt to answer this question, introducing some new evidence on the correlation between extinction and environmental changes in Arctic Siberia. I choose this region not only because of my personal knowledge of it, but also because I am going to forward a suggestion that it was crucial for understanding extinctions in Northern Eurasia. I confine my speculations to this continent only, following Marshall's wise caution that 'extinction on each landmass should be viewed as a discrete event, as it may be unique'.

WERE EARLIER INTERGLACIALS SIMILAR TO THE HOLOCENE ?

Marshall (1984) emphasizes that the human and climatic paradigms are not mutually exclusive; rather they are complementary. He outlines the advantages of a middle-of-the-road view and favours multidimensional paradigms, which are becoming more and more popular. The most recent broad-scale review attempting to be multidimensional is that of Stuart (1991). Although Stuart calls his paradigm a 'combined overkill and environmental change hypothesis', he clearly inclines to the view that even heavily stressed populations could have survived till the next (future) glaciation, had advanced human hunters been absent. Their presence, concludes Stuart, 'is the one critical factor not present earlier in the Pleistocene, whereas rather similar climatic/environmental changes had occurred many times before'. How similar were they in fact?

To answer this question, let us take a look at the last interglacial (Eem-Mikulino-Kazantsevo) faunas in northern Eurasia. In the British Isles the faunas of the optimum phase of the interglacial include hippopotamus *Hippopotamus amphibius*, fallow deer *Dama dama*, straight-tusked elephant *Palaeoloxodon antiquus*, aurochs *Bos primigenius* and hyaena *Crocuta crocuta*. Arctic or continental steppe species were not present; even woolly mammoth and horse *Equus ferus* are not recorded for this phase. Eastward, towards central Europe, many 'southern' species disappear, and the fauna acquires a more continental character (Stuart 1982). Progressive eastward 'steppisation' of the interglacial environment can be traced through Byelorussia and European Russia to West Siberia, where a fauna correlated with the

Kazantsevo is dominated by tundra and steppe species (see references in Sher 1991).

Mammal faunas of an 'interglacial' character (i.e. including some southern species, not known from the faunas of the cold stages in the area) have never been discovered in the north of East Siberia, although they were sought for many years. This does not seem to be in agreement with the fact that several 'interglacial' stages are recognized in the regional pollen record by palynologists who infer a wide distribution of forest communities during these stages. Moreover, on the few occasions when mammal fossils were excavated from units referred to an interglacial by pollen analysis, as in the lower Indigirka (Kaplina *et al.* 1980), they could not be distinguished from those from the underlying and overlying members with typical 'tundra–steppe' pollen assemblages.

Pleistocene intervals of afforestation in the Siberian Arctic are recognized by an increase of arboreal pollen, normally very low in tundra–steppe spectra. However, these arboreal plants are by no means 'exotic' for these regions and are represented most commonly by tree and shrub birch *Betula* spp., shrub alder *Alnus* spp., sometimes shrub pine *Pinus pumila* and minute quantities of larch *Larix*. Although the distinction between tree, shrub and dwarf birches is a well known problem in pollen analysis, the presence of tree birch is usually confirmed by plant macrofossils. Sometimes a few macrofossils of larch are found, but in general plant macrofossil samples include arctic and hypoarctic plants, common in this area at present, in tundra or northern larch taiga. These palaeobotanical data have usually been interpreted as evidence of forest–tundra or northern taiga, quite similar to those existing now in this part of Siberia (full analogue assemblages). However, the insect assemblages, found in the same strata, most commonly did not agree with this interpretation. 'Interglacial' insect faunas were dominated by species of steppe, dry meadow, dry tundra and open ground habitats. Species related to mesomorphic habitats (wet tundra, wet meadows) were much less abundant, while obligatory tree feeders were very rare (Sher *et al.* 1979; Kaplina *et al.* 1980; Kiselyov 1981). These insect assemblages did not differ essentially from the tundra–steppe ones, and were virtually non-analogue (i.e. with no modern equivalent)[2].

The controversy between pollen and palaeoentomological data led me to critically examine the 'interglacial' pollen assemblages (Sher 1988, 1991). Comparison with a large set of surface samples has revealed that the fossil spectra do not represent the modern larch taiga, but a different kind of environment. It has been interpreted as an open birch woodland alternating with grassland communities and relatively few bogs. The forest ground cover had an essentially richer herbaceous component than the present taiga with its moss/lichen/shrublet cover which is absolutely non-supportive for grazers. Of course, larch existed as a species, but never formed a unique monodominant forest formation as it does now. There was moderate degradation of permafrost, but mostly local, since polygonal ice-wedge systems have been preserved unthawed since the preceding 'cold' stages. Climate remained continental, with some increase of humidity.

This 'interglacial' scenario for northeast Siberia requires further multidisciplinary palaeoecological work, especially in re-interpretation of pollen data. It seems, however, that it offers a possible (if not the only) way to solve the existing controversies between the plant and insect records. It helps to explain the persistence of non-analogue (tundra–steppe) insect communities in this region through various Pleistocene climatic intervals (Kiselyov 1981). If insect assemblages only slightly changed their composition (relative weight of different ecological groups) during the 'interglacials', but their general character did not change, we can hardly expect to be able to observe corresponding changes in mammal communities, given their lower palaeoecological resolution and much less abundant fossils. They

---

[2] The non-analogue character of these insect assemblages, as well as of the tundra–steppe ones, is based on the joint occurrence of species presently common in various tundra habitats (mesic to xeric) with large amounts of steppe and open ground species. Although most of the recognized steppe species are present in the modern fauna of northern Siberia, they are restricted to small isolated extrazonal steppe-like habitats south of 68°N, whereas their main ranges lie within true steppe habitats in south Siberia and Mongolia. Among open ground species, one non-flying pill-beetle species, *Morychus viridis,* is especially noticeable. Currently it is restricted to extremely dry and almost bare-ground habitats with an enormous soil temperature range in the Kolyma Highlands (Berman 1990). This beetle has been a marker of tundra-steppe environments in north Siberia since the Early Pleistocene, and is often the dominant species in fossil insect assemblages of that kind.

retained virtually the same non-analogue tundra–steppe character during the 'interglacials' (Sher 1991; Sher & Sulerzhitzky 1991).

All the evidence presented in this chapter allows the following conclusions:

1. In the continental regions of northern Asia *the interglacial environment was essentially different from the modern or the Holocene one.* It differed much less from the 'glacial' environment than it did in Europe, and could support the fauna of large grazing mammals.

2. The observed west–east gradient in the interglacial faunas not only proves that tundra–steppe faunas persisted through the interglacials in Siberia, but allows the deduction that *continental northeast Asia sheltered tundra–steppe mammals intolerant of vegetation changes in the west and in lower latitudes.* Elimination of the westernmost populations did not prevent these mammals from recurrent backward dispersal to the west during the next cold stage, which is well documented in Europe for at least some mammals (Kahlke 1992). This offers a simple answer to the question why no mammals became extinct during the interglacials. Extinction only happened when these huge refugial areas were subjected to a radical environmental change.

THE ENVIRONMENTAL REVOLUTION AND EXTINCTION IN NORTHEAST SIBERIA: AN ATTEMPTED CHRONOLOGY

As an example of megafaunal extinction, the best documented case is the woolly mammoth. Its pattern of extinction is based on a large collection of ^{14}C dates (more than 120) obtained directly from mammoth fossils. This collection was put together by L. Sulerzhitzky (Geological Institute, Moscow) and the writer in 1989; since then, it has been partially published by Stuart (1991), Lavrov and Sulerzhitzky (1992), and finally by Sulerzhitzky (1995). Analysis of this collection (Sher 1992a) showed that prior to 12·5 ka (uncalibrated radiocarbon age) woolly mammoth inhabited the whole of northern Eurasia. At around that time it started to disappear in temperate latitudes, both in Europe and in Siberia. Reliable dates younger than 12 ka are known only from the Siberian High Arctic, north of 70°N – Yamal, Gydan and Taimyr peninsulas, Severnaya Zemlya Islands, Lower Indigirka and Wrangel Island (Figure 1). The upper time limit of these dates elsewhere except Wrangel is around

9·6 ka. That means that by 9·0 ka mammoth was extinct everywhere except the only island where the relic population of dwarf mammoth persisted until at least 3·7 ka (Vartanyan *et al.* 1993). This pattern, which has been called the "retreat to the North extinction model" (Sher 1994), highlights two critical periods in the history of the mammoth's range: around 12·5 and 9·5 ka respectively. At present no evidence is known for saiga antelope, woolly rhino or horse surviving in northeast Siberia after 12·5 ka, but the dates available are in fact very few, and may not reflect the latest occurrences.

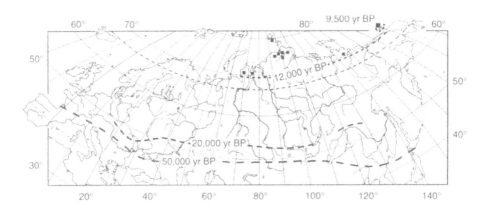

Figure 1. Southern limits of woolly mammoth distribution at certain times during the late Pleistocene. Black squares mark the sites with mammoth fossils dated around 12 ka and younger. The 50 ka and 20 ka boundaries – after Soffer (1993), modified in East Siberia and China.

During the last 'glacial', an extremely continental and cold climate in northeast Siberia was too dry to allow the development of glaciers except on some highlands. All the vast plains remained unglaciated, and were subjected to the accumulation of fine-grained sediments, mostly silt, rich in segregation ice and syngenetic ice wedges ('Yedoma Complex'). Considering the amount of frozen water conserved in Yedoma, the thick cover of this sediment over the vast northern plains can be compared with the continental glaciers. Evidently, release of this water at a time of change in the permafrost régime would have had an enormous effect on the environment. The precise mechanism of Yedoma sedimentation is still debated; currently many students consider it as a polygenetic formation built by fluvial, lacustrine, aeolian, slope and other processes, but under a peculiar climatic régime with intensive cryogenic weathering. The environment corresponding to this régime is reconstructed from various lines of palaeoecological evidence as a dry arctic grassland (tundra–steppe) existing on the frozen substratum. In pollen spectra, it is marked by predominance of grasses and various herbs ('herb zone'), sometimes with high amounts of *Selaginella sibirica* and Bryales spores. Larch probably formed small groves in habitats with higher moisture, as it does today around the relic 'steppe' patches in the extremely continental depressions in northeast Siberia (Yurtsev 1981). Fossil insect assemblages from these sediments are of distinctively non-analogue character (tundra–steppe).

This characteristic environment persisted in northeast Siberia until as late as 13·0 – 12·5 ka, when some very rapid changes started. The subsequent geological and environmental history can be outlined in a few stages, or phases (Figure 2a).

Phase I – ca. 12·5 – 11 ka: The accumulation of Yedoma silts virtually stopped around 12·5 – 12·0 ka, which seems to be an important climatic signal by itself. Initial features of thaw (thermokarst) processes on the surface of ice-saturated deposits are dated in various areas to between 12·3 – 11 ka (see Sher 1992a for references). Pollen spectra display a sharp decline of herbaceous components (the end of the 'herb zone') and an increase in shrubs. No tundra–steppe insect assemblages are known after that time except in some special habitats. Further south in Siberia, the spread of arboreal vegetation, bogs, and moss and lichen ground cover instead of dry grassland must have been even more pronounced. Around 12 ka a significant change in the spatial distribution of large mammals also occurred. Mammoth populations were eliminated in extra-Arctic regions, and this species 'retreated' to the High Arctic (Figure 2B). Some other tundra–steppe species, such as saiga antelope, became locally extinct in northeast Siberia, and lost the northern part of their range ('retreated' to the south).

I suggest that this first phase of environmental restructuring can be seen as an onset of 'normal interglacial' conditions. Mammoth populations still could survive within a huge area of mainland Arctic north of 70°N, and on the shelf islands. Although some populations experienced a notable decrease in body size, like the Berelyokh one, on the whole they were still in good condition and apparently numerous. That means that the conditions of this period were not disastrous for mammoth (and probably some other grazers as well). In other words, it can be quite reasonably assumed that if the general trend had reversed at that time, and tundra–steppe conditions returned, the mammoth could have dispersed back south.

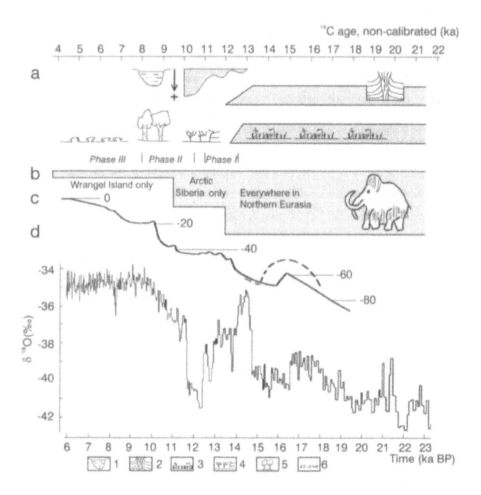

Figure 2. Correlation of environmental events in northeastern Siberia during the Pleistocene – Holocene transition (a), with the dynamics of woolly mammoth range (b), sea level history in northeastern Siberia (c) and the GRIP ice core isotope curve (d). (c after Fartyshev 1993 (figures mark the position of the sea level in m below present; dashed line - local curve for the Laptev Sea). d modified from Dansgaard *et al.* 1993).
1. dynamics of thermokarst activity; 2. accumulation of Yedoma silts; 3. herb zone in pollen spectra; 4. spread of shrubs; 5. spread of tree birch; 6. modern tundra vegetation.

Phase II – ca. 10·5 – 8 ka: Around 10 ka a second major environmental change occurred, and the main feature of it was a catastrophic regional outburst of thermokarst. The dating of this thermokarst event is confirmed by numerous ^{14}C dates from the base of organic sediment infilling former lake kettles (Kaplina & Lozhkin 1982). Most of them range from 9·5 to 8·5 ka, implying that by 9·5 ka many lakes had been already turning into bogs, and from that we infer that the maximum development of lakes took place approximately around 10 ka. By that time myriads of thaw lakes had appeared on formerly rather dry lowlands. Numerous bogs and peatlands spread, leaving little space for grassland communities suitable for mammalian grazers. Mammoth barely survived later than 9·5 ka, except in refugial areas such as Wrangel Island. Soon after, however, many lakes became sealed with the peat accumulation. At that time, tree and shrub birch and alder spread far north in the Arctic, but this sharp spread seems to have been over by 8 ka. That was the last time when the Arctic supported a vegetation different from modern tundra.

Phase III – ca. 8 ka onwards. Although minor fluctuations are observed in some areas, since that time the modern-type vegetation has been generally established in the Arctic.

Thus, in our search for the reasons of extinction we come to a quite short critical period of time (not longer than 1,500 – 2,000 years), marked with important climatic changes. However, that still does not answer the question, why the situation in the Arctic after 11 ka did not follow a 'normal interglacial' scenario, and turned into an irreversible one instead. Looking for an appropriate answer, I believe that we have missed one of the most important components of the environment – the Arctic Ocean.

ARCTIC SEA-LEVEL RISE AND THE ENVIRONMENTAL REVOLUTION

At present, the influence of the ocean on the coastal areas of northern Siberia cannot be overestimated. A huge mass of open cold water in summer is the main source of low temperatures and high humidity during the growing season. Meteorological data show a very rapid increase of continentality towards the inland, e.g. summer temperature increases southward incomparably faster than would be expected due to a latitudinal gradient only. It may even be suggested that the present day lowland tundra – treeless, excessively wet and with impoverished flora and fauna – is climatically supported not only by its latitudinal position, but significantly by the influence of the cold water mass of the Arctic shelf seas.

The lowering of the ocean level during the last glacial was about 100 m. On the very shallow and flat shelf of eastern Siberia even half the magnitude of regression would result in a huge increase of dry land (Figure 3a). Prior to 12 ka the East Siberian Sea ran much further north than the southern limit of permanent pack ice under the

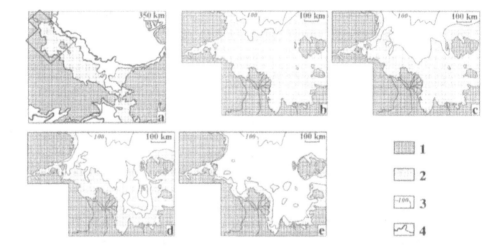

Figure 3. (a) Distribution of sea and land in northeastern Siberia and (b – e) in the Laptev Sea at the end of the Pleistocene and early Holocene.
a. around 15 ka (the sea level stands at -50 m); the frame shows the position of maps b-e; b. around 11·5 ka (–40 – –45 m); c. around 9 ka (–30 – –35 m); d. around 8-9 ka (–20 – –25 m); e. around 7 ka (–10 – –15 m). (a. from Sher 1984. b – e. modified from Fartyshev 1993, Figure 68.)
1. present land; 2. dry areas of the shelf; 3. sea and the present isobath 100 m; 4. the outer boundary of the shelf (approximated by the present isobath 200 m).

present ('interglacial' !) conditions. This means that during the glacial the coastal waters were rarely (if ever) open in summer, which in turn (besides the increase in the land mass) contributed to the continentality of the northeast Siberian climate. The sea level rise at the end of the Pleistocene must have had a tremendous effect on the regional climate and environment. Taking into consideration the flatness of the local shelf, the rate of southward shift of the shore line must have been very high. Combined with the general warming, this process must have resulted very soon in the appearance of open Arctic waters in summer. The rapid intrusion of a huge cold water mass must have drastically changed the most important characteristics of local climate, especially during the growing season, i.e. decreasing summer temperature and increasing precipitation. Together with the increase in summer cloudiness, a sharp drop in effective evaporation would result. Altogether, it provides a classical portrait of modern tundra climate.

To verify this scenario we need a well dated curve of the eustatic sea level rise for this region. The data available, although rather poor (Fartyshev 1993), indicate that prior to 10 ka the sea level rise in the Laptev Sea had, most probably, rather insignificant effect on the mainland (Figs. 2c, 3b). The subsequent transgression took place so rapidly that the data on the age of the submerged shorelines are quite controversial. By 8 ka the sea occupied the major part of the modern Laptev Sea, and by 7 ka the shoreline hardly differed from its present position (Figure 3d,e). Thus, in just 2–3 thousand years, the advent of the sea not only eliminated nearly a million km^2 of land, which presumably had been good grazing habitat, but also introduced a powerful new climatic factor that certainly did not favour preservation of the tundra–steppe biome.

This tremendous regional effect of the end-Pleistocene eustatic sea level rise poses the question whether earlier rises (during previous interglacials) had similar effects. Interestingly, this large sector of the Arctic (between the Taimyr and Chukotka peninsulas) seems to have experienced much less extensive marine transgressions than the western Arctic. Sediments of marine origin are very rarely found on the present mainland, while they are quite widespread west of 110°E and in the Bering Strait area. This difference between the two sectors of the Arctic has been noticed long ago, and assigned to a contrasting tectonic history of the region (Suzdalsky 1971). Recent studies in the eastern Arctic revealed only few outcrops of marine sediments, mostly poorly dated, along the present sea coast and on the islands (Kim 1986; Resolutions... 1987; Makeev et al. 1989). According to geological evidence from the coast and the shelf islands, prior to the end-Pleistocene transgression most of the shelf was covered by non-marine sediments of the Yedoma-type. The formation of this extensive terrestrial sedimentary body started as early as during mid-Quaternary times (Kaplina et al. 1980; Resolutions... 1987). Although these sediments, due to their high ice content, could be easily destroyed by marine and thermal abrasion, their large bodies remained on the shelf until the Holocene transgression; some islands were destroyed in historical times. Thus, it can be reasonably assumed that at least during the second half of the Pleistocene, marine transgressions in this area were more restricted in extent than the modern one; some of them could develop only estuarine ingressions along the major river courses.

WAS THE HOLOCENE UNIQUE REGIONALLY OR WORLDWIDE ?

The fast sea level rise around 10 - 9 ka could be one of the main factors that attached an irreversible character to the environmental changes which started in the Siberian Arctic about 12 ka. A particular combination of global and regional events seems to have made the Pleistocene – Holocene transition *unique in the history of this region*. Although I believe that the Siberian Arctic was of special importance for mammoth survival/extinction, as it supported the latest populations of the species, the suggested explanation remains regional, not general. Is there any evidence for a general unique character of the Holocene? Lister and Sher (1995) see that evidence in the detailed records of past climate from the Greenland ice cores. Both the GRIP (Dansgaard *et al.* 1993) and GISP2 (Grootes *et al.* 1993) cores reveal a unique stability of the Holocene $\delta^{18}O$ curve; such a feature cannot be found elsewhere in the preceding 250 ka record. A totally different pattern, with almost constant, sharp and rapid climatic fluctuations (e.g. temperature changes of up to 10°C during the lifetime of one individual mammoth) was typical for the glacial periods and probably even for the pre-Holocene interglacials. Putting aside a debate on the possible scale of fluctuations during the last interglacial, which has still not led to a certain solution[3], let us concentrate on the $\delta^{18}O$ record of the last 20 – 25 ka, which does not seem to raise any doubts.

It was tempting to compare this North-Atlantic record (strictly speaking, regional, but with probable global implications) with our data on northeast Siberian changes at the

---

[3] Significant fluctuations during the Eemian, revealed by the GRIP core, are not confirmed by some ice and oceanic cores (cf. Zahn 1994 for review), but are supported by the terrestrial pollen records in France (Thouveny *et al.* 1994) and Germany (Field *et al.* 1994).

end of the Pleistocene. To make them comparable, I had to transform the GRIP curve onto a linear time scale, by adjusting the appropriate sections to standard 1,000 yr time intervals. As far as all the environmental and the sea-level curve data are in non-calibrated ^{14}C years, they were plotted against the linear time scale based on correlation of the Younger Dryas event (11,000 - 10,000 yr BP in radiocarbon chronology, 12,500 – 11,500 yr in ice core chronology, and *ca*. 12,200 – 11,000 yr in varve chronology (M. Saarnisto, pers. comm.)), i.e. shifted back 1,500 yr. The earlier time interval of the graph is not accurately correlated, but because the extinction period is just around the Younger Dryas, this adjustment is more or less satisfactory.

The resulting correlation (Figure 2d) permits the following deductions:

a) The initial environmental change in Siberia (Phase I) roughly correlates with Bølling plus Allerød warmings. The lag after the sharp rise of global temperature around 13 ka might represent a timing error, but could be a true response lag, assuming a conservative nature of the biotic system. The sea level rise had not yet reached a position seriously affecting the sea/land ratio.

b) There is no evidence of important changes in northeast Siberia corresponding to the Younger Dryas cold event.

c) The second stage (Phase II) corresponds to the period of a sharp post-Younger Dryas rise in temperature followed by the stage when the isotope curve has acquired virtually a Holocene character (minute oscillations), but is still rising. This period corresponds to a sharp rise of sea level, and very soon more than 50% of the former shelf land became inundated by the sea.

d) The period when fully modern vegetation became established (after 8 ka) corresponds exactly to the stable plateau in the GRIP core (modern phase of climatic stability); the area covered by sea was rapidly approximating the modern shoreline.

Analysis of these correlations is restricted by dating precision, which is far from satisfactory. Both source data and subsequent time rescaling could cause errors as large as ± 500 yr or more. During this most crucial period, the environmental changes were evidently so large-scale and rapid that we need a much higher time resolution to understand what was cause, and what was effect. Even if we succeed in that, we still face a problem of numerous feedbacks in such a complicated system.

DISCUSSION

To return to the climate vs. overkill debate, I am ready to admit that humans could have finally exterminated some individual populations of big game, stressed by environmental deterioration. But it is hard to accept that human hunting alone could result in the total extinction of mammoth or any other mammalian species in Eurasia (except in very recent historical times). Early attempts to simulate extinction by mathematical modeling (Budyko 1967) offered a strong support to the overkill paradigm by concluding that the sudden massive extinctions would occur as soon as the density of human population reached a certain threshold. The modern computer simulations (Mithen 1993) suggest that the hunting impact on mammoth could become dramatic if combined with environmental pressure. Anyway, the fate of extinct large mammals can by no means be viewed separately from the simultaneous biotic turnover. The whole problem of extinction must be considered in a broader, multidimensional aspect.

This kind of a more complex approach is demonstrated by some recent interpretations of the archaeological evidence on extinction. More and more evidence emerges that at least some mass accumulations of mammoth bones in Eurasia are not actually kill sites but rather natural death sites, where Palaeolithic people settled to use this wonderful natural resource (Haynes 1991; Soffer 1993). The Stone Age people may even have created some accumulations by collecting bones from various sites. It seems that the overkill paradigm overestimates the human role, especially their spatial distribution, and subsistence practices, and underestimates the role of environmental revolution, which is especially evident in northern Eurasia.

In the areas between 50 and 55°N, relatively densely populated by humans in European Russia (Dniepr and Don basins) and south Siberia (Yenissey and Angara River valleys), Palaeolithic hunters coexisted with mammoth and other large Pleistocene mammals until as late as around 12·5 ka. Disappearance of large mammals in these areas occurred only at the same time that they became extinct between the Arctic Circle and 70°N, where the human population was much less dense[4]. So, when Stuart (1991 p. 549) writes that further work may prove that mammoth 'survived longest in north-central Siberia, because this area was not colonized by people until near the end of the last Cold Stage', his statement is almost correct except for the word 'because'. As I have shown above, mammoth did survive longest in the Arctic, but for a completely different reason. The way in which mammoths retreat to the Arctic after 12·5 ka correlates with observed environmental change suggests much stronger cause-and-effect relationship than with human dispersal. In short, mammoth and other tundra–steppe mammals shared the fate of the ecosystem with which they had co-evolved for more than a million years (Sher 1992b).

As the palaeoecological record from northeast Siberia shows, the Pleistocene (tundra-steppe) ecosystem was a very stable one, although it seems to have been based on unstable climate. As I have suggested, the constant fluctuations of climate in the Late Pleistocene, indicated by the ice cores, may have been partly responsible for maintaining a mosaic of plant communities by a constant 'stirring' effect favouring a pioneering character of vegetation which supported the grazing megafauna. In contrast, the unique stability of Holocene climate may have contributed to the development of today's strong zonation of climax vegetation types such as tundra and coniferous forest, unsuitable for large grazers like the woolly mammoth (Lister & Sher 1995).

During the first phase of the environmental restructuring, which apparently followed a 'normal interglacial' scenario, the Arctic was still able to support mammoth and probably some other grazers. I suggest that until 10·5 – 10 ka both the tundra–steppe ecosystem and the mammal ranges might still have been capable of recovery. A possible test of this hypothesis would be to find other extinct grazers, such as horse and bison *Bison priscus*, surviving in the Arctic until that time or even

---

[4] The only late Palaeolithic site known beyond the Arctic Circle is Berelyokh in the lower Indigirka Basin. The site is dated around 13 ka. It is associated with one of the richest 'mammoth cemeteries', but real relationship between the human lithic culture and enormous bone accumulation is far from being clear.

as late as 9·5 ka[5]. What happened subsequently, however, can be considered a crisis of ecosystem stability. The events during Phase II were very rapid and most likely interlinked by complex cause-and-effect and feedback relationships. The vegetation was imbalanced, as displayed by a very short-lived peak of tree birch dispersal followed by a rapid destruction of the former ecosystem and the spread of moss/lichen and mesophytic grass/sedge ground cover approaching the recent one. It seems that rather early during this rapid vegetational turnover, the supportive capacity of the Arctic biotopes for mammoth was lost, the populations were fragmented and most of them became extinct except some that survived in particular conditions such as on Wrangel Island.

The onset of the modern vegetation correlates well with both global and regional conditions favouring this new situation (stable climate and enormous role of the Arctic Ocean). The survival of mammoth on Wrangel confirms the intimate relations of this Pleistocene grazer with the tundra–steppe ecosystem: it survived in the place where relics of mosaic grassland were preserved, echoed in the extraordinary modern diversity of herbs and grasses on the island (Vartanyan *et al.* 1993). In addition, there is evidence that the whole period of transition (*ca.* 12 – 8 ka), was difficult even for this population. While more than 40 radiocarbon dates on mammoth from the island represent the time between 8 and 4 ka, and a few are older than 12 ka, no dates have so far been obtained for the transition period. Since mammoth was certainly present on the island during this period (by 8 ka, when the radiocarbon dates begin again, Wrangel Island had already been long separated from the mainland and mammoth could hardly have re-colonized it), I interpret this pattern as evidence that the population was held at very low numbers between 12 and 8 ka. Remarkably, it was during just this period that the Wrangel mammoth dwarfing occurred. If dating is continued, some fossils from the transition period should be found, but they will be predictably rare.

---

[5] At the First International Mammoth Symposium in Saint Petersburg (October 17-21, 1995), after this paper had been submitted, S.Vartanyan and his colleagues announced a new date on bison bone from Wrangel Island which confirmed that this animal survived in the Arctic until at least 9·5 ka. Another new date extended the time of horse survival in the Arctic until about 13·5ka (same source).

Animals of the tundra–steppe ecosystem followed different strategies to cope with these dramatic changes. Some successfully found appropriate habitats in the new ecosystem, either *in situ* (reindeer *Rangifer tarandus*, musk-ox *Ovibos moschatus*, most voles *Microtus* spp. and lemmings *Dicrostonyx* and *Lemmus* spp.), or in a remote (much more southern) part of their formerly huge range (saiga *Saiga tatarica*, horses *Equus* spp.). Others found these habitats on isolated relict patches in the Arctic (steppe insects, ground squirrels *Citellus* spp.). Some large grazers that were most intimately linked with the tundra–steppe grassland, and tried to track it by retreat to the North, were trapped by the Holocene change in the Arctic and failed to alter their adaptations developed during hundreds of thousand of years of coevolution with an environment which did not exist any more. This approaches closely the idea of coevolutionary disequilibrium as the main cause of extinction (Graham & Lundelius 1984). Although some of my particular interpretations are different, I hope that the considerations outlined in this paper will contribute to further development of that idea.

The currently known latest appearances of large mammals other than mammoth in the Arctic fossil record – 14·5 ka for horse, 14·3 ca for woolly rhino, 12·8 for bison – might make an impression that many extinctions happened earlier than the irreversible biotic turnover in the North. I believe that this impression may be erroneous for two reasons. Firstly, the number of dates available for these mammals is so far very low as compared with mammoth. Secondly, some mammals, like saiga for instance, could in fact abandon the Arctic (or become extinct there) at the beginning of the environmental change, during the 'normal interglacial' Phase I, i.e. between 13 and 12 ka. For the other extinct mammals the real dead-line must have been close to that for mammoth, i.e. 10 – 9 ka. This correlates with the beginning of the period of stability documented by the GRIP curve as a rise at 10 ka and, reaching a plateau at around 8·5 ka. I believe that the further dating will demonstrate the approximation of the true extinction times to one of these critical periods.[6]

---

[6] The most recent evidence seems to corroborate this hypothesis. See footnote 5 for the latest dates for horse and bison. Harington and Cinq-Mars (1995) have just shown that saiga survived in the North American Arctic until 13·4 ka. Unpublished evidence of R.D. Guthrie (pers. comm.) may extend this period by a further *ca.* 1,000 yr.

In conclusion, I would note that, although at first sight the suggested model of mammoth extinction may seem *regional,* taken together with the suggestion that northern Siberia was *the key region*, that supported the survival of the tundra–steppe biome throughout the earlier interglacials, the model acquires much broader implications.

ACKNOWLEDGEMENTS

The main ideas in this paper were presented at several meetings in 1994 – 95 (Neogene and Quaternary Mammals of the Palaearctic, Krakow, May 1994; IGCP Project 253 "Termination of the Pleistocene" Symposium, Stockholm, October 1994; NSF Workshop "Research Priorities for Russian Arctic Land–Shelf Systems", Columbus, Ohio, January 1995; NATO ARW "Past and Future Rapid Environmental Changes", Crieff, Scotland, June 1995), and I thank numerous colleagues for useful discussion and comments. The latter meeting eventually triggered the paper, and I thank its Director, Brian Huntley, for the opportunity to attend. I appreciate very much the help of Matti Saarnisto with the Younger Dryas problem. I thank Adrian Lister for long and useful discussions of the subject and invaluable help in preparing this paper, as well as Russ Graham, Olga Soffer and Igor Krupnik for their comments. The final work was completed with the support of the BBSRC Research Grant No. 602228 on the evolution of Quaternary mammals and the Russian Foundation for Basic Research Grant No. 95-04-12816.

REFERENCES

Berman DI (1990) Ecology of *Morychus viridis* (Coleoptera, Byrrhidae), a mass beetle from Pleistocene deposits in the northeastern USSR. *in* Kotlyakov V, Sokolov V (eds) Arctic Research, Advances and Prospects, Volume 2, 281-287. Nauka Moscow

Budyko MI (1967) On the causes of the extinction of some animals at the end of the Pleistocene. Soviet Geogr Rev & Transl 8:783-793

Dansgaard W, Johnsen SJ, Clausen HB, Dahl-Jensen D, Gundestrup NS, Hammer CU, Hvidberg CS, Steffensen JP, Sveinbjörnsdottir AE, Jouzel J, Bond G (1993) Evidence for general instability of past climate from a 250-kyr ice-core record. Nature 364:218-220

Fartyshev AI (1993) Features of the coastal/shelf cryolithozone in the Laptev Sea. Nauka Novosibirsk:1-136 (In Russian) (Фартышев АИ Особенности прибрежно-шельфовой криолитозоны моря Лаптевых. ВО Наука, Новосибирск)

Field MH, Huntley B, Müller H (1994) Eemian climate fluctuations observed in a European pollen record. Nature 371:779-783

Graham RW, Lundelius EL Jr (1984) Coevolutionary disequilibrium and Pleistocene extinctions. *in* Martin PS, Klein RG (eds) Quaternary Extinctions: a Prehistoric Revolution, 223-249. Univ Arizona Press Tucson

Grayson DK (1984) Explaining Pleistocene extinctions: thoughts on the structure of a debate. *in*: Martin PS, Klein RG (eds) Quaternary Extinctions: a Prehistoric Revolution, 807-823. Univ Arizona Press Tucson

Grootes PM, Stuiver M, White JWC, Johnsen S, Jouzel J (1993) Comparison of oxygene isotope records from the GISP2 and GRIP Greenland ice cores. Nature 366:552-554

Harington CR, Cinq-Mars J (1995) Radiocarbon dates on saiga antelope (*Saiga tatarica*) fossils from Yukon and the Northwest Territories. Arctic 48:1-7

Haynes G (1991) Mammoths, mastodonts, and elephants: biology, behavior and the fossil record. Cambridge University Press Cambridge New York

Kahlke RD (1992) Repeated immigration of Saiga into Europe. Courier Forsch-Inst Senckenberg 153, 187-195. Frankfurt/Main

Kaplina TN, Lozhkin AV (1982) History of vegetation development in Maritime Lowlands of Yakutia during the Holocene. *in* Development of environment in the USSR during the late Pleistocene and Holocene, 207-220. Nauka Moscow (In Russian) (Каплина ТН, Ложкин АВ История развития растительности Приморских низменностей Якутии в голоцене. В кн.: Развитие природы территории СССР в позднем плейстоцене и голоцене. Наука, Москва)

Kaplina TN, Sher AV, Giterman RE, Zazhigin VS, Kiselyov SV, Lozhkin AV, Nikitin VP (1980) The key section of the Pleistocene deposits on the Allaikha River (the Indigirka lowstream). Bull Commis Quat Res 50:73-95 Moscow: (In Russian) (Каплина ТН, Шер АВ, Гитерман РЕ, Зажигин ВС, Киселев СВ, Ложкин АВ, Никитин ВП Опорный разрез плейстоценовых отложений на реке Аллаихе (низовья Индигирки). Бюлл. Комиссии по изучению четвертичного периода АН СССР, Наука, Москва)

Kim BI (1986) Cenozoic history of development of the East-Arctic shelf and paleoshelf. *in* The Structure and Evolution of the Arctic Ocean, 105-119. USSR Ministry of Geology, PGO "Sevmorgeologiya" Leningrad (In Russian) (Ким БИ Кайнозойская история развития восточно-арктического шельфа и палеошельфа. В кн.: Структура и история развития Северного Ледовитого океана. Министерство геологии СССР, ПГО "Севморгеология", Ленинград.)

Kiselyov SV (1981) Late Cenozoic Coleoptera of North-East Siberia. Nauka Moscow (In Russian) (Киселев СВ Позднекайнозойские жесткокрылые северо-востока Сибири. Наука, Москва)

Kurtén B, Anderson E (1980) Pleistocene mammals of North America. Columbia University Press, New York

Lavrov AV, Sulerzhitzky LD (1992) Mammoths: radiocarbon data on the time of existence. *in* Secular dynamics of biogeocenoses, 46-51. 10th Sukachev Chteniya. Nauka, Moscow (In Russian) (Лавров АВ, Сулержицкий ЛД Мамонты: радиоуглеродные данные о времени существования. В кн.: Вековая динамика биогеоценозов. Х Чтения памяти академика В.Н.Сукачева, Наука, Москва)

Lister AM, Sher AV (1995) Ice cores and mammoth extinction. Nature 378:23-24

Makeev VM, Arslanov KhA, Baranovskaya OF, Kosmodamiansky AV, Ponomaryova DP, Tertychnaya TV (1989) Late Pleistocene and Holocene stratigraphy, geochronology, and paleogeography of Kotelny Island. Bull Commis Quat Res 58:58-69 Moscow (In Russian) (Макеев ВМ, Арсланов ХА, Барановская ОФ, Космодамианский АВ, Пономарева ДП, Тертычная ТВ Стратиграфия, геохронология и палеогеография позднего плейстоцена и голоцена о-ва Котельного. Бюлл. Комиссии по изучению четвертичного периода АН СССР, Наука, Москва)

Marshall LG (1984) Who killed Cock Robin?; an investigation of the extinction controversy. in Martin PS, Klein RG (eds) Quaternary Extinctions: a Prehistoric Revolution, 785-806. Univ Arizona Press, Tucson

Martin PS, Klein RG (eds) (1984) Quaternary Extinctions: a Prehistoric Revolution Univ Arizona Press, Tucson

Mithen S (1993) Simulating mammoth hunting and extinction: implications for the Late Pleistocene of the Central Russian Plain. in Petersen GL, Bricker HM Mellars P (eds) Hunting and Animal Exploitation in the Palaeolithic and Mesolithic of Eurasia. Archaeol Paps Amer Anthrop Assoc 4:163-178

Resolutions of the Interdepartmental Stratigraphic Conference for the Quaternary System in the East of the USSR (Magadan, 1982) (1987). SVKNII DVO AN SSSR Magadan:1-241 (In Russian) (Решения Межведомственного стратиграфического совещания по четвертичной системе Востока СССР (Магадан, 1982 г.). СВ КНИИ ДВО АН СССР, Магадан)

Sher AV (1984) The role of Beringian Land in the development of Holarctic mammalian fauna in the Late Cenozoic. in Kontrimavichus VL (ed.) Beringia in the Cenozoic Era, 296-316. Amerind Publ Co Pvt Ltd, New Delhi

Sher AV (1988) Environment of Plio-Pleistocene mammals in Northeast Siberia. in Stratigraphy and Correlation of Quaternary Deposits of Asia and Pacific Region (Abstr. Intern. Symp.) pt 2, Vladivostok:78-79 (In Russian) (Шер АВ Среда обитания плио-плейстоценовых млекопитающих северо-восточной Сибири. В кн.: Стратиграфия и корреляция четвертичных отложений Азии и Тихоокеанского региона. Тезисы Международного симпозиума (Находка, 9-18 октября 1988 г.) Владивосток)

Sher AV (1991) Problems of the last interglacial in Arctic Siberia. Quat Intern 10-12:215-222

Sher AV (1992a) Biota and climate in Arctic Northeast Siberia during the Pleistocene/Holocene transition. in The 22nd Arctic Workshop, 125-127. Boulder

Sher AV (1992b) Beringian fauna and early Quaternary mammalian dispersal in Eurasia: ecological aspects. Courier Forsch-Inst Senckenberg 153:125-133 Frankfurt/Main

Sher AV (1994) Pleistocene extinctions: how dwarf mammoths escaped the net. in Neogene and Quaternary Mammals of the Palaearctic, 63-65. Conf in Honour Prof. K.Kowalski, Krakow

Sher AV, Kaplina TN, Giterman RE, Lozhkin AV, Arkhangelov AA, Kiselyov SV, Kouznetsov YuV, Virina EI, Zazhigin VS (1979) Late Cenozoic of the Kolyma Lowland: XIV Pacific Science Congress, Tour Guide XI, Moscow

Sher AV, Sulerzhitsky LD (1991) Did mammoth escape from Arctic Siberia during the "bad times": a review of the ^{14}C dating. Abstr. 21st Arctic Workshop, 90-91. Fairbanks, Alaska

Soffer O (1993) Upper Paleolithic adaptations in central and eastern Europe and man-mammoth interactions. *in* Soffer O, Praslov ND (eds) From Kostenki to Clovis: Upper Paleolithic - Paleo-Indian adaptations, 31-49. Plenum Press, New York

Stuart AJ (1982) Pleistocene vertebrates of the British Isles. Longman, London

Stuart AJ (1991) Mammalian extinctions in the late Pleistocene of northern Eurasia and North America. Biol Rev 66:453-562

Sulerzhitzky LD (1995) Features of radiocarbon chronology of the woolly mammoth (*Mammuthus primigenius*) in Siberia and north of Eastern Europe. Proc Zool Inst 265, 163-183. Russ Acad Sci St-Petersburg (In Russian) (Сулержицкий ЛД Черты радиоуглеродной хронологии мамонтов (Mammuthus primigenius) Сибири и севера Восточной Европы. Труды Зоол. ин-та РАН 265, ЗИН РАН, Санкт Петербург)

Suzdalsky OV (1971) Pattern of the recent tectonic movements as the basic reason for peculiarities of the Anthropogene sediments in northern Eurasia. *in* The problems of correlation of the Plio-Pleistocene deposits in Northern Eurasia. Materials of Symposium, Leningrad March 1971, 137-143. USSR Geogr Soc Pleist Commis, Leningrad (In Russian) (Суздальский ОВ Режим новейших тектонических движений - первопричина особенностей строения антропогена на севере Евразии. В кн.: Проблемы корреляции новейших отложений севера Евразии. Материалы симпозиума (Март 1971, Ленинград). Географическое о-во СССР, Плейстоценовая Комиссия, Ленинград)

Thouveny N, de Beaulieu J-L, Bonifay E, Creer KM, Guiot J, Icole M, Johnsen S, Jouzel J, Reille M, Williams T, Williamson D (1994) Climate variations in Europe over the past 140 kyr deduced from rock magnetism. Nature 371:503-506

Vartanyan SV, Garutt VE, Sher AV (1993) Holocene dwarf mammoths from Wrangel Island in the Siberian Arctic. Nature 362:337-340

Yurtsev BA (1981) Relic steppe complexes in northeast Asia (problems of reconstruction of cryoxerotic landscapes of Beringia). Nauka, Novosibirsk (In Russian) (Юрцев БА Реликтовые степные комплексы Северо-Восточной Азии (проблемы реконструкции криоксеротических ландшафтов Берингии). Наука, Новосибирск)

Zahn R (1994) Core correlations. Nature 371:289-290

Section 5

Mechanisms enabling evolutionary responses

Variation in plant populations: History and chance or ecology and selection?

Honor C. Prentice

Department of Systematic Botany

Lund University

Östra Vallgatan 18-20

S-223 61 Lund

Sweden

GENETIC DIVERSITY WITHIN SPECIES AND POPULATIONS IS THE RAW MATERIAL FOR NATURAL SELECTION AND EVOLUTIONARY CHANGE

More than a hundred years ago, and before the results of Mendel's genetic experiments were available, Charles Darwin argued that a species' ability to exploit a wide range of habitats in subsequent generations was related to levels of infraspecific diversity: "...natural selection can do nothing until favourable individual differences and variations occur" (Darwin 1890, p.137).

Within the framework of modern quantitative genetics, the response to natural selection can be related to the level of genetic variation in the previous generation by the following equation:

$$R = h^2 S$$

where R is the response to selection (the difference in the mean phenotypic value between the offspring of the selected parents and the whole of the parental generation before selection), S is the selection differential (the mean phenotypic value of the selected parents, expressed as a deviation from the population mean) and h^2 is the narrow sense heritability (a measure of the genetic variability that is available for selection to act upon). Information on the amount of heritable variation in a population, together with knowledge of the genetic correlation between

NATO ASI Series, Vol. I 47
Past and Future Rapid Environmental Changes:
The Spatial and Evolutionary Responses of
Terrestrial Biota
Edited by Brian Huntley et al.
© Springer-Verlag Berlin Heidelberg 1997

characters, can thus be used to predict the response to selection over a few generations (Falconer 1981).

A population may show low levels of genetic variation but be well-adapted to its present environment. However, theory suggests that low reserves of genetic diversity will limit the ability of populations and species to respond, via natural selection, to future environmental change and will increase the chance of extinction (Falconer 1981; Lande & Arnold 1983; Hoffmann & Blows 1993).

LEVELS OF GENETIC DIVERSITY IN NATURAL POPULATIONS ARE DETERMINED BY STOCHASTIC AS WELL AS SELECTIVE PROCESSES

If we are to be able to predict evolutionary responses to climate change, we need to understand the mechanisms that control the spatial patterning of genetic variation and the amounts of genetic diversity within populations and species. We are well-equipped with elegant theoretical models that can explain the loss or maintenance of genetic diversity. Unfortunately, we have only a superficial understanding of the way in which different selective and non-selective processes interact on different spatial and temporal scales to determine the levels of genetic diversity that we can observe in natural populations at the present day.

The levels of genetic diversity within natural populations are not simply a reflection of ecology and natural selection. Factors that influence population size and patterns of gene flow on different spatial scales may also have a profound influence on population genetic structure – levels of genetic diversity within populations and the degree of between-population differentiation. Populations in which the effective number of breeding individuals is low will be exposed to the effects of genetic drift, and undergo random changes in allele frequencies between generations as a result of sampling error. The impact of genetic drift will increase with decreasing population size. Extreme, prolonged or repeated reductions in population size will lead to the random loss of allelic variation within populations and to increased divergence between populations (Nei et al. 1975; Barrett & Kohn 1991).

A number of recent studies of fragmented plant populations have examined the relationship between present-day population size and levels of genetic diversity within populations or the degree of between-population differentiation (e.g. Ouberg et *al.* 1991, van Treuren *et al.* 1991). However, demographic inertia may mean that populations of perennial herbs that have been only recently fragmented have not yet been exposed to the effects of genetic drift and random changes in allele frequencies (cf. Widén & Andersson 1993). The high levels of genetic diversity and low degree of population divergence in disjunct populations of the extremely long-lived conifer, *Pinus longaeva*, suggest that the species' demographic and genetic structure has remained essentially unaltered since the last glacial period, when *P. longaeva* was widespread south of the North American ice sheets (Hiebert & Hamrick 1983). On the other hand, the cumulative effects of demographic accidents and genetic stochasticity in the distant past may still be reflected in low levels of genetic diversity in present-day populations (e.g. Fowler & Morris 1977; Ledig & Conkle 1983; Sage & Wolff 1986).

MAJOR MIGRATIONAL EVENTS INFLUENCE THE STRUCTURING OF GENETIC VARIATION AND LEVELS OF GENETIC DIVERSITY WITHIN SPECIES

The trace gas-induced climatic change that is predicted over the next century, with a rise in the global mean temperature of 2 to 4°C (cf. Cramer & Steffen, 1996), is of a magnitude that is comparable to the major climatic events associated with glacial–interglacial cycles. Such major climatic events have previously been associated with massive episodes of plant and animal migration on a continental scale (cf. Huntley & Birks 1983; Davis 1976).

The geographic pattern of intraspecific genetic differentiation within a number of widespread European animal and plant species can be clearly interpreted in terms of their histories of northward immigration during the postglacial period (e.g. Tegelström 1987; Ferris *et al.* 1993). As well as determining geographic differentiation patterns, postglacial immigration history may also have had an impact on levels of genetic diversity within populations and species that is still observable at present (Critchfield 1984; Lewis & Crawford 1995). Range-expansion and range-contraction are inevitably associated with recurrent episodes of population disjunction. Populations of retreating species may undergo fragmentation and isolation as a result of successive loss of suitable habitats, while populations on the expanding range-front of immigrating species may be subjected to repeated episodes of low effective size during founder events.

The perennial shrub *Hippocrepis emerus* (Leguminosae) provides an example of a plant species in which both the geographic structure of genetic differentiation and the low levels of within-population genetic diversity are plausibly explained in terms of the species' postglacial migrational history. The main distributional range of *H. emerus* is centred on southern Europe and the Mediterranean region, and the species is represented in Scandinavia by three, highly isolated, regional populations. The genetic structure in the native Scandinavian populations and an introduced population – derived from the central area of the species' range – was assessed using DNA fingerprinting and enzyme electrophoresis (Lönn *et al.* 1995). In contrast to the newly-introduced population, the native populations showed a high degree of allozyme fixation, with the Norwegian and Swedish isolates of the species fixed for alternate alleles at most loci. Levels of DNA diversity were low in the native populations and high in the introduced population (Figure 1). The allozyme and DNA fingerprint data showed closely-congruent patterns of geographic differentiation. Evidence from both allozymes and DNA suggests a historical explanation for the levels of diversity and the geographic structure of genetic variation in native Scandinavian *H. emerus*. In Scandinavia, *H. emerus* is only found growing in close-association with hazel (*Corylus avellana*) and it is reasonable to speculate that the immigration history of *H. emerus* followed that of hazel. While *H. emerus* is not represented in the pollen record, there is abundant pollen analytical evidence on the spatiotemporal variation in the range-extent of hazel throughout the postglacial period (Huntley & Birks 1983). Hazel immigrated into Scandinavia in two separate waves, colonizing Norway from the west around 10,500 yr B.P. and entering southern Sweden along a broader front a thousand years later. It is likely that the loss of alleles and gene diversity from the native Norwegian and Swedish regional populations of *H. emerus* reflects the effects of genetic drift during founder events and repeated periods of small population size during postglacial range expansion. The reciprocal fixation of alternate alleles in the Norwegian and Swedish populations and the close similarity of the populations within each of the two areas suggest that allelic fixation occurred during the process of immigration, rather than resulting from local, *in situ*, population fragmentation.

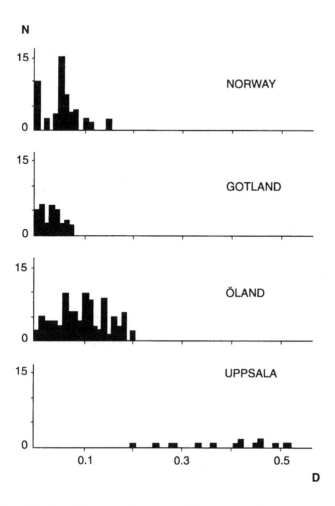

Figure1: Scandinavian *Hippocrepis emerus*. Histograms of pairwise DNA fingerprint dissimilarities between individuals within the three native regional populations (Norway, Öland and Gotland) and a recently-introduced population (Uppsala) derived from the central area of the species' range.
D = dissimilarity (based on a band-sharing statistic); N = number of pairwise comparisons. For further background, see Lönn *et al.* (1995).

On a local scale, within the Öland (Sweden) regional population of *H. emerus*, population isolation and marginality, but not population size, are related to levels of differentiation between populations. Marginal populations show a higher degree of inter-population divergence than the central Öland populations (Lönn & Prentice 1995). Such peripheral populations are likely to be particularly vulnerable to variation in local habitat availability and to have had a history of size fluctuations.

Random genetic divergence between populations during successive episodes of low population size is less likely to be counteracted by gene flow in isolated, marginal populations than in range-centre populations.

THE SELECTIVE MAINTENANCE OF GENETIC DIVERSITY: MULTIPLE NICHE SELECTION IN LOCAL POPULATIONS

The reserves of genetic variation that are available for the operation of natural selection are generated by mutation (Lande 1995) and may be augmented by gene flow from other species or populations (Ellstrand 1992). Loss of genetic diversity may occur as a result of random genetic drift (Barrett & Kohn 1991). Directional or stabilizing selection may also reduce levels of intrapopulation variation (Falconer 1981).

While selection may lead to the loss of genetic variation, there are also various models that suggest that selection may contribute to the maintenance of high levels of polymorphism within populations (see Ennos, 1983, and Ledig, 1986, for useful reviews). Elegant, classical models, such as that of single-locus overdominance (in which heterozygous individuals are fitter than either type of homozygous individual), which apply to a constant environment, may be of limited relevance in natural populations. However, there are a number of models (see Ennos 1983) which apply to environments that are heterogeneous in space or time. Such models are intuitively attractive to researchers working with variation in field populations. One such model is the 'niche variation' model of Van Valen (1965).

The niche variation model proposes that, if different genotypes have different relative fitnesses in different habitats or niches, multiple niche selection, or habitat selection (*sensu* Bazzaz 1991), will contribute to the maintenance of genetic polymorphism within populations. Tests of the niche variation hypothesis have been problematic. Most studies of associations between genetic variation and habitat variation have been carried out on animals, and the results of such studies have been interpreted as both supporting and rejecting the hypothesis (see e.g. Soulé & Stewart 1970; Van Valen & Grant 1970; Hedrick 1986; Smith 1987). The study of niche variation in plants should, in principle, be simpler than in animals. Except in the case of clonal species (cf. Bazzaz 1991), established plant individuals cannot move to a more favourable microhabitat. The habitat experienced by plant individuals may vary throughout their lifetime, but the realized niche of the individuals is fixed in space. If habitat selection is operating in local plant populations, it should be possible to detect significant associations between genotypic or allelic variation and habitat variation. However, the spatial structuring of genetic variation within plant populations may reflect the effects of inbreeding, localized gene flow and population history as well as the effects of niche differentiation. Thus it may be difficult to separate the effects of habitat selection from raw spatial autocorrelation of genetic variation in investigations of inbreeding species (cf. Nevo *et al.* 1986, 1994).

The steppe-like 'alvar' grasslands on the Baltic island of Öland (Sweden) are characterized by high levels of species diversity and by a diverse and spatially-repeating fine-scale mosaic of habitat types (Bengtsson *et al.* 1988) Studies of genetic variation in populations of outbreeding species in these habitats have revealed significant associations between allozyme variation and gradients of edaphic and plant community variation (Prentice & Cramer 1990; Prentice *et al.* 1995; Lönn *et al.* 1996), suggesting that niche-differentiation may be contributing to the maintenance of genetic diversity within populations. There has been considerable debate over whether allozyme–habitat associations result from direct selection on the allozyme loci in question, or whether such associations reflect the effects of selection on other, linked, loci that confer differential fitness in different microhabitats (cf. Hedrick 1986). There is an increasing amount of evidence from both plant and animal populations that suggests that selection may act directly on allozyme genotypes at particular loci (e.g. Brown *et al.* 1976; Powers *et al.* 1991; Riddoch 1993). Studies that link information on the *in vitro* kinetics of different allozymes to experimental tests of the relative fitness of different genotypes under controlled conditions, followed by tests of the predicted patterns of genotype–habitat associations in field populations, provide convincing evidence of direct selection on allozymes (e.g. Brown *et al.* 1976; Simon 1987; Lönn 1993).

LEVELS OF GENETIC DIVERSITY ON DIFFERENT SPATIAL SCALES IN PRESENT-DAY POPULATIONS REFLECT THE EFFECTS OF BOTH STOCHASTIC, NON-SELECTIVE, PROCESSES AND NATURAL SELECTION – INTEGRATED OVER TIME

In practice, it may be extremely difficult to assess the relative roles of selection and genetic drift as determinants of the levels of genetic diversity in natural plant populations. Natural selection may act locally, within plant populations, to create a spatial mosaic of genetic variation that is fine-tuned to different microhabitats and plant communities. The response to selection in such situations may be rapid, with significant differences in fitness-related characters being detectable as little as six years after changes in habitat conditions (see e.g. Snaydon & Davies 1982). On the other hand, restricted gene flow may lead to a non-adaptive, fine-scale structuring of genetic variation within populations of self-pollinating species (e.g. Heywood 1991) and even of outcrossing species (e.g. Schaal 1980). Spatial structuring of variation as a result of restricted gene flow may reinforce the effects of selection, but genetic drift may also lead to the loss of genetic diversity from substructured populations which have poor gene flow between genetic neighbourhoods.

On a larger geographic scale, and on a longer temporal scale, selection may lead to, for example, clinal variation that is associated with climatic variation along altitudinal or latitudinal gradients (Ledig & Korbobo 1983; Roy & Lumaret 1987). However, clinal variation may also have non-adaptive origins and reflect historical migration patterns (Endler 1986).

PREDICTIONS FOR THE FUTURE

Genetic diversity within species and populations is the raw material for evolutionary change. Populations or species with low reserves of genetic diversity may not experience problems with survival and reproduction under a particular range of environmental conditions. However, the ability to respond to future environmental change will depend on the availability of genetic variation upon which natural selection can operate (Hoffmann & Blows 1993, Lynch & Lande 1993). Human-induced environmental changes, such as atmospheric pollution and the contamination of soils by heavy-metals, already constitute an evolutionary challenge to many plant species (Antonovics *et al.* 1971, Gregorius 1989, Geber & Dawson 1993).

Many widespread perennial plant species contain ample reserves of genetic diversity that should permit an evolutionary response to climate change (cf. Geber & Dawson 1993, Mátyás, 1996). An evolutionary response to climate change may be the only option available to species that are confined to disjunct habitat fragments (Geber & Dawson 1993). However, even species that have the option of migrating in response to climate change will be exposed to novel environments. The increase in temperature predicted over the next century will be accompanied by changes in other abiotic factors, such as seasonality, precipitation and evapotranspiration. Migrating plant populations will also experience novel biotic environments, with changes in, for example, competitive interactions, pollinator guilds and predators. Hence, a capacity for adaptive change will be necessary in species that are able to respond to climate change by migration (Geber & Dawson 1993). Natural selection is expected to continue to contribute to both large-scale geographic patterns of variation and the fine-tuning of intrapopulation genetic variation to local habitats.

The trace gas-induced global warming that is predicted over the next century can be expected to trigger distributional changes, comparable to those that followed the end of the last glaciation, in many plant species. However, the assumption that migration is likely to be the most common response to climate change (cf. Geber & Dawson 1993) may be unrealistic in late-successional landscapes with closed vegetation cover. For example, pollen analytical studies in Europe suggest that most tree types (apart from *Betula*, *Pinus* and *Salix*) may have died out in northern Europe during cold stages, rather than retreating southwards as the climate deteriorated at the end of interglacial periods (Bennett *et al.* 1991). The southern European 'glacial refugia' for tree species are likely also to have served as refugia during warm periods. The pollen analytical data thus suggest that there may have been relatively little genetic replenishment of the refugial isolates that have served as a source for the reforestation of northern Europe during successive interglacial periods. Levels of genetic diversity and the structure of geographic differentiation within the northern range of European tree species may, to a large extent, reflect local evolutionary processes which have taken place within small and isolated southern European refugial populations.

At the present day, large tracts of northern Europe and North America are occupied by intensive agriculture while less intensively exploited areas are covered by closed forest vegetation. The possibilities for species to respond to future climate change by migrating are likely to be restricted. Levels of genetic diversity and the structuring of genetic variation presently observed in widespread species may be of little relevance as predictors of species' capacity for evolutionary response to climate – unless we are able to foresee which areas of a species' current range will contribute to migrating populations in the future. As with the major range-shifts during the early Holocene, it is likely that chance events during migration will have a disproportionate impact on the subsequent genetic structure in many species. Local reserves of genetic diversity will be lost as populations become small and disjunct, and founder effects and genetic bottlenecks during range expansion or contraction will lead to random population divergence and loss of genetic diversity on a more extensive geographic scale.

REFERENCES

Antonovics J, Bradshaw AD, Turner RG (1971) Heavy metal tolerance in plants. Advances in Ecological Research 7:1-85

Barrett SCH, Kohn JR (1991) Genetic and evolutionary consequences of small population size in plants: implications for conservation. in Falk DA, Holsinger KA (eds) Genetics and conservation of rare plants, 3-30. Oxford University Press, Oxford

Bazzaz FA (1991) Habitat selection in plants. American Naturalist 137:S116-S130

Bengtsson K, Prentice HC, Rosén E, Moberg R, Sjögren E (1988) The dry alvar grasslands of Öland: ecological amplitudes of plant species in relation to vegetation composition. Acta Phytogeographica Suecica 76:21-46

Bennett KD, Tzedakis PC, Willis KJ (1991) Quaternary refugia of north European trees. Journal of Biogeography 18:103-115

Brown AHD, Marshall DR, Munday J (1976) Adaptedness of variants at an alcohol dehydrogenase locus in Bromus mollis L. (Soft bromegrass). Australian Journal of Biological Science 29:389-396

Cramer W, Steffan W (1996) Forecast changes in the global environment: What they mean in terms of ecosystem responses on different time scales. in Huntley B, Cramer W, Morgan AV, Prentice HC, Allen JRM (eds) Past and future rapid environmental changes: The spatial and evolutionary responses of terrestrial biota, 415-426. Springer-Verlag, Berlin

Critchfield WB (1984) Impact of the Pleistocene on the genetic structure of North American conifers. in Lanner RM (ed) Proceedings of the eighth North American forest biology workshop, 77-118. Utah State University, Logan

Darwin C (1890) Origin of species (6th edn, with additions and corrections to 1872). John Murray, London

Davis MB (1976) Pleistocene biogeography of temperate deciduous forests. Geoscience and Man 13:13-26

Ellstrand NC (1992) Gene flow by pollen: implications for plant conservation genetics. Oikos 63:77-86

Endler JA (1986) Natural selection in the wild. Princeton University Press, Princeton

Ennos RA (1983) Maintenance of genetic variation in plant populations. Evolutionary Biology 16:129-155

Falconer DS (1981) Introduction to quantitative genetics (2nd edn) Longman,London

Ferris C, Oliver RP, Davy AJ, Hewitt GM (1993) Native oak chloroplasts reveal an ancient divide across Europe. Molecular Ecology 2:337-344

Fowler DP, Morris RW (1977) Genetic diversity in red pine: evidence for low genic heterozygosity. Canadian Journal of Forest Research 7:343-347

Geber M, Dawson TE (1993) Evolutionary responses of plants to global change. in Kareiva PM, Kingsolver JG, Huey RB (eds) Biotic interactions and global change, 179-197. Sinauer, Sunderland, Mass.

Gregorius H-R (1989) The importance of genetic multiplicity for tolerance of atmospheric pollution. in Scholz F, Gregorius H-R, Rudin D (eds) Genetic effects of air pollutants in forest tree populations, 163-172. Springer-Verlag, Berlin

Hedrick PW (1986) Genetic polymorphism in heterogeneous environments: a decade later. Annual Review of Ecology and Systematics 17:535-566

Heywood JS (1991) Spatial analysis of genetic variation in plant populations. Annual Review of Ecology and Systematics 22:335-355

Hiebert RD, Hamrick JL (1983) Patterns and levels of genetic variation in Great Basin bristlecone pine, *Pinus longaeva*. Evolution 37:302-310

Hoffmann AA, Blows MW (1993) Evolutionary genetics and climate change: will animals adapt to global warming? in Kareiva PM, Kingsolver JG, Huey RB (eds) Biotic interactions and global change, 165-178. Sinauer, Sunderland, Mass.

Huntley B, Birks HJB (1983) An atlas of past and present pollen maps for Europe, 0-13,000 years ago. Cambridge University Press, Cambridge

Lande R (1995) Mutation and conservation. Conservation Biology 9:782-791

Lande R, Arnold SJ (1983) The measurement of selection on correlated characters. Evolution 48:1460-1469

Ledig FT (1986) Heterozygosity, heterosis, and fitness in outbreeding plants. in Soulé ME (ed) Conservation biology: the science of scarcity and diversity, 77-104. Sinauer, Sunderland, Mass.

Ledig FT, Conkle MT (1983) Gene diversity and genetic structure in a narrow endemic, Torrey pine (*Pinus torreyana* Parry ex Carr). Evolution 37:79-85

Ledig FT, Korbobo DR (1983) Adaptation of sugar maple populations along altitudinal gradients: photosynthesis, respiration, and specific leaf weight. American Journal of Botany 70:256-265

Lewis PO, Crawford DJ (1995) Pleistocene refugium endemics exhibit greater allozymic diversity than widespread congeners in the genus *Polygonella* (Polygonaceae). American Journal of Botany 82:141-149

Lönn M (1993) Genetic structure and allozyme-microhabitat associations in *Bromus hordeaceus*. Oikos 68:99-106

Lönn M, Prentice HC (1995) The structure of allozyme and leaf shape variation in isolated range-margin populations of the shrub *Hippocrepis emerus* (Leguminosae). Ecography 18:276-285

Lönn M, Prentice HC, Tegelström H (1995) Genetic differentiation in *Hippocrepis emerus* (Leguminosae): allozyme and DNA fingerprint variation in disjunct Scandinavian populations. Molecular Ecology 4:39-48

Lönn M, Prentice HC, Bengtsson K (1996) Genetic structure, allozyme–habitat associations and reproductive fitness in *Gypsophila fastigiata* (Caryophyllaceae). Oecologia 106:308-316

Lynch M, Lande R (1993) Evolution and extinction in response to environmental change. *in* Kareiva PM, Kingsolver JG, Huey RB (eds) Biotic interactions and global change, 234-250. Sinauer, Sunderland, Mass.

Mátyás Cs (1996) Genetics and adaptation to climate change: A case study of trees. *in* Huntley B, Cramer W, Morgan AV, Prentice HC, Allen JRM (eds) Past and future rapid environmental changes: The spatial and evolutionary responses of terrestrial biota, 357-370. Springer-Verlag, Berlin

Nei M, Maruyama T, Chakraborty R (1975) The bottleneck effect and genetic variability in populations. Evolution 29:1-10

Nevo E, Krugman T, Beiles A (1994) Edaphic natural selection of allozyme polymorphisms in *Aegilops peregrina* at a Galilee microsite in Israel. Heredity 72:109-112

Nevo E, Beiles A, Kaplan D, Golenberg EM, Olsvig-Whittaker L, Naveh Z (1986) Natural selection of allozyme polymorphisms: a microsite test revealing ecological genetic differentiation in wild barley. Evolution 40:13-20

Ouberg NJ, van Treuren R, van Damme JMM (1991) The significance of genetic erosion in the process of extinction. II. Morphological variation and fitness components in populations of varying size of *Salvia pratensis*. Oecologia 86:359-367

Powers D, Lauerman T, Crawford D, DiMichele L (1991) Genetic mechanisms for adapting to a changing environment. Annual Review of Ecology and Systematics 25:629-659

Prentice HC, Cramer W (1990) The plant community as a niche bioassay: environmental correlates of local variation in *Gypsophila fastigiata*. Journal of Ecology 78:313-325

Prentice HC, Lönn M, Lefkovitch LP, Runyeon H (1995) Associations between allele frequencies in *Festuca ovina* and habitat variation in the alvar grasslands on the Baltic island of Öland. Journal of Ecology 83:391-402

Riddoch BJ (1993) The adaptive significance of electrophoretic mobility in phosphoglucose isomerase (PGI). Biological Journal of the Linnean Society 50:1-17

Roy J, Lumaret R (1987) Associated clinal variation in leaf tissue water relations and allozyme polymorphism in *Dactylis glomerata* L. populations. Evolutionary Trends in Plants 1:9-19

Sage RD, Wolff JO (1986) Pleistocene glaciations, fluctuating ranges, and low genetic variability in a large mammal (*Ovis dalli*). Evolution 40:1092-1095

Schaal BA (1980) Measurement of gene flow in *Lupinus texensis*. Nature 284:450-451

Simon J-P (1987) Thermal adaptation and acclimation of higher plants at the enzyme level: malate dehydrogenase in populations of the legume *Lathyrus japonicus*. Evolutionary Trends in Plants 1:78-83

Smith TB (1987) Bill size polymorphism and intraspecific niche utilization in an African finch. Nature 392:717-719

Snaydon RW, Davies TM (1982) Rapid divergence of plant populations in response to recent changes in soil conditions. Evolution 36:289-297

Soulé M, Stewart BR (1970) The 'niche-variation' hypothesis: a test and alternatives. American Naturalist 104:85-97

Tegelström H (1987) Transfer of mitochondrial DNA from the Northern red-backed vole (*Clethrionomys rutilus*) to the Bank vole (*C. glareolus*). Journal of Molecular Evolution 24:218-227

van Treuren R, Bijlsma R, van Delden W, Ouberg NJ (1991) The significance of genetic erosion in the process of extinction. I. Genetic differentiation in *Salvia pratensis* and *Scabiosa columbaria* in relation to population size. Heredity 66:181-189.

Van Valen L (1965) Morphological variation and width of ecological niche. American Naturalist 99:377-390

Van Valen L, Grant PR (1970) Variation and niche width reexamined. American Naturalist 104:589-590

Widén B, Andersson S (1993) Quantitative genetics of life-history and morphology in a rare plant, *Senecio integrifolius*. Heredity 70:503-514

Genetics and adaptation to climate change: A case study of trees

Csaba Mátyás
Department of Plant Sciences
University of Sopron
P.O.B. 132
H - 9401 Sopron
Hungary

FOREST TREES AS OBJECTS OF STUDY OF CLIMATIC ADAPTATION

Most species are ephemeral features of the Earth system - an average species exists for about 10 million years (Vitousek 1992). Forest trees, especially evolutionarily ancient conifers, certainly belong to the longer lasting class of species. Trees have not only successfully survived changing geological periods as species, but also endure during an individual's lifetime considerable fluctuations of environment without the chance of migration or short-term genetic adaptation on the population level as in case of annual plants and many animals.

The present interglacial period, the Holocene, has lasted for approximately 10,000 years; radiocarbon dating confirmed a relatively recent migrational past of forest trees in the temperate-boreal forest zone of a few thousand years only (Davis 1980). This timespan is very short with regard to possibilities of climatic adaptation when compared to the lifespan of trees. In the extreme, for the oldest bristlecone pines (*Pinus aristata*) of the Californian White Mountains, reaching ages of over four thousand years, the whole Holocene might appear as a period of only three generations.

The adaptive strategy of trees, buffering relatively faster environmental changes as compared with shorter-lived organisms, is therefore an interesting subject to study, especially in the perspective of the expected uncertainty of future climate conditions. Another reason to give trees special consideration is that trees are keystone species

NATO ASI Series, Vol. I 47
Past and Future Rapid Environmental Changes:
The Spatial and Evolutionary Responses of
Terrestrial Biota
Edited by Brian Huntley et al.
© Springer-Verlag Berlin Heidelberg 1997

in every ecosystem where present; their stability and vitality determines the presence and vitality of most other species at all trophic levels.

Recent years have yielded an increasing wealth of information on the genetic diversity of forest trees; the majority of these studies, however, deal with diversity on the molecular level. Although molecular polymorphism may be a conditionally useful marker for loci of adaptive significance (Ledig *et al.* 1983, Devey *et al.* 1995), the importance of quantitative (growth) traits as measured in long-term field tests cannot be stressed enough as these traits are the most reliable, quantitative expressions of long-term plant–environment interactions. Investigations on the relation of environment and intraspecific variability of growth traits have a long history in forestry, especially in connection with so-called provenance experiments (Langlet 1971; Mátyás 1996).

Provenance research is the expression used in forestry for the analysis of common garden plantations of 'wild' tree populations originating from geographically different locations. The original goal of provenance tests was to identify stands, populations or areas of origin which provide progenies with the most desirable traits and commercially best results in the test area. After the development of forest tree breeding programs, provenance tests were regarded as being redundant. The upcoming issues of genetic diversity and of expected climate instability, and the growing importance of the adaptive potential, will hopefully direct research priorities back to these experiments (demonstrating that trees may live longer than research concepts).

The tracing of between-provenance variation probably represents the most powerful available tool for testing hypotheses of climatic adaptation in trees. Instead of analysing genetic changes in subsequent generations, an unthinkable task in forestry, the observed variation is interpreted as an adaptive response to changes in climate conditions. The outplanting of the tests may be regarded as a simulation of environmental change over time and may be modeled (Mátyás & Yeatman 1992; Mátyás 1994). This approach offers direct applications in forecasting climate change effects on trees and forests.

Compared to common-garden experiments with short-lived plants, an advantage might be the long duration of the forest tree tests; effects of annual weather fluctuations and rare anomalies are integrated in the end results. The longevity of the objects themselves binds the organisms much more strongly to the given environment than in the case of ephemeral plants. It may be expected therefore that the interaction between climate and genetic makeup of tree populations yields general patterns of variation which exhibit adaptive effects.

These distinct evolutionary and ecological implications make provenance tests important and interesting objects to study beyond direct silvicultural applications. Although provenance research might be among the most important contributions of forestry to biological sciences, up to now its results have failed to capture much attention outside the forestry community. Even the fact that much of the climatic adaptation research has been initiated and studied on forest trees is not generally known (Langlet 1971; Mátyás 1996).

ROLE OF NATURAL SELECTION IN INCREASING THE ACTUAL FITNESS OF THE POPULATION

The fitness of a species is determined by fitness at the individual, population and species levels, covering both the genetic, morphological and physiological adjustments of the species to its environment. On the population level, the maintenance of fitness is mainly influenced by the effects of natural selection, favouring genotypes with advantageous traits – therefore adaptability is strongly determined by the available genetic diversity. In the course of *genetic adaptation* the population undergoes a change in gene frequencies, as less fit individuals are removed by natural selection. The genetic change in the makeup of the population is directed toward a temporary optimum in a given ecological situation. Genetic adaptation is expected to increase the average fitness of the population. Covering relatively short periods in time, adaptational changes in populations may be regarded as reversible.

In the following discussion the term adaptive variation is used to describe climate-related genetic variation within species, although it is clear that other forms of variation support adaptation (and fitness) as well; however, climate-related variation is the most prudent interpretation of adaptive genetic variation.

To evaluate or even quantify adaptedness is certainly a question of interpretation. Without disregarding conflicting opinions in this respect, the concept of Ayala (1969) is followed: to assess adaptedness, the ability of the plant to transform available nutrients and energy into its own living matter is used. This corresponds to growth and dry matter production; in forestry terms to volume (timber) yield. Amongst the components of yield of a forest stand, height growth is under the strongest genetic control; only this growth trait will be discussed. The genetic differences between populations manifest themselves as differences in the mean (phenotypic) performance at the various test locations. The observed or measured phenotype is interpreted as a response of the genotype to the given environment.

To investigate the effect of natural selection on adaptive traits is especially difficult considering the overwhelming complexity of biotic and abiotic interactions in a forest stand. Provenance tests however, also offer interesting insights in this respect. Observations on the phenological behaviour (e.g. timing of budset) of populations of various origins reveal marked differences between population means in relation to the climatic conditions at the location of origin, but the within-population variation is also remarkably broad even for traits under very strong adaptive pressure such as length of annual growing cycle (Mátyás 1981; Mikola 1982). Analysis of the between-population differences in provenance tests shows, however, that these differences clearly tend to diminish toward maturity. This effect leads to decreasing relative differences in the productivity of provenances as illustrated by century-old experiments (Timofeyev 1975). The levelling-off of experimental data may be partially explained by the selective effect exerted on transplanted populations by the local site conditions; the least fit genotypes are removed, which reduces the variability between populations. Beuker and Koski (1995) demonstrated the same effect with phenological data from old provenance experiments. The directional genetic change in the population over time is supported by allozyme frequency analyses (Müller-Starck 1991).

In addition to the above-described effect of natural selection and genetic adaptation, there exist other phenomena offering 'shortcut solutions' to the adaptation problem within the lifetime of the same generation. With age, there is a certain change in growth and phenological traits in forest trees, which is developmentally controlled; e.g. budburst occurs at a later date in older trees (Ununger et al. 1988). There are also indications that certain gene regulation changes activated through environmental conditions at the time of sexual reproduction may cause lasting effects on the behaviour and performance of the progenies. In Norway spruce *Picea abies*, progenies from controlled crosses carried out on mother plants transferred into altered environments (into warmer or longer-day climates) surprisingly perform as if adapted to the new environment and maintain their altered traits even in their original, native environment. The most likely explanation is the existence of a regulatory mechanism affecting the expression of genes controlling adaptive traits. In certain phases of the life cycle the future gene expression might be influenced by thermal or photoperiodic effects (Bjørnstad 1981; Skrøppa & Johnsen 1994). Similar phenomena, affecting both juvenile mortality and growth have been found in Scots pine *Pinus sylvestris* (Lindgren & Wei 1994). If these after-effects persisted over the whole lifetime, this would offer a mechanism for fast adaptation to changing environmental conditions and could open completely new perspectives in interpreting genetic mechanisms of climatic adaptation. At present neither the inheritance of after-effects, nor their occurrence in other species, are known.

HOMEOSTASIS AND PHENOTYPIC STABILITY OF POPULATIONS

Individual or 'physiological' homeostasis appears on the population level as *phenotypic stability*. Phenotypic stability or plasticity stands for the change in the mean expressed phenotype as a function of changing environmental effects. Stable populations maintain certain phenotypic traits, such as relative growth vigour, across different environments. Although clearly determined by hereditary factors, phenotypic stability does not involve genetic change in the population when planted into a new environment (therefore the often synonymously used term 'genotypic flexibility' is somewhat misleading). Phenotypically stable populations are specifically valuable because of their potential to adjust better to extreme environmental fluctuations. Therefore, stability as a trait should receive more attention in the future.

The most striking feature of provenance tests is the generally low sensitivity of tree populations to changing environments, even with regard to highly adaptive traits such as height growth. Provenances transferred over large distances into very different environments are able to grow and sometimes even compete with the locally adapted, native populations. This indicates a very high level of individual homeostasis and of phenotypic stability of tree populations.

While the value of phenotypically stable populations is very great, the evolutionary causes for their appearance are not clear. Phenotypic stability *per se* need not be adaptive. However, various theoretical models agree upon the fact that environmental heterogeneity in time favours the evolution of stability by increasing individual homeostasis (Mitton & Grant 1984; Powers *et al.* 1991; Scheiner 1993). Indications that higher climatic instability (e.g. in areas of climatic transition) may be responsible for the evolution of phenotypically more stable populations has been found for some coniferous species (Mátyás 1986).

As a consequence of homeostasis and stability, the response of individual populations to changing conditions is generally more-or-less linear across the sites of the common garden tests. Significant interactions, i.e. rank changes of populations, appear usually more toward the limits of the area of distribution, such as in the boreal zone (Eriksson 1980), but also at the lower/southern limits where the water régime plays an important role (Giertych 1995). In most tests, growth differences between populations from adjacent areas are difficult to detect in the absence of distinctly unfavourable climatic effects. In planar, contiguous distributional areas, the distance between populations with measurable growth differences may exceed 50 to 100 km. Under mountainous conditions, however, already a few hundred metres of altitudinal difference cause clear genetic differentiation between populations, mainly due to the elevational temperature gradient (Rehfeldt l988; Mátyás 1996).

FORECASTING CLIMATE CHANGE EFFECTS: THE EXAMPLE OF PONDEROSA PINE *PINUS PONDEROSA*

The assessment of effects of climate change with the help of common garden experiments has been proposed by the author (Mátyás 1994). The use of existing provenance or family tests to predict such effects is based upon the concept that between-provenance genetic variation patterns represent a response to transfer into altered environments, and may be interpreted as a simulation of responses to climate change over time. According to the concept, growth (=phenotypic response) at a given location is determined by the genetic constitution of the population, the site potential and the effect of transfer from the original environment to the given site. The transfer variable is defined as ecological distance or ecodistance (Mátyás & Yeatman 1992; Mátyás 1994) and expressed as the difference value of ecological variables (e.g. temperature, precipitation) or of combined variables (e.g. principal components of the ecological variable set) between the locations of origin and that of the test. At any test site, the locally adapted population has an ecodistance value of zero. Depending on the direction and distance of transfer, a tested population will have different ecodistance values in different tests. The site (growing) potential is characterised relative to the populations in the test by the average ecodistance at every test location. The averages correlate strongly with mean test heights.

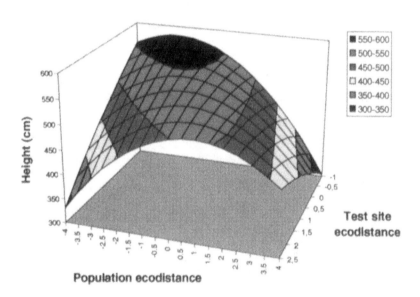

Figure 1. Response regression model of open-pollinated ponderosa pine families for 12-year height, measured in 4 tests in the Sierra Nevada range (California). Calculated from unpublished data of the US Forest Service PSW Expt. Sta. (Colours indicate height categories)

The effect of changing temperature conditions can be modeled utilizing growth response to changing temperature ecodistances. A nonlinear multiple regression analysis technique was developed to define phenotypic response across an ecological space. In Figure 1 an example of a ponderosa pine (*Pinus ponderosa*) family test in California is presented. A subset of an extensive parent tree testing program (Kitzmiller 1983), including 45 families at 4 locations has been analyzed.

Twelve-year height served as indicator of adaptive response. The response surface for temperature ecodistances displays in a simply interpretable way the effect of changing annual average temperature on height growth. In general, local populations (0 ecodistance) do not perform best in the test series and are mostly outperformed by populations moved into cooler environments. That growth increases until a certain limit of transfer into cooler environments is reached, partly may be explained by the constraints on adaptedness (see Gray 1996). More favourable ecological conditions (decrease of drought stress on cooler sites) may contribute to this. The characteristic shape of the response surface is determined by heritable performance differences between populations; provenances originating from mid-elevation sites perform best across all tested sites.

On the other hand, the increase of temperature affects growth significantly. Within reasonable ecological distance, the majority of populations react linearly to changing conditions; accordingly, the average effect of temperature increase on growth appears fairly linear within the analyzed ecocline. Relative to the changes in environment, the response is comparatively weak. For the above mentioned experiment, the height loss amounts to approximately 5 % $°C^{-1}$ temperature increase. A scenario of 3°C temperature increase would trigger a 14·2 % growth depression at juvenile age if precipitation conditions remain roughly the same. The calculated decline in height growth is only an indication of the actual volume (and assimilated C) loss, as a similar decline in diameter and basal area has a quadratic effect on volume.

Regarding the elevational significance of temperature change, the temperature lapse rate for the Sierras was established as 5·4°C km^{-1} (Mátyás 1995). A temperature increase of 1·62 °C is equal to an elevational change of 300 m, which is generally proposed as the critical limit of safe genetic adaptation by genecological studies in some western conifers (Rehfeldt 1989; Ying & Liang 1994; Sorensen & Weber 1994), even though the actual loss in growth is not yet dramatic at this point. In addition to growth loss, other consequences of maladaptation increase in significance with growing transfer distances, such as susceptibility to snow break; or the loss of resistance to diseases and pests, which are otherwise often harmless or of minor importance; e.g. saprophytic fungi may turn into real parasites. A direct link between ecodistance and mortality has been demonstrated for *Pinus banksiana* (Mátyás & Yeatman 1992).

At low elevations, close to the lower limit of the species, the temperature increase certainly would lead to the thinning out and disappearance of the species, losing its competitive ability against other species. At the upper limits, temperature increase will bring an improvement of site conditions, although its utilisation will be hindered by the fact that soil conditions (usually shallow, less developed soils) will remain the same, as soil development is a slow process. Thus, the site potential will not follow the improvement in climate.

CONSTRAINTS ON ADAPTATION; REASONS FOR AND VALUE OF HIGH DIVERSITY

The presented features of adaptive genetic variation in trees illustrate very effective constraints on 'perfect' adaptation. The population average of a fitness-related trait for a locally adapted population is often lower than the average of other populations introduced from other environments; a phenomenon termed *adaptation lag* (Mátyás 1990). Parallel to a weakly responsive, constrained adaptedness to the environment (=high stability), a high level of genetic diversity may be expected in trees both at the individual and population levels. In fact, genic diversity of trees seems to be relatively high compared to other organisms, as assessed by biochemical marker studies (Hamrick & Godt 1990). The highest levels of allozymic variation are observed in widely distributed, strictly outbreeding, wind-pollinated and wind-dispersed species, such as most conifers and many of the broad-leaved tree species.

In the case of highly adaptive traits, which display a clearly climate-related pattern of variation at the between-population (provenance) level, diversity still can be substantial among individuals, in spite of obviously high selection pressure. A prominent example is the diversity in the length of the growing cycle; phenological phases within one population may vary considerably and may be manifested in form of phenologic polymorphisms (e.g. late- and early-flushing forms). In a Scots pine test, the range of variability in termination of growth in autumn was more than 70 days within populations, while the difference between the means of the two most distant populations in the test, North Russia (lat. 60° N) and Turkey (lat. 40° N), was 89 days (Mátyás 1986). Similarly, a single elevational cline at one location may encompass nearly all the adaptive genetic variation observed within a range of distribution (Rehfeldt 1988).

The importance of genic diversity itself for the survival and reproductive success of a species or population, although generally recognized, should not be overestimated (Ledig 1986). Some studies propose a link between protein heterozygosity and consistency of performance (Mitton & Grant 1984). Considering the limited predictive value of allozymic diversity on heterozygosity of loci determining adaptive traits, these results cannot be generalized. Homeostasis is not a function of average observed heterozygosity, it is likely regulated both through changing allelic expression and changing interactions among loci in relation to environmental conditions (Scheiner 1993), which means that it is the genic diversity of certain loci only which matters.

The role of the environment in adaptation is generally understood in two ways: first, by setting the fitness function and selecting the fittest individuals, and second, affecting the developmental process of the individual by determining the phenotype. Both the genetic system of the species and random effects play a role in maintaining the adaptation lag. The genetic system of temperate-boreal tree species is characterised by usually very effective gene flow through wind-pollination, an outbreeding mating system with strong mechanisms to prevent inbreeding, and a high genetic load (Ledig 1986, Koski 1991). These genetic factors act against the development of genetically related neighborhoods and local adaptation. Peculiarities of reproductive ecology (e.g. differential fertility, polyembryony, non-random mating; Müller-Starck 1991) further prevent a straightforward action of abiotic selective forces.

Another reason for observed non-optimality of growth characters could be 'trade-off effects'; the adaptive value of (height) growth is certainly relative, and depends on the competition conditions and the on the environment. Reproductive ability as an antagonistic trait to growth gains importance on extreme sites with open vegetation. While partitioning of assimilates between vegetative and reproductive organs is strongly influenced by the environment (mainly by plant density), the hereditary relation between the two trait-groups is less than clear for forest trees.

Randomness in the occurrence of disturbances, and other stochastic events may have a very important role in shaping diversity. 'Rare events' are irregularly occurring disturbances (e.g. extreme drought years, severe frosts) which contribute to a high level of temporal heterogeneity on time scales which are difficult to follow. On an evolutionary scale, the effect of migration, fragmentation, isolation and drift induced by climatic shifts across geologic times, can be demonstrated on the genetic structure of populations and species, and contributes to the presently observed patterns of diversity (Hamrick & Godt 1990; Ledig 1986). On the scale of the lifespan of trees, regeneration success and biotic interactions have strongly random components. The possibility for regeneration through appearance of suitable niches is unpredictable, depending on some catastrophic event or mortality of neighbors. Biotic interactions include long-lasting competition with the same and other species, and incidental interactions with consumers, symbionts, parasites, pathogens etc. In the entireness of ecosystems, it seems that evolution promotes rather the complexity of biotic interactions at the expense of the 'proper fitting' of the organism-environment relation.

In summary, it seems that both the genetic processes characteristic for the investigated tree species (outbreeding, strong gene flow etc.) and certain biotic and abiotic effects of the environment strengthen the maintenance of a high level of genetic variability at the individual, population and species levels, which is in turn responsible for their evolutionary success.

LIMITATIONS AND FUTURE TASKS

Common garden tests are carried out in forestry with easy-to-propagate, widely distributed species. The majority of species in large-scale, often range-wide test networks up to now are conifers; these predominantly occupy the initial phases of succession in temperate forest communities. It is therefore not certain that the described phenomena apply to all widely distributed tree species. The described modeling method has its limitations as well; as the effect of climate variables at a given test location cannot be evaluated in its full complexity. In the model presented the effects of latitudinal transfer have been neglected, although changes in the photoperiod conditions might have an effect on light-sensitive species.

Because of the conservative nature of the genetic adaptation process, and of the relative speed of expected changes, long-lived, immobile organisms, such as trees will presumably need human interference in order to enhance gene flow and adaptation to altered conditions, in spite of an impressive capacity to adjust to changes. National forest policies have to incorporate this task into the agenda of the next decades. To counteract genetic erosion and extinction, populations along the southern (or low-elevation) limits of species distribution areas need special attention.

REFERENCES

Ayala RJ (1969) An evolutionary dilemma: fitness of genotypes versus fitness of populations. Canad Journ Genet Cytol 11:439-456

Beuker E, Koski V (1996) Adaptation of tree populations to climate as reflected by aged provenance tests. in Mátyás Cs (ed), Forest genetics and tree breeding: perspectives and challenges. IUFRO World Series, IUFRO Secretariat ed. Vienna (in press)

Bjørnstad Å (1981) Photoperiodical after-effect of parent plant environment in Norway spruce seedlings. Medd Nors Inst Skogforsk 36:1-30

Davis MB (1980) Quaternary history and the stability of forest communities. in West DC, Shugart HH, Botkin DB (eds) Forest succession: Concepts and Application, 132-153. Springer-Verlag, New York

Devey ME, Delfino-Mix A, Kinloch BB, Neale DB (1995) Random amplified polymorphic DNA markers tightly linked to a gene for resistance to white pine blister rust in sugar pine. Proc Natl Acad Sci USA 92:2066-2070

Eriksson G (1980) Severity index and transfer effects on survival and volume production of *Pinus silvestris* in Northern Sweden. Stud For Suec Nr 156

Giertych M (1996) Effect of thinnings on the evaluations of provenance experiments as exemplified by a Norway spruce (*Picea abies* Karst.) trial. in Mátyás Cs (ed) Forest genetics and tree breeding: perspectives and challenges. IUFRO World Series, IUFRO Secretariat ed. Vienna (in press)

Gray AJ (1996) Climate change and the reproductive biology of higher plants. in Huntley B, Cramer W, Morgan AV, Prentice HC, Allen JRM (eds) Past and future rapid environmental changes: The spatial and evolutionary responses of terrestrial biota, 371-380. Springer-Verlag, Berlin

Hamrick JL, Godt MJ (1990) Allozyme diversity in plant species. in Brown AHD, Clegg MT, Kahler AL, Weir BS (eds), Plant population genetics, breeding, and genetic resources, 43-63. Sinauer Ass,

Kitzmiller J (1983) Progeny testing - objectives and design. in Charleston, SC (ed) Progeny testing. Proc Servicewide Genetics Workshop Dec. 5-9 1983, 231-247. USDA For Serv Timber Management

Koski V (1991) Generative reproduction and genetic processes in nature. in Giertych M, Mátyás Cs (eds) Genetics of Scots pine, 59-72. Elsevier Sci Publ

Langlet O (1971) Two hundred years of genecology. Taxon 20:(5/6) 653-722

Ledig FT, Guries RP, Bonefield BA (1983) The relation of growth to heterozygosity in pitch pine. Evolution 37:1227-1238

Ledig FT (1986) Heterozygosity, heterosis, and fitness in outbreeding plants. *in* Soulé M (ed) Conservation biology: the science of scarcity and diversity, 77-104. Sinauer Assoc Sunderland, Mass

Lindgren D, Wei RP (1994) Effects of maternal environment on mortality and growth in young *Pinus silvestris* in field trials. Tree Physiol 14:323-327

Mátyás Cs (1981) Kelet-európai erdeifenyo származásak fenológial változékonyság [Phenologic variability of East European Scots pine populations; in Hungarian with English summary] Erdészeti Kutat 74: 71-79

Mátyás Cs (1986) Nemesített szaporítóanyag gazdálkodás [Improved planting stock in forestry; in Hungarian] Akadémia Publ Budapest 136

Mátyás Cs (1990) Adaptation lag: a general feature of natural populations. Proc Jt Meet WFGA and IUFRO WP Olympia, WA, USA Aug 20-25, 1990 Pap No 2.226

Mátyás Cs (1994) Modeling climate change effects with provenance test data. Tree Physiology 14:797-804

Mátyás Cs (1995) Climate of the Central Sierras. Progr. Report, Inst. of Forest Genetics, Placerville (manuscript)

Mátyás Cs (1996) Climatic adaptation of trees: rediscovering provenance tests. Euphytica, Kluwer, Dordrecht (in press)

Mátyás Cs, Yeatman CW (1992) Effect of geographical transfer on growth and survival of jack pine (*Pinus banksiana* Lamb.) populations. Silvae Gen 43:6,370-376

Mikola J. (1982) Bud-set phenology as an indicator of climatic adaptation of Scots pine in Finland. Silvae Fennica 16:178-184

Mitton JB, Grant MC (1984) Associations among protein heterozygosity, growth rate, and developmental homeostasis. Ann Rev Ecol Syst 15:479-499

Müller-Starck G (1991) Genetic processes in seed orchards. *in* Giertych M, Mátyás Cs (eds), Genetics of Scots pine, 147-162. Elsevier Sci Publ

Powers DA, Lauerman T, Crawford D, DiMichele L (1991) Genetic mechanisms for adapting to a changing environment. Ann Rev Genet 25:629-659

Rehfeldt GE (1988) Ecological genetics of *Pinus contorta* from the Rocky Mountains (USA): a synthesis. Silvae Gen 37:131-135

Rehfeldt GE (1989) Ecological adaptations in Douglas-fir (*Pseudotsuga menziesii* var. *glauca*): a synthesis. For Ecol Manage 28:203-215

Scheiner SM (1993) Genetics and evolution of phenotypic plasticity. Ann Rev Ecol Syst 23:1-14

Skrøppa T, Johnsen Th (1994) The genetic response of plant populations to a changing environment: the case for non-Mendelian processes. *in* Boyle TJB, Boyle CEB (eds), Biodiversity, temperate ecosystems, and global change, 183-199. Springer Verlag, Berlin

Sorensen FC, Weber JC (1994) Genetic variation and seed transfer guidelines for ponderosa pine in the Ochoco and Malheur National Forests of Central Oregon. USDA For Serv Res Pap PNW-RP-468 26

Timofeyev VP (1975) Stareishij opyt geograficheskikh kultur sosny obykhnovennoj [Oldest provenance test with Scots pine in the USSR at the TSHA experimental station; in Russian] *in* Materiali soveshchaniya o rabote uchebno-opytnykh leskhozov, 11-28. Tartu

Ununger J, Ekberg I, Kang H (1988) Genetic control and age-related changes of juvenile growth characters in *Picea abies*. Scand Journ For Res 3:55-66

Vitousek PM (1992) Global environmental change: an introduction. Ann Rev Ecol Syst 23:1-14

Ying ChC, Liang Q (1994) Geographic pattern of adaptive variation of lodgepole pine within the species' coastal range: field performance at age 20 years. Forest Ecol Manage 67:281-298

Climate change and the reproductive biology of higher plants

Alan J. Gray
ITE Furzebrook Research Station
Wareham
Dorset BH20 5AS
U.K.

INTRODUCTION

If there is to be an evolutionary response among plant populations to climate change, three necessary conditions must apply. First, climate, or some aspect of climate, must be capable of exerting a selective effect. Secondly, there must be variation in traits which affects the fitness of their possessors when this selective effect is applied. Thirdly, this variation must be heritable. These conditions will empower, but not guarantee, an evolutionary response.

In this contribution I will discuss aspects of the reproductive biology and ecology of flowering plants under three headings broadly equivalent to the three issues above. The discussion is confined to plants of temperate climates and to aspects of their reproductive biology.

CLIMATE AS A DETERMINANT OF PLANT DISTRIBUTIONS

It hardly seems necessary to ask whether climate can be an agent of selection since on a geographical scale climate and vegetation are quite clearly correlated. That plant distributions are controlled by climate seems axiomatic – if, in a particular instance, we cannot see *how*, we assume it must be because we have failed to detect the appropriate climatic variable (July temperature, frost-free days in March, rainfall in May, and so on) or combination of variables. However, and despite the broad correlations between plant distribution and climate, it is extremely difficult in

NATO ASI Series, Vol. I 47
Past and Future Rapid Environmental Changes:
The Spatial and Evolutionary Responses of
Terrestrial Biota
Edited by Brian Huntley et al.
© Springer-Verlag Berlin Heidelberg 1997

practice to find evidence that the distribution of a particular species is limited by climate.

One place to look for evidence of limiting effects is among the populations at the edge of the species' geographical (climatic) range. Acceptable evidence of signs that the species is struggling would be reduced fecundity, failure to set seed, or even failure to flower. In fact, using as examples species with northern limits in the Northern Hemisphere temperate zone, the actual number of species revealing indications of reproductive failure at their northern boundaries is very limited. Some well-known cases include the classic work on the tree *Tilia cordata*, which only produces fertile fruit in north-western England in exceptionally warm years (Pigott & Huntley 1981, Pigott 1992) and the herb *Cirsium acaule*, increasingly confined to south-facing slopes at the northern edge of its range and setting seed there only in years with warm, dry, late summers (Pigott 1970). The early study of Iversen (1944) related reproductive success in *Hedera helix*, *Ilex aquifolium* and *Viscum album* to average winter and summer temperatures, again pointing out the reduction in ripe fruit production at the plants' limits. He also documented the clinal pattern of death of individual *Ilex* trees and *Hedera* plants in Denmark during the severe winters of 1939-1942.

Mostly, however, the evidence for climatic control of species' distributions is not obvious. Plants seem simply to stop, apparently performing as well at the edge of their range as in the centre. *Tamus communis* provides a good example of a species with a sharply-defined range, the reproductive output of which has been carefully monitored and found to be high at its northern limit (Pigott 1992).

It is perhaps not surprising that the effects of climatic factors are not commonly found at the range edge of natural populations. The distribution of many species will not be in equilibrium with the climate - some species may be spreading northwards, for example - and historical factors unrelated to climate will determine the distribution of others. Distributional boundaries may coincide with geological discontinuities, with barriers imposed by modern agriculture, or with the southern edge of a competitor or pathogen's distribution. The asymmetric nature of any explanation of species distributions (*i.e.* the reason for a plant's *absence* from a site or area is not strictly amenable to scientific explanation) demands that long-term studies or, ideally, transplant experiments, are required to establish whether climatic factors have a selective rôle.

Importantly, too, populations at the edge of the species' range may display very different dynamics to those at the centre for several reasons. These include their escape from specialist predators (enemy-free space), the effect of low relative-growth rate (low r) due to resource limitation, and the reduction in the amount of suitable and stable habitat. It may be that failure to spread (say northwards) can only be understood at the scale of the population (birth rates being consistently less than death rates : $r < 1$), or even the metapopulation (populations establishing at a lower rate than that at which they go extinct). Carter and Prince (1981, 1988) provide a model for relating metapopulation dynamics to distribution limits which demonstrates that small changes in (i) the number of available habitats, (ii) the fecundity, dispersal and establishment of seedlings (governing the rate at which habitats can be occupied), and (iii) the rate at which populations go extinct locally, can produce sharp distribution boundaries which do not correspond to the physiological tolerance limits of individuals.

All of this suggests that the selective effects of climate on reproductive biology will be difficult to detect, and may be small, subtle and not necessarily related to the physiological response of individual plants to individual climate variables. It indicates also that clues to the ways in which climate affects distribution may best be sought in comparing the performance of plants and populations across the whole climate gradient, including edge-of-range populations. A recent example reinforces this point.

In a study of the winter annual grass *Vulpia ciliata*, Carey et al. (1995) observed that populations at the northern edge of the species' range were smaller in area, size and density, and that plants flowered later and produced fewer seeds in these northern populations than in those in the centre of the plant's distribution. The decline in the basic reproductive rate (R_o) towards the edge of the range culminates in populations at the extreme range-margin in which, during the year of study, this rate fell below 1. Although there was no physiological failure, the decline in fecundity (mortality was increased only slightly) was sufficient for the finite rate of population increase to fall below the level of persistence (and to 'explain' a potential northern limit).

CLIMATE AND PATTERNS OF VARIATION IN REPRODUCTIVE BIOLOGY

If it is difficult to establish which aspects of climate limit the distribution of particular species, what clues about the selective effects of climate can be gathered from latitude-related variation in reproductive biology?

Although the number of studies is small, they point towards one or two generalisations (perhaps they should be called hypotheses since they are largely unsupported!). The most interesting studies have been with monocarpic species with variable life-spans (annual → triennial or more), although some data are available for polycarpic perennials.

For these species there is a tendency for populations at the northern edge of a species' range to comprise individuals that delay reproduction in favour of vegetative growth, producing a cline from south to north in the length of the non-reproductive phase.

At its extreme, this trend produces annuals in the south and perennials in the north, as is found in *Aster tripolium* (Gray 1971 and Figure 1). Plants from Scandinavian and north Scottish saltmarsh populations took at least two years to flower when grown in central England, whereas those from Portugal flowered within a year from seed. Some populations included mixtures of first- and second-year flowerers, and not all first-year-flowering plants died after flowering (but most did). Such contrasting life-histories were also found on a very local scale in the centre of the species' range in eastern England where annuals occurred on the edges of brackish pools and long-lived perennials were found on the pioneer zones of saltmarshes less than 500 m away (Gray 1987). Nevertheless, the overall geographical trend is clear.

The same type of pattern has been observed in the monocarpic perennial *Verbascum thapsus* (Reinartz 1984) in which the delay in flowering to a third year was most common among northern (Canadian) genotypes, and annuals were found only in southern (Texas and Georgian) populations. In *Daucus carota*, also a short-lived monocarpic species, a similar latitudinal cline in age at first reproduction has arisen in North America since the species' introduction in the 17th century (Lacey 1988). Other species which display the same trend include *Melilotus alba* (Smith 1927), *Prunella vulgaris* (Bocher 1949) and *Holcus lanatus* (Bocher & Larsen 1958).

Figure 1. Variation in the year of first flowering in European populations of *Aster tripolium* when grown in a common garden experiment in central England (University of Keele). Numbers of plants population[1] varied from 16 to 73. From Gray (1971).

Recent unpublished work in our laboratory on *Pulicaria dysenterica* reveals a similar pattern (Daniels & Moy, pers. comm.).

It should be noted that such trends in intraspecific variation are paralleled by a general trend towards species with annual life histories being more common in southern temperate, and especially Mediterranean, zones. This trend is confounded by clines in polyploidy and the breeding systems - there being more polyploids and higher levels of apomixis in northern climates - clines which themselves may be related to colonisation following the repeated advances and retreats of the Pleistocene ice sheets (Gray 1986).

This is not the place to consider the mechanisms of selection involved. The latitudinal cline may be driven entirely by resource availability, plants in short days and low temperatures requiring longer to garner the resources to reproduce. Variation in the threshhold age or size at which individuals can be induced to flower may also be important (see below) as well as environmental variation confounded by latitude such as the degree of density-independent selection and of habitat permanence and the relationship between adult and juvenile mortality.

Accompanying the variation in age-to-first-reproduction there may be variation in the proportion of a plant's resources allocated to reproduction and in the time of flowering during a particular year (season). The first of these is, in most species, an autocorrelate of year of first flowering but can be separated from it among perennial plants which flower in the same year. In *Pulicaria dysenterica*, plants from more northerly populations devoted relatively more resources to rhizome (and overwintering shoot) production than did those from southern populations – suggesting a trade-off between reproduction and the ability to overwinter (Daniels & Moy unpublished). Interestingly, plants from the north in this species flowered later than those from the south, against the trend that "populations from higher latitudes flower earlier than do populations from lower latitudes when planted in low latitudes" (Rathke & Lacey 1985). Relatively few *Pulicaria* from the north flowered in the south in this experiment and these late-flowering individuals may have been eliminated in the north by their failure to overwinter. Some grass species from higher latitudes also form unseasonal (vulnerable) inflorescences in the autumn when grown in low latitudes. This is because normally they initiate inflorescence primordia in the autumn, an adaptation which facilitates early heading and flowering in the shorter, cooler, northern summer (Heide 1994).

CLIMATE, GENETIC VARIATION AND REPRODUCTIVE BIOLOGY

Finally, it is important to note that many studies have established that the traits discussed above are under genetic control. Such control is usually exercised through differential requirement for the type and intensity of the environmental cues that induce flowering and reproduction. It can be measured as variation in the requirement for vernalisation, in sensitivity to photoperiod, or in the threshhold size or age at which a response to such stimuli can be made. Genetic variation in induction requirements has been analysed in several, predominantly crop, species and generally found to be under polygenic control by vernalisation genes or photoperiod-response genes (e.g. Cooper 1963; Murfet 1977; Rathke & Lacey 1985; Rees 1987; Heide 1994). However, major genes have also been found. One example from a natural population is that of *Beta vulgaris* ssp. *maritima*, in which the major 'bolting' gene controls the requirement for vernalisation (allele *B*, no vernalisation requirement, being dominant to *b* – *bb* plants requiring vernalisation) (Boudry *et al.* 1994). The dominant *B* allele has a high frequency in southern, especially Mediterranean, populations but appears to be absent from northern Europe (van Dijk pers. comm.). Quantitative trait loci also control the vernalisation requirement in *bb* plants in this species.

Range-related variation in reproductive biology is found in populations of *Agrostis curtisii*, a perennial grass confined to acid soils, principally heathland, in south-west Britain. In one experiment (Gray & Moy unpublished), plants from three populations in each of five regions were grown (as 15 half-sibs from 10 families population[-1]) in a common environment in the centre of their range. Figure 2 summarises the flowering performance of these populations in relation to their position along a cline from Cornwall in the west eastwards to the outlier populations on the Bagshot sands in Surrey and northwards to the extreme limit of the species' world distribution in South Wales. This cline, which is mirrored by a gene-frequency cline at several allozyme loci (Gray 1988), is difficult to interpret. The populations from which the plants were sampled (as seed) experience virtually no difference in daylength, and the main climate variables are winter temperature (milder in the west) and rainfall (higher by 250 mm in the west). Since the differences in flowering performance appear to be genetic (the heritabilities of these and several other flowering traits have yet to be calculated), it suggests that conditions at the edge of the species' range may have led to the differential selection of plants which allocate fewer resources to flowering , at least in the first year of flowering.

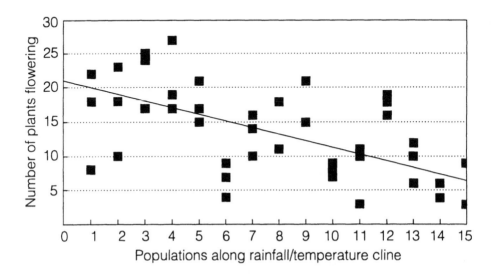

Figure 2. Variation in the number of plants flowering in the first flowering year in 15 populations of *Agrostis curtisii* collected along a south-west– north-east cline in SW Britain and grown in a common environment. Each point is the number of plants to flower (out of 50) in a single block, each population being split across three fully-randomised blocks. (from Gray & Moy unpublished).

Although the mode of inheritance may vary (and sometimes be complex or little understood), and extreme latitude-related or climate-related ecotypic differentiation may be widespread, it is essentially at the level of the local population that we might expect an evolutionary response to climate change. Speculations about future plant species' distributions in relation to future climates beg several important questions about dispersal and establishment. Initially at least, the response of individual species to climate warming or increasing CO_2 levels will be seen as shifts in performance, often comprising directional selection for heritable traits, within individual populations. Indeed, plant community changes will result from the different abilities of the constituent species to make such *in situ* responses. Furthermore, evidence of genetic differentiation in relation to latitude or climate, whilst indicating past selection, by no means indicates that further genetic response is possible.

For this reason, detailed population level studies are extremely important. There are several of these, demonstrating that genetic variation is an important component of variation in reproductive biology and life-history in local populations (e.g. in local populations of *Plantago major* (Lotz 1990), *Daucus carota* (Lacey 1986) and *Bouteloua rigidiseta* (Miller & Fowler 1994).

A recent exemplary study is that of Wesselingh and de Jong (1995) on *Cynoglossum officinale* in which genetic variation for threshhold size for flowering was quantified by performing a two-way artificial selection experiment using as the parental generation seed from a natural population. Narrow-sense heritabilities of 0.32 and 0.35 for low and high selection lines, which are high for a fitness-associated trait, indicate that there is ample genetic variation in this population to enable a response to selection. Whatever the mechanism maintaining this variation (environmental heterogeneity, some form of trade-off, a loose coupling between rosette size in the first year and eventual seed production), its presence suggests that an evolutionary response is possible in this species to changing climatic conditions.

REFERENCES

Bocher TW (1949) Racial divergences in *Prunella vulgaris* in relation to habitat and climate. New Phytologist 48:285-314

Bocher TW, Larsen K (1958) Geographical distribution of initiation of flowering, growth habit and other characters in *Holcus lanatus*. Botaniska Notiser 111:289-300

Boudry P, Weiber R, Saumitou-Laprade P, Pillen K, van Dijk H, Jung Chr. (1994) Identification of RFLP markers closely linked to the bolting gene B and the significance for the study of the annual habit in beets (*Beta vulgaris* L.). Theoretical and Applied Genetics 88:852-858

Carey PD, Watkinson AR, Gerard FFO (1995) The determinants of the distribution and abundance of the winter annual grass *Vulpia ciliata* ssp. *ambigua*. Journal of Ecology 83:177-187

Carter RN, Prince SD (1981) Epidemic models used to explain biogeographical limits. Nature 293:644-645

Carter RN, Prince SD (1988) Distribution limits from a demographic viewpoint. *in* Davy AJ, Hutchings MJ, Watkinson AR (eds) Plant population ecology, 165-184. Blackwell Scientific Publications, Oxford

Cooper JP (1963) Species and population differences in climatic response. *in* Evans LT (ed) Environmental control of plant growth, 381-403. Academic Press, New York

Gray AJ (1971) Variation in *Aster tripolium* L. with particular reference to some British populations. PhD Thesis, University of Keele

Gray AJ (1986) Do invading species have definable genetic characteristics? Philosophical Transactions of the Royal Society of London B 314:655-674

Gray AJ (1987) Genetic change during succession in plants. *in* Gray AJ, Crawley MJ & Edward PJ (eds) Colonisation, succession and stability, 273-293. Blackwell Scientific Publications, Oxford

Gray AJ (1988) Demographic and genetic variation in a post-fire population of *Agrostis curtisii*. Oecologia Plantarum 9:31-41

Heide OM (1994) Control of flowering and reproduction in temperate grasses. New Phytologist 128:347-362

Iversen J (1944) *Viscum, Hedera* and *Ilex* as climate indicators. Geologiske Föreningen i Stockholm Förhandlingar 66:463-483

Lacey EP (1986) The genetic and environmental control of reproductive timing in a short-lived monocarpic species, *Daucus carota* (Umbelliferae). Journal of Ecology 74:73-86

Lacey EP (1988) Latitudinal variation in reproductive timing of a short-lived monocarp, *Daucus carota* (Apiaceae). Ecology 69:220-232

Lotz, LAP (1990) The relation between age and size at first flowering of *Plantago major* in various habitats. Journal of Ecology 78:757-771

Miller RE, Fowler NL (1994) Life history variation and local adaptation within two populations of *Bouteloua rigidiseta* (Texas grama). Journal of Ecology 82:855-864

Murfet IC (1977) Environmental interaction and the genetics of flowering. Annual Review of Plant Physiology 28:253-278

Pigott CD (1970) The response of plants to climate and climatic change. *in* Perring FH (ed) The flora of a changing Britain, 32-44. Classey, Faringdon.

Pigott CD (1992) Are the distributions of species determined by failure to set seed? *in*: Marshall C, Grace J (eds) Fruit and seed production, 203-216. Cambridge University Press, Cambridge.

Pigott CD, Huntley JP (1981) Factors controlling the distribution of *Tilia cordata* at the northern limits of its geographical range. 3. Nature and causes of seed sterility. New Phytologist 87:817-839

Rathke B, Lacey EP (1985) Phenological patterns of terrestrial plants. Annual Review of Ecology & Systematics 16:179-214

Rees AR (1987) Environmental and genetic regulation of photoperiod - a review. *in* Atherton JG (ed) Manipulation of flowering, 187-202. Butterworths, London.

Reinartz JA (1984) Life history variation of common mullein (*Verbascum thapsus* L.). I. Latitudinal differences in population dynamics and timing of reproduction. Journal of Ecology 72:897-912

Smith, HB (1927) Annual *versus* biennial growth habit and its inheritance in *Melilotus alba*. American Journal of Botany 14:129-146

Wesselingh RA, de Jong TJ (1995) Bidirectional selection on threshold size for flowering in *Cynoglossum officinale* (hound's-tongue). Heredity 74:415-424

Space and time as axes in intraspecific phylogeography

John C. Avise
Department of Genetics
University of Georgia
Athens, GA 30602
U.S.A.

Except in rare instances where DNA might be extracted from a well-preserved and well-dated fossil series, molecular genetic assays normally are confined to living organisms sampled from the present-day horizon in time. Nevertheless, temporal aspects of evolution can be recovered from extant organisms using genomic differences accumulated since shared ancestry. This temporal, phylogenetic dimension of evolution has traditionally been the province of macroevolutionary studies that deal with relationships among species and higher taxa. Phylogenetic perspectives were rarely applied, or even perceived as relevant, at the intraspecific level because of a lack of empirical approaches to assess historical relationships, and because of a widespread perception that phylogeny had no real meaning for potentially interbreeding assemblages of populations. In the last two decades, studies of mitochondrial (mt) DNA have changed this view dramatically by demonstrating that the matriarchal component of intraspecific phylogeny can be estimated empirically (Avise 1989).

By virtue of its cytoplasmic housing within mitochondria, mtDNA is normally inherited by progeny only from their mothers, and only female offspring can pass mtDNA to future generations. For purposes of phylogeny reconstruction, this maternal mode of mtDNA transmission has the favourable property of circumventing complicating aspects of genetic recombination that can plague similar attempts to recover the phylogenies of nuclear genes at the intraspecific level. Furthermore, mtDNA has proved to evolve rapidly in nearly all organisms studied to date, such that numerous markers of matriarchal ancestry can usually be recovered from laboratory assays that involve sampling variable restriction sites or nucleotide sequences from the mtDNA molecule. For simplicity, different mtDNA genotypes can be thought of as

NATO ASI Series, Vol. I 47
Past and Future Rapid Environmental Changes:
The Spatial and Evolutionary Responses of
Terrestrial Biota
Edited by Brian Huntley et al.
© Springer-Verlag Berlin Heidelberg 1997

providing female family names that are relatable to one another by consideration of the nature and pattern of mutational differences that distinguish them.

Almost by definition, conspecific populations of most species are distributed geographically. Thus, a contemporary spatial dimension is associated with whatever temporal dimension of evolution might be estimated from the molecular genetic makeup of extant organisms. *Phylogeography* refers to the study of the principles and processes governing the geographic distributions of genealogical lineages, including those at the intraspecific level (Avise *et al.* 1987). The purpose of this brief report is to consider whether phylogeographic perspectives might contribute to an understanding of populational responses to past (and future) rapid environmental changes. In this context, ideally we would like to know the precise timeframes associated with changes in the size or configuration of a species' range, which requires that an observable spatial axis of population genetic structure be translated to an absolute temporal axis.

Such space-time conversions are certainly plausible. Consider, for example, Figure 1, which shows genealogical relationships among conspecific individuals from two geographically separated regions, within each of which gene flow is also constrained. As drawn, the isolation-by-distance characterizing each region results in a hierarchy of temporal depths in the gene genealogy for local populations over increasingly large spatial scales. Furthermore, the two regional populations are assumed to have been isolated for a sufficiently long period of time for this genealogy to have achieved the status of reciprocal monophyly, wherein all of the lineages in region 1 are phylogenetically closer to one another than to any lineages in region 2, and *vice versa* (Neigel & Avise, 1986). Empirical genetic analyses of these extant populations would be expected to yield the space-time patterns presented in Figure 2A. For example, if successive pairs of spatially adjacent individuals were sampled along a linear transect traversing the species' range, genetic distances should display a sharp spike at the boundary between the two regions (Figure 2A, left), and should otherwise be of lower but variable magnitudes for the local populations and clades (monophyletic groups) within each region. If random pairs of individuals were to be compared from these same extant populations, a positive slope in the space-time regression should characterize individuals within regions, whereas the major phylogeographic gap should be evidenced by much larger but rather uniform genetic distances in interregional comparisons (Figure 2A, right).

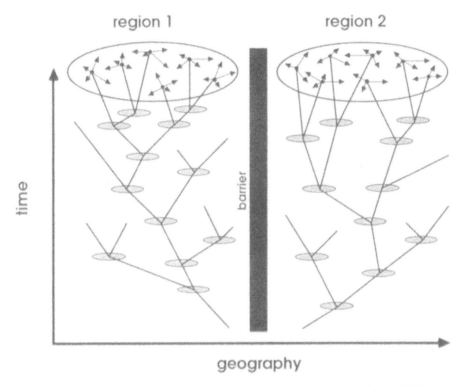

Figure 1. Hypothetical gene genealogy (such as that provided by mtDNA) for a species displaying restricted gene flow within each of two physically-separated regional populations. Vectors (arrows) in the extant populations indicate magnitudes of individual dispersal per generation from natal sites.

Several empirical mtDNA surveys of animal populations have revealed phylogeographic patterns closely similar to those presented in Figures 1 and 2A (reviews in Avise, 1994, 1996). For example, the matriarchal genealogy of the musk turtle *Sternotherus minor* exhibited pronounced but shallow spatial structure regionally, plus a deeper phylogenetic gap that distinguished all individuals in the southeastern portion of the species' range (peninsular Florida, eastern Georgia) from those to the north and west (western Georgia, central Alabama, Tennessee and

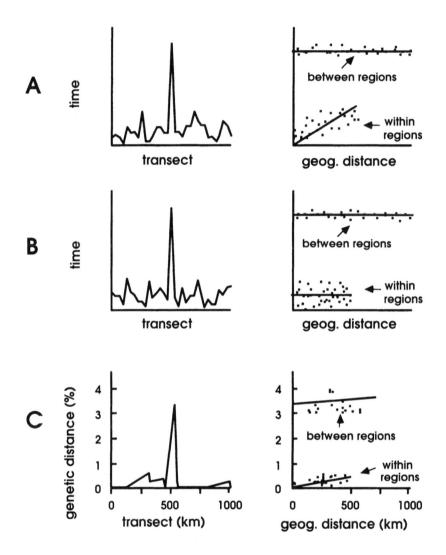

Figure 2. Examples of possible relationships between geography and intraspecific phylogeny. Left: coalescence times between successive pairs of adjacent alleles along a linear transect through extant populations with phylogeny similar to that shown in Figure 1. Right: coalescence times for random pairs of alleles drawn from such populations. **A**, restricted gene flow *within* regions; **B**, extensive gene flow *within* regions; **C**, empirical results from mtDNA restriction site data in the musk turtle *Sternotherus minor* (Walker *et al.* 1995). In this latter case, the transect was from central Florida to Alabama and then north (at a right angle) to west Virginia; the calculated regressions on the right were based on distances between random pairs of individuals, sampled without replacement from the data base.

western Virginia) (Walker *et al.* 1995). The empirical data for musk turtles, when analyzed according to the format of Figure 2A, are presented in Figure 2C.

In principle, any specified pattern of historical demography should carry phylogeographic consequences for extant populations, such that empirical assessments of the latter may be employed to estimate the demographic histories of populations (Templeton *et al.* 1995). For example, exceptionally high dispersal within the regional populations of Figure 1 should diminish the positive slope between space and time in the within-region regression that is otherwise expected under a restricted gene flow scenario (compare Figures 2A and 2B). In general, coalescence theory (Hudson 1990; Takahata,1995) provides a conceptual and mathematical framework for relating historical population demography to the genealogical structures registered in particular genes. From the vantage of the present looking backward in time, allelic lineages must eventually come together or coalesce in a common ancestor. In other words, extant alleles are phylogenetically related.

Within this framework, several distinct analytical approaches have been introduced to formalize historical demographic predictions from observed phylogeographic data such as that provided by mtDNA. One class of approaches involves estimation of the gene flow parameter Nm, which can be interpreted as the mean number of individuals exchanged between populations per generation (Slatkin & Barton 1989; Slatkin & Maddison 1989). A second approach involves estimation of single-generation dispersal distances based on expected spatial distributions of lineages of various ages, assuming that lineage dispersal has occurred via a multigenerational "random-walk" process from specifiable centres of origin for each molecular clade (Neigel *et al.* 1991). A third approach involves nested cladistic appraisals of gene-genealogical structure in formal statistical tests for alternative phylogeographic patterns (Templeton 1993; Templeton *et al.* 1995; Templeton & Georgiadis 1996). These and similar approaches as applied to mtDNA data sets from numerous terrestrial (and other) animal species indicate that demographic and phylogeographic patterns have been strongly impacted by changing environmental circumstances of the Quaternary (Avise 1994).

How far can such phylogeographic perspectives be taken to assess prospects for populational responses to anthropogenically-altered environments over yearly or decade long timeframes? In some respects, conventional molecular assays are unlikely to be particularly informative at these levels for the following reasons. First, the rate of *de novo* nucleotide substitution in mtDNA of vertebrate animals, although higher than that of most nuclear genes, remains far too low (ca. 2% sequence divergence between lineages per million years) to permit meaningful assessments of *de novo* mutational change over extreme microevolutionary scales (at the "conventional" pace quoted above, only about one new mutational difference per kilobase is expected to arise between two lineages that coalesce to 50,000 years ago). Second, extreme evolutionary stochasticity weakens any single-gene inferences about evolutionary processes that may have led to the phylogeographic state of an extant species (Weir 1990; Templeton 1993). This stochasticity has many sources, including randomness of the underlying mutational process, the great diversity of probable factors (e.g. genetic drift, selection, migration) influencing a gene genealogy, inherent vagaries in the nature of genic transmission and syngamy (fertilization producing a zygote) such that multiple nuclear gene genealogies within one-and-the-same pedigree can differ considerably in topology (Ball *et al.* 1990), and uncertainties in molecular rate calibration and phylogeny estimation *per se*.

Nonetheless, general lessons from phylogeographic analyses may add some useful perspectives to discussions of spatial and temporal responses of biotas to rapid environmental change. First, it is now abundantly clear that most species exist not as monolithic entities, but rather as a spatially assorted mosaic of populations with varying and sometimes deep temporal structure. Thus, different conspecific populations (particularly those that have been separated for a long time) may respond differently to environmental challenges because of accumulated genetic differences, which may well include alterations of possible adaptive relevance. Second, methods now exist for distinguishing the "deeper" from "shallower" population separations, based on genealogical concordance principles which involve searches for congruent molecular genetic patterns across multiple characters within a gene, across multiple independent loci, across co-distributed species, and between genetic and distributional classes of data (Avise & Ball 1990; Avise 1996). Thus, keynote intraspecific population units worthy of special attention in studies of response to rapid environmental modification may be identified. Third, phylogeographic analyses and coalescense theory have made evident the close connection between population demography and genetics, and this can have conservation implications particularly in periods of swift environmental change. For example, because of a special relationship between female reproduction and population viability, significant population structure as registered by mtDNA (even in the absence of nuclear gene differentiation) can indicate that populations compromised or extirpated by humans or other causes will be unlikely to recover via natural recruitment from foreign sources, at least over the ecological timeframes relevant to human interests (Avise 1995).

Finally, phylogeographic studies have shown that extant species often carry genetic records of idiosyncratic historical-demographic events. Pronounced expansions, contractions, and fragmentations of populations and their geographic distributions have all been a natural part of the evolutionary process long before human influence on the planet. However, whether the unusual magnitude and tempo of environmental impact by humans fall within a range that can be accommodated by most species via range shifts and adaptational changes remains to be seen.

REFERENCES

Avise JC (1989) Gene trees and organismal histories: a phylogenetic approach to population biology. Evolution 43:1192-1208

Avise JC (1994) Molecular Markers, Natural History and Evolution. Chapman and Hall, New York

Avise JC (1995) Mitochondrial DNA polymorphism and a connection between genetics and demography of relevance to conservation. Cons Biol 9:686-690

Avise JC (1996) Toward a regional conservation genetics perspective. in Avise JC & Hamrick JL (eds) Conservation Genetics: Case Histories from Nature, 431-470. Chapman and Hall, New York.

Avise JC, Arnold J, Ball RM Jr, Bermingham E, Lamb T, Neigel JE, Reeb CA, Saunders NC (1987) Intraspecific phylogeography: the mitochondrial DNA bridge between population genetics and systematics. Annu Rev Ecol Syst 18:489-522

Avise JC, Ball RM Jr (1990) Principles of genealogical concordance in species concepts and biological taxonomy. Oxford Surv Evol Biol 7:45-67

Ball RM Jr, Neigel JE, Avise JC (1990) Gene genealogies within the organismal pedigrees of random mating populations. Evolution 44:360-370

Hudson RR (1990) Gene genealogies and the coalescent process. Oxford Surv Evol Biol 7:1-44

Neigel JE, Avise JC (1986) Phylogenetic relationships of mitochondrial DNA under various demographic models of speciation. in Nevo E & Karlin S (eds) Evolutionary Processes and Theory, 515-534. Academic Press, New York.

Neigel JE, Ball RM Jr, Avise JC (1991) Estimation of single generation migration distances from geographic variation in animal mitochondrial DNA. Evolution 45:423-432

Slatkin M, Barton NH (1989) A comparison of three indirect methods for estimating average levels of gene flow. Evolution 43:1349-1368

Slatkin M, Maddison WP (1989) A cladistic measure of gene flow inferred from the phylogenies of alleles. Genetics 129:555-562

Takahata N (1995) A genetic perspective on the origin and history of humans. Annu Rev Ecol Syst 26:343-372

Templeton AR (1993) The "Eve" hypotheses: A genetic critique and reanalysis. Amer Anthropol 95:51-72

Templeton AR, Georgiadis NJ (1996) A landscape approach to conservation genetics: conserving evolutionary processes in the African bovidae. *in* Avise JC, Hamrick JL (eds) Conservation Genetics: Case Histories from Nature, 398-430. Chapman and Hall, New York

Templeton AR, Routman E, Phillips CA (1995) Separating population structure from population history: a cladistic analysis of the geographical distribution of mito-chondrial DNA haplotypes in the tiger salamander, *Ambystoma tigrinum.* Genetics 140:767-782

Walker D, Burke VJ, Barák I, Avise JC (1995) A comparison of mtDNA restriction sites vs. control region sequences in phylogeographic assessment of the musk turtle (*Sternotherus minor*). Molec Ecol 4:365-373

Weir B (1990) Genetic Data Analysis. Sinauer, Sunderland MA

Migratory birds: Simulating adaptation to environmental change

Paul M. Dolman
School of Environmental Sciences
University of East Anglia
Norwich
Norfolk NR4 7TJ
U.K.

INTRODUCTION

In this paper I consider the possible response of migratory birds to environmental change. Elsewhere in this workshop the term 'migration' has been used in relation to changes in the range of a population; here I use the term to describe the predictable seasonal movements of individuals between specific wintering and breeding grounds. Environmental change can have serious consequences for such migratory species. A change in quality or loss of habitat in the breeding grounds or wintering grounds may cause a population to decline or even become extinct, but this may be avoided if they are able to migrate to alternative areas of habitat in a different geographical region.

This workshop has contrasted two mechanisms by which terrestrial biota may respond to environmental change; the geographic range of a species may shift or, alternatively, there may be genetic adaptation to the changed conditions. In many migratory birds these two themes combine, as a change in range or the use of new geographic areas for breeding or wintering may require adaptive evolutionary changes in the migration strategy. In this paper I use a simulation model to show that whether migration routes are flexible enough for such changes to occur may depend on details of the biology of particular species.

NATO ASI Series, Vol. I 47
Past and Future Rapid Environmental Changes:
The Spatial and Evolutionary Responses of
Terrestrial Biota
Edited by Brian Huntley et al.
© Springer-Verlag Berlin Heidelberg 1997

SIMULATION MODELS OF MIGRATORY BIRD POPULATIONS

I have developed a behaviour-based model that has applications for predicting the response of populations to habitat loss. The size of each population depends on the dynamic interaction between breeding productivity and mortality (Fretwell 1972; Williamson 1972).

Breeding productivity is often density-dependent. This frequently arises through competition for good quality territories (Ens *et al.* 1992; Holmes & Sherry 1992; Newton 1992), but may also occur through effects of density on frequency of breeding, individual fecundity or neonatal mortality (Coulson *et al.* 1982; Coulson 1984; Arcese & Smith 1988; Owen & Black 1989). Density-dependent breeding productivity (P) is incorporated using the equation of Maynard Smith and Slatkin (1973):

$$P = NF\left(1 + (aN)^b\right)^{-1}$$

where:

N   is the number of individuals

F   the per capita fecundity in the absence of density-dependence

a   is a scaling constant

b   determines the strength of the inverse relation between density and productivity.

Mortality is considered by simulation modelling of a wintering population, consisting of individuals that differ in competitive ability, foraging for food in a heterogeneous environment. Foraging individuals often suffer interference (a reversible decline in food intake rate due to the presence of competitors), through disturbance of prey, antagonistic encounters, kleptoparasitism, displacement or avoidance (Goss-Custard 1980). Interference is incorporated using the function:

$$a'_{(S,i)} = QP_i^{-mR_{(S,i)}}$$

where:

$a'_{(S,i)}$ is the searching efficiency of all individuals of phenotype S in patch i

P_i is the density of birds in patch i

Q   is the Quest constant (the searching efficiency achieved by a solitary individual)

m   is the coefficient of interference

$R_{(S,i)}$ is the relative competitive ability of phenotype S in patch i, expressed as the mean competitive ability of all other individuals in patch i divided by the competitive ability of an individual of phenotype S (Parker & Sutherland 1986).

The intake rates and distribution of individuals between patches of differing resource density are determined using game theory, assuming a Type II functional response (Holling 1959). Mortality is then determined by assuming that individuals have a minimum intake requirement for survival (Sutherland & Dolman 1994; Dolman & Sutherland 1995). As the density of birds increases, interference and displacement to poorer quality feeding areas reduce the intake rates of individuals, so that the resulting mortality is density-dependent. It is then possible to consider how behavioural parameters, such as the strength of interference or the rate of depletion, interact with density-dependent breeding productivity to determine population size (Sutherland & Dolman 1994).

This population model may be extended to consider complex migration systems consisting of a series of breeding populations with a choice of potential wintering sites. I assume that there is a cost of migration that increases with the length of the journey. By considering the relative fitness of individuals that use different migration routes it is possible to predict the evolutionarily stable migration strategies using game theory simulation (Sutherland & Dolman 1994; Dolman & Sutherland 1995). The effect of environmental change may be considered by removing habitat from the model and determining the new evolutionarily stable migration strategies and population sizes. This shows that 'knock-on' effects may occur, with changes in the migration strategy and size of populations that did not originally use the affected site.

This game theory approach predicts the new optimal migration system that may evolve in response to environmental change. However, it is important to ask whether migration strategies actually are flexible enough to adapt to the changing environment or whether they may instead fail to respond so that populations experience severe declines.

In some long-lived species that migrate in family groups, such as storks or wildfowl, the migration route is largely passed on by cultural learning (Gwinner 1971; von

Essen 1991). Recent events such as the change in wintering grounds of the Red-breasted Goose *Branta ruficollis* (Sutherland & Crockford 1993) and colonisation of islands in the Baltic by the Barnacle Goose *Branta leucopsis* (Larsson & Forslund 1994) suggest that such systems may change rapidly once a few innovative individuals have pioneered a new route or site. However, there is now a large body of evidence that many other species, such as waders and passerines, have a genetically determined spatial and temporal program of migration (Gwinner 1990; Wiltschko & Wiltschko 1991; Berthold 1993). Breeding experiments using small passerines have shown genetic variation, strong heritability and a response to selection in various aspects of migration (see review in Berthold 1993). This genetic component is important in determining the broad geographic region that birds migrate to. Local patterns of settlement of naïve migrants may then be facultative, influenced by weather, food availability or the density of competitors (Pienkowski & Evans 1984, 1985; Townshend 1985); the geographic scale at which it is appropriate to consider genetic processes rather than behavioural decisions will vary between different species.

I will now use a simple genetic model to show that whether the migration strategy evolves in response to environmental change may depend on details of the biology of a particular species.

I consider a system consisting of two breeding populations, A and B, and two wintering sites, 1 and 2. The bulk of population A winters in site 1, but part of this population adopt an alternative strategy and winter in site 2. I assume that the choice of wintering site is controlled by two alleles at a single locus with the allele for the alternative site recessive. I consider the response of population A to habitat loss in its primary wintering site, site 1. I assume that population B is constrained to winter in site 2. Population size is modeled as before by considering density-dependent breeding productivity and migration costs, and incorporating density-dependent winter mortality by simulation of foraging behaviour. For simplicity I assume that individual reproductive success, competitive ability in winter and migratory phenotypes are independent. Differences in the level of winter mortality experienced by the two migratory phenotypes determine genotype frequencies in the breeding population. I initially assume that mating is random with respect to migratory phenotype.

RESULTS

Figure 1 shows the response of population A to gradual habitat loss in its primary wintering site occurring over a period of 100 years, for simulations that differ in the initial frequency of the alternative allele. This shows that the migration route only tracks the changing environment if there is sufficient initial genetic variation.

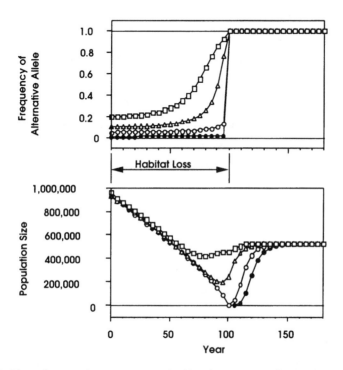

Figure 1. The effect on the response to habitat loss in the primary wintering site as a result of varying the initial frequency of the alternative allele. Habitat loss takes place from year zero to year 100. Results are shown for the changing frequency of the alternative allele and the post-breeding size of population A (see text), after the operation of density-independent mortality. Breeding sites are equal in area and quality; area = 2,000 km^2, fecundity - F = 5; a = 0·1; b = 0·3. Density-independent mortality (10%) acts on the post breeding population. Migration costs are a further density-independent mortality of 1% on each journey. Wintering sites are initially equal in size. Handling time = 5 s; Q = 0·01 m^{2}s^{-1}; interference - m = 0·2. Simulations consider 20 competitive phenotypes, with competitive ability normally distributed over ±3 standard deviations around a mean of 10 with variance = 4. Threshold intake for survival = 0·04 items s^{-1}. Both winter sites consist of 10 patches whose initial resource density ranges from 10 to 50 items m^{-2}. (Initial frequency of alternative allele: ● 1%, o 5%, ∆ 10%, ☐ 20%.)

As the primary wintering site is reduced in area, competition and density-dependent mortality increase, resulting in selection for the alternative allele. Where the frequency of the alternative allele is initially high it increases in response to this selection. As habitat loss proceeds population A declines, but as an increasing proportion of the population are migrating to the alternative site the numbers do not fall to a very low level. After the primary wintering site has been completely lost population A increases to a new equilibrium determined by competition with population B in the remaining wintering site. However, where the alternative allele is initially rare it fails to increase in frequency despite selection. The bulk of population A continue migrating to the primary wintering site, consequently as this site declines population A falls to a very level. Although in this deterministic model I show population A recovering to the new equilibrium, populations that reach such a low level may be vulnerable to stochastic extinction.

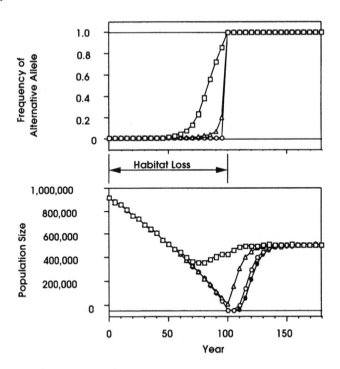

Figure 2. The effect of varying the degree of assortative mating on the response to habitat loss in the primary wintering site. The initial frequency of the alternative allele = 1%. Other density-dependent breeding, demographic, site and foraging parameters as in Figure 1. (Frequency of assortative mating: ● 0·0; o 0·4; △ 0·8; □ 1·0.)

However, other details of the biology of a species may also affect the response to habitat loss; as an example I consider the effect of the mating system. In the simulations presented in Figure 1, mating is random with respect to migratory phenotype. In Figure 2, I show the response of population A to habitat loss in the primary wintering site, for simulations that vary in the degree of assortative mating. In each simulation the initial frequency of the alternative allele is low, only 1%. Where mating is largely random then the alternative allele fails to respond to selection and the population declines catastrophically, as in Figure 1. But if individuals adopting the alternative migration strategy mate assortatively, then the migration route tracks the changing environment and severe population declines do not occur.

DISCUSSION

Some bird species that rely on highly variable food supplies may disperse opportunistically and, in any one year, may breed or winter in different geographic regions depending on food availability and population density (Newton 1970; Gauthreaux 1982; Korpimäki 1994). Such nomadic species may be expected to respond rapidly to environmental change. In contrast, many other bird species undertake regular migration between traditional wintering and breeding grounds; for populations of such species to respond to environmental change may require adaptive changes in the migration strategy.

There are a number of cases where migration routes have changed during the last century, suggesting that migration strategies may respond to environmental change. In a number of species northward expansion of breeding ranges within Europe has been accompanied by changes from partial to full migration, in other cases northern populations derived from resident populations have become migratory (Berthold 1993). This suggests that, at least in some resident populations, genes controlling migratory behaviour have been 'turned off' rather than lost. Similarly, both experiments and field evidence suggest that a shift from partial migration to residency may be relatively rapid if there is amelioration of winter climate (Schwabl 1983; Berthold 1984, 1993). That migratory birds have the potential to evolve such changes is not surprising, as current migration routes must have arisen since the last ice age. It is likely that migratory birds have tracked the changing distribution of biomes during the Holocene by gradually extending or re-orienting their migratory journey.

However, current environmental change is unlike anything seen before. Although the expected rate of climate change may not be unprecedented, it is combined with the destruction and modification of biotopes through intensive agriculture and other forms of human land-use. For some species, adapting to the loss of suitable breeding or wintering habitat may involve more than progressive shifts in orientation or journey length; instead populations may need to radically change their migration route to reach a completely different but suitable region. The simulation model presented here shows that whether or not the migration route of a particular species is flexible enough to allow this to happen may depend on details of its biology.

If the extent and scale of environmental change mean that a population must change its migration strategy beyond the range of existing genetic variation then extinctions may be likely. For example, during the last 30 years increasing numbers of Blackcaps *Sylvia atricapilla* from western central Europe have adopted a new migration strategy, wintering in Britain instead of flying south-west to the Mediterranean. This is thought to be a response to increasingly favourable wintering conditions in Britain and may also result from increased competition in the traditional wintering grounds (Berthold 1993). Breeding experiments have shown that this change in migration route has a genetic basis (Berthold *et al.* 1992). Experiments in orientation cages using Blackcaps from southern Germany show that, although the mean autumn orientation is south-west, there is a scatter in preferred directions between individuals, from north-west to south (Berthold *et al.* 1992). This suggests that the shift to a westerly or north-westerly orientation may have arisen relatively easily in response to selection. In contrast, if in order to adapt to environmental change southern German Blackcaps had to migrate east then the migration route may fail to respond, as no individuals showed such orientation.

There is anecdotal evidence to suggest that some bird species are unable to radically re-orientate their migration route. The breeding range of the Wheatear *Oenanthe oenanthe* is now circum-polar but all populations, from as far west as eastern Canada and as far east as Alaska, winter in Africa. It may be that marginal populations undertake the journey to Africa, not because this is the optimal strategy, but because they have not yet evolved new routes south to what appear to be suitable and closer wintering grounds. Edge-of-range populations of a number of other species such as the Ruff *Philomachus pugnax*, Red-backed Shrike *Lanius collurio*, Red-breasted Flycatcher *Ficedula parva*, Wood Warbler *Phylloscopus sibilatrix* and Lesser Whitethroat *Sylvia curruco* also follow apparently anomalous migration routes. It has been suggested that such populations have not yet evolved new routes as there may be a lack of suitable genetic variation on which selection can act (Berthold 1993).

The simulations presented here have shown that assortative mating may allow the migration route to evolve even where the amount of genetic variation is initially low. Assortative mating may occur if individuals wintering in a new region return to the breeding grounds at a different time, as occurs with British wintering Blackcaps (Berthold *et al.* 1992), or if there is low natal dispersal. Assortative mating may also occur if birds pair-up on the wintering grounds, as long as these wintering grounds are not also used by other conspecific breeding populations. In the simulations presented here the simplifying assumption is made that individual reproductive success and performance on the wintering grounds are independent, so that the only selection on the migration strategy arises from differential winter survival. However, additional selection and faster change may occur if individuals returning from the alternative site have a higher breeding success, for example, through greater accumulation of physiological reserves important for subsequent breeding success (Black *et al.* 1991), or earlier return migration allowing settlement in a better quality territory.

The simulation model I have presented considers single species in isolation; however, birds interact in complex communities. Environmental change may cause changes in the range of predators or competitors that may have unexpected consequences for populations of migratory birds. For example, it is known that late-returning migratory species may be excluded from potential breeding habitat by resident or early-returning species (O'Connor 1981; Garcia 1983). If environmental change allows partially migratory populations to become resident, as may occur for example with milder winters, then populations of long-distance migrants may decline (Berthold 1993).

The simple genetic model presented in this paper shows that, whether the migration route of a species responds to environmental change may depend on details of its biology, so that we cannot assume that all migratory birds will successfully shift their range to track the changing distribution of biomes. This work has shown that it may be possible to predict how both the demography and the migratory strategy of a population respond to large scale environmental change, using behaviour based models that explicitly consider biological and ecological processes underlying the population dynamics. However, more information is needed before quantitative empirical predictions can be made for particular migratory populations. It is important to refine our understanding of the genetic mechanisms controlling the migration strategy; for example, breeding experiments using Blackcaps suggest that orientation is probably a quantitative character under multi-locus control (Helbig 1991; Berthold 1993). It is also important to quantify the extent of genetic variation within migratory populations. Selection acting on the migration strategy arises from the demography of a population. However, except for a few well studied species such as the Oystercatcher *Haematopus ostralegus* (Goss-Custard *et al.* 1995), only limited estimates are currently available for many critical demographic parameters, such as the strength of density-dependence in the breeding grounds.

REFERENCES

Arcese P, Smith JNM (1988) Effects of population density and supplemental food on reproduction in the song sparrow. J Anim Ecol 57:119-136

Berthold P (1984) The control of partial migration in birds: a review. The Ring 10:253-265

Berthold P (1993) Bird migration: a general survey. Oxford University Press, Oxford

Berthold P, Helbig AJ, Mohr G, Guerner U (1992) Rapid microevolution of migratory behaviour in a wild bird species. Nature 360:668-670

Black J M, Deerenberg C, Owen M (1991) Foraging behaviour and site selection of barnacle geese *Branta leucopsis* in a traditional and newly colonised spring staging habitat. Ardea 79:349-358

Coulson JC (1984) The population dynamics of the eider duck *Somateria mollisima* and evidence of extensive non-breeding by adult ducks. Ibis 126:525-543

Coulson JC, Duncan N, Thomas C (1982) Changes in the breeding biology of the herring gull (*Larus argentatus*) induced by reduction in the size and density of the colony. J Anim Ecol 51:739-756

Dolman PM, Sutherland WJ (1995) The response of bird populations to habitat loss. Ibis 137:S38-S46

Ens BJ, Kersten M, Brenninkmeijer A, Hulscher JB (1992) Territory quality, parental quality and reproductive success of Oystercatchers (*Haematopus ostralegus*). J Anim Ecol 61:703-715

Fretwell SD (1972) Populations in a Seasonal Environment. Princeton University Press, Princeton

Garcia EFJ (1983 An experimental test of competition for space between blackcaps *Sylvia atricapilla* and garden warblers *Sylvia borin* in the breeding season. J Anim Ecol 52:795-805

Gauthreaux SA Jr (1982) The ecology and evolution of avian migration systems. Avian Biology 6:93-168

Goss-Custard JD (1980) Competition for food and interference among waders. Ardea 68:31-52

Goss-Custard JD, Clarke RT, Briggs KB, Ens BJ, Exo K-M, Smit C, Beintema AJ, Caldow RWG, Catt DC, Clark NA, Durell SEA, Le V dit, Harris MP, Hulscher JB, Meininger PL, Picozzi N, Prys-Jones R, Safriel U, West AD (1995) Population consequences of winter habitat loss in a migratory shorebird. I. Estimating model parameters. J Appl Ecol 32:320-336

Gwinner, E (1971) Orientierung. *in* Schütz E (ed) Grundriß der Vogelzugskunde, 299-348. Parey, Berlin

Gwinner, E (1990) Bird Migration: The Physiology and Ecophysiology. Springer-Verlag, Berlin

Helbig AJ (1991) Inheritance of migratory direction in a bird species: a cross-breeding experiment with SE- and SW-migrating Blackcap (*Sylvia atricapilla*) Behav Ecol Sociobiol 28:9-12

Holling CS (1959) Some characteristics of simple types of predation. Can Ent 91:385-398

Holmes RT, Sherry, TW (1992) Site fidelity of migratory warblers in temperate breeding and Neotropical wintering areas: Implications for population dynamics, habitat selection, and conservation. *in* Hogan JN, Johnston DW (eds) Ecology and Conservation of Neotropical Migrant Landbirds, 563-575. Smithsonian Institution Press, Washington DC

Korpimäki E (1994) Rapid or delayed tracking of multi-annual vole cycles by avian predators. J Anim Ecol 63:619-628

Larsson K, Forslund P (1994) Population-dynamics of the barnacle goose *Branta leucopsis* in the Baltic area: density-dependent effects on reproduction. J Anim Ecol 63:954-962

Maynard Smith J, Slatkin M (1973) The stability of predator-prey systems. Ecology 54:384-391

Newton I (1970) Irruptions of crossbills in Europe. *in* Watson A (ed) Animal Populations in Relation to their Food Resources, 337-357. Blackwell, Oxford

Newton I (1992) Experiments on the limitation of bird numbers by territorial behaviour. Biol Rev 67:129-173

O'Connor RJ (1981) Comparisons between migrant and non-migrant birds in Britain. *in* Aidley DJ (ed) Animal Migration, 167-195. Cambridge University Press, Cambridge

Owen M, Black JM (1989) Factors affecting the survival of barnacle geese on migration from the breeding grounds. J Anim Ecol 58:603-617

Parker GA, Sutherland WJ (1986) Ideal free distributions when individuals differ in competitive ability: phenotype limited ideal free models. Anim Behav 34:1222-1242

Pienkowski MW, Evans PR (1984) Migratory behaviour of shorebirds in the Western Palearctic. *in* Burger J, Olla BL (eds) Behaviour of Marine Animals Volume 6 Shorebirds: Migration and Foraging Behaviour, 73-123. Plenum Press, New York

Pienkowski MW, Evans PR (1985) The role of migration in the population dynamics of birds. *in* Sibly RM, Smith RH (eds) Behavioural Ecology: Ecological Consequences of Adaptive Behaviour, 331-352. Blackwell, Oxford

Schwabl H (1983) Ausprägung und Bedeutung des Teilzugverhaltens einer südwestdeutschen Population der Amsel *Turdus merula*. J Ornithol 124:213-217

Sutherland WJ, Crockford NJ (1993) Factors affecting the feeding distribution of red-breasted geese *Branta ruficollis* wintering in Romania. Biol Conserv 63:61-65

Sutherland WJ, Dolman PM (1994) Combining behaviour and population dynamics with applications for predicting consequences of habitat loss. Proc R Soc Lond B 255:133-138

Townshend DJ (1985) Decisions for a lifetime: establishment of spatial defence and movement patterns by juvenile grey plovers (*Pluvialis squatarola*). J Anim Ecol 54:267-274

von Essen L (1991) A note on the lesser white-fronted goose *Anser erythropus* in Sweden and the result of a re-introduction scheme. Ardea 79:305-306

Williamson M (1972) The Analysis of Biological Populations. Edward Arnold, London

Wiltschko R, Wiltschko W. (1991) Magnetic orientation and celestial cues in migratory orientation. *in* Berthold P (ed) Orientation in Birds, 16-37. Birkhäuser, Basel

Terrestrial invertebrates and climate change:
Physiological and life-cycle adaptations

Jennifer E.L. Butterfield and John C. Coulson
University of Durham
Department of Biological Sciences
South Road, Durham DH1 3LE
U.K.

INTRODUCTION

Invertebrates are ectotherms and their growth rates are directly dependent on environmental temperatures; being small, they are also highly susceptible to desiccation (Andrewartha & Birch 1954). Changes in temperature and rainfall régimes thus are likely to have major direct effects on invertebrate distributions. Equally important, climate change will have indirect effects through its effect on food plant distributions and phenologies. Many specialist herbivores have shifted their distributions in response to changes in plant distributions in the past (Hengeveld 1990), others must have adapted to changed timing of budburst and early leaf growth in their host plant (Dixon 1985; Murray *et al.* 1989).

Apart from tropical areas and man-made food storage depots, where there is little annual variation in temperature and precipitation, most invertebrate species experience seasonal changes, such as winter cold or summer drought, which render parts of the year inimical to growth. These species live through the unfavourable period in a stage of suspended animation, known as diapause in insects. The length of the diapause stage may be temperature independent or even negatively related to temperature (Tauber *et al.* 1986) and thus the length of a life-cycle which includes diapause will not necessarily be related to temperature. In addition, some winter diapausing species such as the silkworm *Bombyx mori* or the emperor moth *Saturnia pavonia* have a chilling requirement, with the favourable temperature range for diapause being well below that for active growth (Danilevskii 1965).

NATO ASI Series, Vol. I 47
Past and Future Rapid Environmental Changes:
The Spatial and Evolutionary Responses of
Terrestrial Biota
Edited by Brian Huntley et al.
© Springer-Verlag Berlin Heidelberg 1997

In general, species which have widespread latitudinal or altitudinal ranges already exist under a wide range of climatic conditions and are, in effect, preadapted to climate change. However, the diversity of invertebrates and especially insects, which make up some 75% of all animal species, means that there is a large number of invertebrate species adapted to specialised habitats throughout the globe. Many of these species are confined within relatively narrow climatic ranges and their response to climate change depends on whether individuals survive the changed conditions and whether there is sufficient genetic variability within the populations for appropriate adaptation to occur. In Britain, the northern and mountain plateaux species will be especially vulnerable to global warming. In the event of temperature rise, southern and low altitude species can migrate northwards or upward but the northern and high altitude species are at the limits of their distributions already and are likely to disappear.

In this paper, we draw on examples of differing life-cycle strategies shown by invertebrates over an altitude transect of 700 m in the northern Pennines of England. The mean annual temperature difference encountered over the transect, which takes in the summit of Great Dun Fell (847 m), is approximately 5°C, and is equivalent to the higher figures quoted for global warming over the next 50 years (Schneider 1992). By analogy, comparison of the breadths of altitudinal distributions and the life-cycle adaptations of species encountered over the transect allows prediction of their future response to global warming. In addition, it provides the opportunity to compare the life-history strategies of some of the restricted high altitude species, which are likely to be more vulnerable to climate warming, with those of widespread species.

BROAD SCALE EFFECTS OF CLIMATE CHANGE ON INVERTEBRATE DISTRIBUTION

Invertebrate diversity declines from the tropics to the poles (Begon *et al.* 1986) and shows a similar decline with increase in altitude (Coulson 1988). It has long been known that the northern limits of many insect species' distributions coincide with the isotherms of winter minima (Uvarov 1931). Figure 1 shows the numbers of arthropod species caught at a series of sites on the altitude transect and it is apparent that, in general, diversity declines with increasing altitude and hence with decreasing temperature. When the relationships of individual taxa with altitude are examined spiders (Araneae), harvest spiders (Opiliones), moths (Lepidoptera), bees (Apidae), grasshoppers (Acrididae) and ground beetles (Carabidae) all show significant decreases in species numbers with increasing altitude. The numbers of cranefly species (Tipulidae) and fungus gnats (Mycetophilidae), however, are positively related to altitude (Coulson 1988). This suggests that global warming will favour most invertebrate taxa, the higher temperatures allowing distributions to expand to higher altitudes and latitudes in the northern hemisphere. The two taxa where species number is positively related to altitude have larvae that are particularly sensitive to desiccation (Coulson 1962) and are probably responding to the increased precipitation at higher altitudes rather than the decrease in temperature. Despite the difficulty of predicting future rainfall, higher temperatures will increases the rate of evapotranspiration so it is likely that most areas will become drier. The distributions of species within taxa such as the Tipulidae, that have stages that are particularly vulnerable to desiccation, will contract.

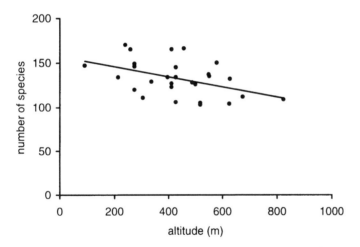

Figure 1. Relationship between the number of invertebrate species (y) at a site and altitude (x); $y = 158.5 - 0.06x$, $r_{26} = -0.44$, $p < 0.01$

LIFE-CYCLE STRATEGIES

1. The growth stage

Some species, especially invertebrate herbivores, which are adapted to live in cool climates and maintain annual life cycles, have growth rates which show a relatively small response to rise in temperature, i.e. they have a low Q_{10} over the favourable temperature range for growth. Figure 2 compares the rate response to temperature of the larval period of the arctic/alpine cranefly *Tipula subnodicornis*, which has a univoltine life-cycle throughout its distribution, with that of the multivoltine large white butterfly *Pieris brassicae*, which has a distribution extending from North Africa to northern Russia. Both the favourable temperature range for development and the response to temperature rise are appreciably less in the cranefly than in the butterfly. A mean temperature rise from 10 to 20°C would have virtually no effect on the length of the larval growth period in *T. subnodicornis* whereas it would cause an 80% shortening in that of *P. brassicae* (Danilevskii 1965; Butterfield 1976).

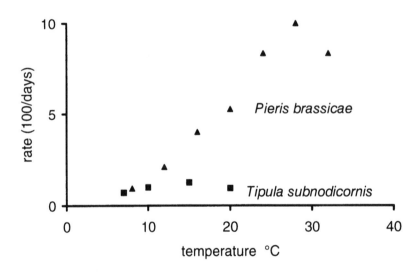

Figure 2. Comparison of the growth rate response to temperature in *Tipula subnodicornis* and *Pieris brassicae* larvae

The low growth rate response to temperature rise in *T. subnodicornis*, and other species, is an adaptation that favours the retention of an annual life-cycle over a relatively wide climatic range. As long as mean temperatures remain within the favourable range, climate change will have little effect on the life-cycle. However, the shallow response curve is also an adaptation to low assimilation rates. Schramm (1972) showed that the capacity of herbivorous invertebrates, or their symbiotic micro-organisms, to digest cellulose is very low at temperatures below 13°C and Remmert (1980) has pointed out that only very slow growing organisms can use plant material as food at temperatures below this. The slow growth strategy tends to be found in species with restricted cold climate distributions rather than in widespread species which have distributions that extend to cold areas. This can be illustrated by a comparison between the widespread emperor moth *Saturnia pavonia*, which has a distribution that extends from southern Europe to the Arctic Circle, and the northern eggar *Lasiocampa quercus callunae*, a sub-species of the oak eggar moth, which is restricted to upland and northern Britain. Both feed on heather *Calluna vulgaris* and at 500 m in the Pennines both hatch in June. The emperor grows rapidly over the summer period and pupates by late summer, overwintering as a diapausing pupa, while the larvae of the northern eggar grow extremely slowly and overwinter as small caterpillars. The latter complete the growth period during the following summer and pupate before overwintering for a second time. The northern eggar is apparently unable to complete its life-cycle in one year, even when kept at higher temperatures in the laboratory.

At lower altitudes the northern eggar is replaced by the oak eggar *Lasiocampa quercus* with an annual life-cycle. There are many examples where high altitude species with biennial life-cycles are replaced by annual species at lower altitude, not only among the herbivores such as *Tipula rufina* and *Strophingia ericae* (Coulson & Whittaker 1978), but also among predatory species such as the ground beetles. For instance, the biennial, fell top species *Nebria gyllenhali* is replaced by the annual *Nebria salina* on grassland habitats at altitudes below 500 m. Other species, such as *Carabus problematicus*, are able to switch from biennial cycles at high altitude to annual cycles at lower altitudes (Butterfield 1986). In general, the more highly adapted the species is to low temperature the less likely it is to be able to adapt to a warmer environment. Species which have low favourable temperature ranges will be vulnerable to temperature rise as will those that cannot depart from a biennial cycle.

2. The diapause stage

Figure 3 compares the favourable temperature ranges for larval growth and pupal diapause stages in the emperor moth and shows that the favourable temperature range for diapause development is adapted for overwintering. The combination of the two developmental stages within an annual life-cycle allows the species to exist

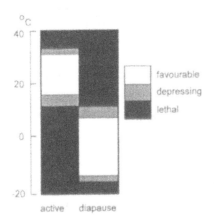

Figure 3. Comparison of the favourable temperature ranges for growth and diapause in *Saturnia pavonia* (redrawn from Danilevskii 1965)

over a wide range of climatic conditions. In species with an annual life-cycle, diapause is usually an obligatory part of the life-cycle, entered spontaneously at a fixed stage (stages) in the life-history.

The presence of the diapause stage allows a wide variation in the larval growth rate response to temperature among annual species, with *T. subnodicornis* spending many months in the larval stage, while the emperor grows rapidly, passing through the vulnerable larval stages within weeks, and entering a prolonged diapause. Other species which respond to temperature elevation with greatly increased growth rates become multivoltine and have several generations in the warmer parts of their range. These species do not automatically enter diapause at the end of the growth phase yet still need to overwinter in the inactive diapause stage. This means that multivoltine species must have some method of predicting the onset of unfavourable conditions. Figure 4 shows the response of developing larvae of the large white butterfly to a series of photoperiod regimes, diapause occurring on daylengths of 15 h or less. Daylength may induce, terminate or moderate diapause development in both univoltine and multivoltine species and, as it is a more reliable indicator of season than temperature, its use as an environmental trigger for the onset of diapause is widespread among multivoltine species.

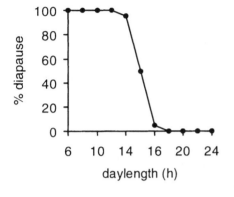

Figure 4. Diapause response to daylength in *Pieris brassicae* (redrawn from Danilevskii 1965)

At higher latitudes and altitudes the proportion of invertebrates that have annual or biennial life-cycles increases. As the summer period gets shorter at higher latitudes, the selective pressure on ectotherms to synchronise their active stages with this favourable development period increases. Diapause, instead of being a stage which allows multivoltine invertebrates to survive an unfavourable period, becomes an obligatory stage which occurs during an annual life-cycle and synchronises it with season. In annual species, summer diapause is usually terminated by photoperiod but winter diapause often ends spontaneously (Tauber *et al.* 1986). For instance, the larval diapause in *Tipula subnodicornis* is maintained in its early stages by short photoperiod but ends within the winter period when temperatures are below the developmental threshold, with pupation and emergence being synchronised by spring rise in temperature (Butterfield 1976). This strategy can be compared with that of another cranefly *Molophilus ater* which also pupates in spring. When the diapausing larvae of *M. ater* and *T. subnodicornis* are kept at 10°C (well above the threshold temperature for pupation) and 6 h daylength, pupation in *M. ater* is inhibited by the short photoperiod. In *T. subnodicornis* pupation is delayed and asynchronous, compared to larvae exposed to an 18 h daylength, but does eventually occur (Coulson *et al.* 1976). Species which rely on a winter diapause followed by spring rise in temperature to synchronise emergence or hatching may lose this synchrony if winter temperatures are not low enough to prevent development. In the case of *T. subnodicornis* this would have adverse effects on the mating system, which depends on high densities of both sexes being present over a short time span. More generally, the presence of larvae may no longer coincide with their optimal food supply.

GENETIC HETEROGENEITY WITHIN POPULATIONS

Within all species there is a degree of heterogeneity in growth rate and in many species diapause response varies between individuals. The asynchronous emergence, recorded above for *Tipula subnodicornis,* indicates a wide range of developmental response on short photoperiod and Figure 5 shows a similar pattern for hatching in eggs of another cranefly, *Tipula pagana. T. pagana* lays its eggs in the autumn and short autumn daylengths slow development so that hatching does not occur before winter (Butterfield & Coulson 1988). With this type of variation there is considerable scope for natural selection and rapid adaptation to climate change (e.g. Masaki (1978) selected a strain of cricket *Pteronemobius fascipes* with 70% diapausing eggs from a strain with 10% within 15 generations). Such adaptation is apparent in many widespread species that show latitudinal clines in growth rates or photoperiod responses. For instance the knot grass moth *Acronicta rumicis* has a distribution extending from north of St. Petersburg (60°N) to the Black Sea (42°N) and overwinters in pupal diapause. The onset of development towards this diapause is triggered by response to short autumn daylength in the developing caterpillar. Figure 6 shows that, when the caterpillars were collected from the north (55°N) and the southern edge of the range (43°N) and kept at a constant temperature of 23°C, the northern population entered diapause on daylengths less than 19 h while the southern population delayed entry until the daylength had decreased to below 15 h (Danilevskii 1965). Here there has been selection to respond to the different daylengths that signal approaching winter at the two latitudes and, as a result, populations with the appropriate responses have evolved.

Species that are highly specialised may be expected to show a decrease in genetic heterogeneity and this must apply to species which exhibit adaptations that allow them to live at low temperatures. However, we know of at least one case where a species, which has hitherto been considered as a high altitude specialist, has adapted to warmer conditions. The cranefly *Tipula montana* is biennial at 1000 m on

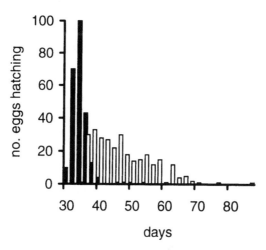

Figure 5. The distribution of hatching of *T. pagana* eggs kept at 10°C under L:D 18:6 (filled columns) and L:D 6:18 (open columns)

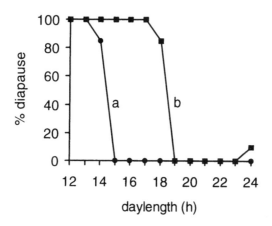

Figure 6. Geographical variation in the photoperiodic response of *Acronicta rumicis* collected from 43°N (a) and 55°N (b) (redrawn from Danilevskii 1965)

the Cairn Gorm plateau in Scotland and annual at 350 m in the north of England. It needs a photoperiod of more than 14 h for the fully grown larvae to pupate. At 1000 m, larvae grow so slowly that development is prolonged into a second calendar year. Autumn pupation is restrained by the photoperiod requirement and there is a synchronised spring emergence in the following year. At low altitude in the north of England the main emergence period is in August and the life-cycle has become annual to a large extent. The high altitude requirement for slow growth, however, is reflected in part of the population not having completed growth by August in the year following hatching. These individuals pupate the following spring (Coulson in prep.).

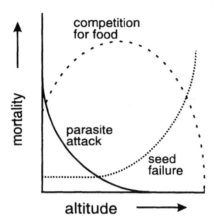

Figure 7. Changing intensity in three mortality factors in *Coleophora alticolella* over a gradient of increasing altitude (redrawn from Randall 1982)

CONCLUSION

In general, global warming will increase invertebrate diversity in temperate latitudes, even on islands such as Britain where the sea acts as a barrier to immigration. Species that will benefit from global warming will tend to be species that are widespread at present and exhibit a high degree of genetic heterogeneity. Northern and high altitude species, which have been highly selected through time for existence at low temperatures, will lack the genetic flexibility to adapt to temperature change. Despite this generalisation, most of the invertebrate species that have been used as examples above are able to complete development over a wider temperature range than that of the climatic range which they occupy in the field. Equally, the species that we have studied show heterogeneity in the length of the growth and diapause periods allowing scope for evolution. In the presence of global warming, one of the major factors that will affect future distribution ranges, especially of the Arctic–Alpine species, will be the expansion of the ranges of thermophilous species to higher latitudes and altitudes. These immigrant species may be competitively superior members of the same taxa or predators and parasites. Using the altitude transect as a model, Figure 7 shows how the comparative importance of parasites as a mortality factor in the life-cycle of the rush moth *Coleophora alticolella* increases with increasing temperature. At higher altitudes, the failure of its food supply becomes progressively more likely with the moor rush *Juncus squarrosus* failing to set seed at low temperature. At low altitude parasites exact a heavy mortality, while in the middle of the altitude range intraspecific competition regulates the population (Randall 1982). This example can be taken as a simple model of the influence that climate change has on species interactions and thus on their population dynamics.

REFERENCES

Andrewartha HG, Birch LC (1954) The Distribution and Abundance of Animals. University of Chicago Press, Chicago

Begon M, Harper JL, Townsend CR (1986) Ecology: Individuals, Populations and Communities. Blackwell Scientific Pubications, Oxford

Butterfield JEL (1976) The response of development rate to temperature in the univoltine cranefly *Tipula subnodicornis* Zetterstedt. Oecologia 25:89-100

Butterfield JEL (1986) Changes in the life-cycle strategies of *Carabus problematicus* over a range of altitudes in Northern England. Ecol Entomol 11: 17-26

Butterfield J, Coulson JC (1988) The rate of development in the overwintering eggs of *Tipula pagana* Meigen. J Insect Physiol 34:53-57

Coulson JC (1962) The biology of *Tipula subnodicornis* Zetterstedt with comparative observations on *Tipula paludosa* Meigen. J Anim Ecol 31: 1-21

Coulson JC (1988) The structure and importance of invertebrate communities on peatlands and moorlands and the effects of environmental and management changes. *in* Usher MB & Thompson DBA (eds) Ecological Change in the Uplands. Blackwell Scientific Publications Oxford

Coulson JC, Horobin JC, Butterfield J, Smith GRJ (1976) The maintenance of annual life-cycles in two species of Tipulidae (Diptera); A field study relating development, temperature and altitude. J Anim Ecol 45: 215-233

Coulson JC, Whittaker JB (1978) Ecology of moorland animals. *in* Heal OW & Perkins DF (eds) Production Ecology of British Moors and Montane Grasslands. Springer, Berlin

Danilevskii AS (1965) Photoperiodism and Seasonal Development of Insects. Oliver & Boyd, Edinburgh

Dixon AFG (1985) Aphid Ecology. Blackie, Glasgow

Hengeveld R (1990) Dynamic Biogeography. Cambridge University Press, Cambridge

Masaki S (1978) Seasonal and latitudinal adaptations in life cycles of crickets. *in* Dingle H (ed) Evolution of Insect Migration and Diapause. Springer Berlin

Murray MB, Cannell MG, Smith RI (1989) Date of budburst of 15 tree species in Britain following climatic warming. J Appl Ecol 26:693-700

Randall MGM (1982) The dynamics of an insect population throughout its altitudinal distribution. Ecol Entomol 7:177-185

Remmert H (1980) Arctic Animal Ecology. Springer Berlin

Schneider SH (1992) The climatic response to greenhouse gasses. *in* Woodward FI (ed) The Ecological Consequences of Global Climate Change. Advances in Ecological Research 22:1-32. Academic Press London

Schramm U (1972) Temperature-food-interaction in herbivorous insects. Oecologia 9:399-402

Tauber MJ, Tauber CA, Masaki S (1986) Seasonal Adaptations of Insects. Oxford University Press Oxford

Uvarov BP (1931) Insects and climate. Trans R Entomol Soc Lond 79:1-247

Section 6

Predicted future environmental changes and simulated responses

Forecast changes in the global environment:

What they mean in terms of ecosystem responses on different time-scales

Wolfgang Cramer and Will Steffen[1]

Potsdam Institute for Climate Impact Research

P.O. Box 60 12 03

D-14412 Potsdam

Germany

INTRODUCTION

The nature of anticipated global environmental change

If seen from a human perspective, the global environment of the future in, say, several decades from now *will* be different from that of today – the best-known reason for this is climate change. 'Climate' is a description of the *average* conditions of a notoriously variable atmospheric system. Relatively small changes in the radiation balance, due to changes in atmospheric composition or due to changing surface characteristics, such as sea surface temperatures or terrestrial albedo, can trigger major changes in circulation pattern and, hence, the regional pattern of temperature and precipitation. The longer term history of Earth, on the other hand, is a history of changes too, but these are dominated by the slow but powerful endogenous changes in the lithosphere and by changes in incoming solar radiation due to changes in the Earth's orbit or the Sun's activity. Between the lithosphere and the atmosphere, the biosphere is a thin, rather fragile home for plants, animals and humans; it can be greatly affected by significant changes in the abiotic environment. Changes with the greatest relevance to the *terrestrial* biosphere occur over a range of time-scales, from rapid changes in radiation (minutes) to changing temperatures

[1] GCTE core project office, CSIRO, Div. Wildlife and Ecology, Lyneham ACT, Australia

NATO ASI Series, Vol. I 47
Past and Future Rapid Environmental Changes:
The Spatial and Evolutionary Responses of
Terrestrial Biota
Edited by Brian Huntley et al.
© Springer-Verlag Berlin Heidelberg 1997

and/or amounts of available water (days to months) to changing locations of major climatic zones (decades to centuries).

Rates and pattern of anticipated climate change

The most recent estimates about rates of global warming are between 0·1 and 0·2°C per decade (IPCC Working Group I, unpublished). Around the middle of the 21st century, this rate is likely to rise to be between 0·2 and 0·35°C, and global annual mean temperature is expected to be between 2 and 4°C warmer than today. It is also expected that warming will be more intense at high latitudes in winter, and in the interior of continents. The diurnal range of temperatures is likely to decrease, i.e. nights may be warming more than days. The hydrological cycle is likely to become intensified, i.e. rainfall may increase in many areas, but so will evapotranspiration – the end result this could have on the terrestrial water balance is likely to vary widely between regions and is therefore difficult to predict.

The current limits to predictions of the regional pattern of climate change do not preclude the development of meaningful biospheric impact scenarios at that scale. A broad range of downscaling techniques exists that allows the assimilation of knowledge from local climate patterns and from broad-scale circulations. These do not by themselves provide more accurate predictions of regional change, but they allow for the local to regional investigation of implications to specific climate change scenarios (Gyalistras et al. 1994). Moreover, given the current uncertainty of climate predictions, it is clearly important to be able to investigate the *sensitivity* of the biosphere to a range of *possible* climatic changes. Knowledge of this sensitivity not only allows for the improved planning of abatement or mitigation strategies, but it may also enhance the capacity of ecosystem models to be diagnostic tools for climate models (Claußen 1994).

Impacts on and feedbacks from land surface processes

General circulation models (GCM's) deal primarily with patterns of atmospheric flow and attempt to balance energy inputs and outputs on a broad scale. Their lower boundary conditions need to be set from information about the surface of the globe (Noblet, 1996). Notwithstanding recent progress in ocean and atmosphere GCM coupling, the assignment of appropriate ocean sea surface temperatures (SSTs) remains one of the most critical factors in driving climate models. On the terrestrial surface, however, the importance of realistic boundary conditions has recently received greater awareness (Claußen 1994; Bonan 1995; Henderson-Sellers & McGuffie 1995). As a result, there now is a demand for a range of more appropriate land surface models that can provide the necessary inputs for climate models. Conventional parameters that are required for these process descriptions are albedo, surface roughness and stomatal conductance – more will have to be included as surface parameterizations become more sophisticated. In addition to these direct feedbacks from the land surface to the climate, a changing land surface also interacts with the atmosphere through the exchange of CO_2 and trace gases.

This demand for improved simulation of feedback processes coincides with an increased demand for models that are capable of assessing likely impacts of climate change on the biosphere in its own right. The life support system for humans depends totally on the biosphere, and impact assessments such as those of the Intergovernmental Panel on Climate Change (IPCC) increasingly base their scenarios on simulation models of ecosystem processes. To simulate both kinds of changes requires tools (models) that can cover a wide range of ecological processes, from physiology and biophysics to ecology, and these need to have known and plausible sensitivities to the major driving variables of climate.

Short and long term dynamics in current terrestrial land surface models

Of the slow to fast processes mentioned above, most terrestrial land surface models now address the fast processes (from daily to monthly changes in surface characteristics and energy and mass fluxes) fairly well (Parton *et al.* 1993; Lüdeke *et al.* 1994). Longer term changes, however, such as the redistribution of vegetation types (either anthropogenic or natural) or the degradation of vegetation greatly affect the parameters of these models. Ecosystem structure models have been shown to be capable of predicting the *equilibrium* characteristics of the biosphere (Prentice *et al.* 1992), and several land surface models are now initialized by them (e.g. PLAI – Plöchl & Cramer (1995)). None of these models, however, has the capability to assess the successional time lags or the dynamics of other processes that are involved in changing from one state of vegetation into another.

TYPES OF ECOSYSTEM RESPONSES AND WAYS TO SIMULATE THEM ON A
GLOBAL SCALE

Types of response

If environmental conditions, such as climate, change beyond a certain tolerance limit,
then the basic response strategies of plants change accordingly. Responses can
roughly be classified into several different categories, depending on the nature of the
response and on their position along the time axis:

- *Physiological* responses (e.g. opening or closing of stomata, $\Delta t \approx$ min - h),

- *Morphological (or phenotypic)* responses occur at short time-scales ($\Delta t \approx$ days to
  weeks), e.g. by shedding leaves when drought conditions require this, or by
  enhanced growth when conditions are more favourable.

- Responses of *stand structure* usually occur over time spans of years to decades,
  e.g. when certain plant types disappear because of unfavourable environmental
  conditions or because of enhanced competition, and are possibly replaced by
  others. Disturbances may often be an important element in these changes and
  hence produce relatively rapid disruptions that are later followed by slower
  regrowth processes where this is possible.

- *Genetic* responses occur when certain plant types (ecotypes, species) change into
  types that can operate under conditions where their ancestors could not. The
  temporal aspect of these responses is dependent on the level of genetic
  differentiation under consideration—it may vary from a few generations (on the
  level of ecotypic differentiation) to centuries or millennia (on the level of species).

Most major responses of stand structure and/or genetic characteristics to
environmental change seem to occur less rapidly than the currently forecast changes
in the environment. Therefore, it is necessary to investigate the resulting time lags by
estimating the nature and (maximum) rate of responses of the biosphere to
environmental change. The majority of current ecosystem models assume some kind
of equilibrium (no dynamic response) for some or all of these strategies—they have
therefore only been capable of predicting the equilibrium condition of the Earth
system. This has been recognized both by critics and by some of the modelling
teams themselves. To improve upon this, however, it is necessary to more

specifically consider both the nature of expected environmental change, and the kind of response strategies that may need to be accounted for by the ecosystem models.

Multi-scale considerations

For each of the four basic response classes (physiological, morphological, structural and genetic), a wide range of plant processes needs to be considered, and only some of these are covered by each of the currently available simulation models. Moreover, usually only one of the four classes is adequately described by a particular model, while the others are implicitly kept constant.

For example, broad-scale (i.e. continental to global) assessments of the likely response of the distribution of terrestrial vegetation to changes in the global environment usually involve a considerable amount of smoothing with respect to shorter term changes in the environment (such as daily or hourly changes in insolation) and they usually disregard interannual variability of weather altogether (Box 1981; Woodward 1987; Prentice *et al.* 1992). Hence, the responses are those that deal with seasonal dynamics, such as the sensitivity of plants to the (average) seasonality of rainfall and temperature (phenology). A realistic transient behaviour with respect to structural or genetic responses cannot be expected from these models.

Physiological response models, on the other hand, mainly involve shorter-term processes such as the daily fluctuations of energy and water balances (e.g. Pinter *et al.* 1994) and would produce wrong results if their environmental driving variables were smoothed over periods longer than a few hours. This usually disconnects these models from being representative for larger heterogeneous areas (i.e. away from the weather stations), because the spatial distributions of the high-resolution time series of driving variables are not known. Also, inevitably, they too cannot describe longer term structural and/or genetic responses.

Lastly, successional models (for forests) usually need to rely rather heavily on the response of plants (trees) to each other and they thus describe the competition for resources such as light, water and nutrients over longer time spans. Temporal (over several years or decades) and spatial (landscape) variability is often dealt with by involving carefully parameterized stochastic functions. These models typically resolve time lags in species composition changes of specific forest types (Prentice *et al.* 1993a; Bugmann & Fischlin 1994), but they are unable to demonstrate physiological or morphological responses to varying local weather conditions, and they are also unlikely to produce plausible results for changing genetic characteristics.

MODELLING MEDIUM-TERM DYNAMICS (DECADES TO CENTURIES): PERSPECTIVES FOR A TRANSIENT GLOBAL ECOSYSTEM MODEL

Current state-of-the-art

In the remainder of this contribution, we explore one of several possible ways to bridge the gap between transient behaviour of models on different time-scales. For global assessments of biospheric change, including impacts on ecosystems and feedbacks to the atmosphere, it has become apparent that potential transient shifts of ecosystem structure need to be known for the physiological behaviour of the biosphere, and hence the total amounts of trace gas fluxes, to be extrapolated into the future (Prentice 1993; Solomon & Cramer 1993). After the first, speculative suggestions of a 'major carbon pulse' which would result from broad-scale dieback of North American forests once conditions became unsuitable for them due to climate change (Neilson 1993), it has been postulated that the magnitude and temporal evolution of this pulse could be predicted from simple assumptions of the processes of carbon release and sequestration (Smith & Shugart 1993). The rationale of both approaches is that the structure of ecosystems, as predicted by rule-based models, is likely to change from one (assumed) equilibrium state into another along only a few fundamental pathways, such as dieback, slow compositional change or invasion. It is assumed that each of these pathways can be translated into carbon release or sequestration rates which commence immediately after the change has occurred. From balancing these rates against each other, total estimates of the net release (or gain) of carbon can be made.

Essentially, this approach retains the strengths and weaknesses of global equilibrium models in being spatially explicit and globally comprehensive and building on the wealth of biogeographical knowledge and data. Several of the major assumptions about temporal dynamics, however, are extremely difficult to justify, such as the step-wise change of climate from pre-industrial to 'post-industrial' conditions, or the monotonous rates of either carbon release or sequestration in various subsystems.

Investigating the nature of ecosystem change

If environmental conditions change beyond a certain limit of tolerance, then plants fail to complete one or several important parts of their life cycle. The result for the ecosystem is primarily that these plants eventually die and may be replaced by others. The range of possible time-scales for processes associated with such shifts is very broad. At the fast end are dieback processes with relatively rapid responses – e.g. occurring after a catastrophic incident, or after a few very unfavourable seasons, and releasing carbon with the rate of decaying biomass immediately after that. The other extreme are slow (successional) replacement processes where the main limitation to a species may be the inability to produce offspring under certain changed conditions – in these cases, nothing may be visible before the normal life cycle of the present individuals has been completed, and they may then be slowly replaced by other species that are able to invade a habitat that only then becomes available.

The class of model that deals with these changes is the so called gap model (named so because its fundamental assumption is that forest dynamics are characterized by the opening and re-filling of canopy gaps). Most gap models have been developed for forests of specific geographic regions or types. It is currently not possible to apply a gap model globally, because they require a set of parameters for every major plant species that could potentially occur at a site, and these parameters are not available for all areas of the globe. To reduce this burden, research is in progress to generalize gap models into plant functional types dynamic models—they would allow the total set of species to be classified into a set of easily distinguishable types (Bugmann in press). This generalization is matched by developments towards subdividing the 'green slime' ecosystem description of global ecosystem models into combinations of similar plant functional types (Box 1981; Prentice *et al.* 1992). Ideally, these two developments would converge into a common key of plant types (Steffen *et al.* 1992) that would allow the application of gap models within the framework of global ecosystem models.

An intermediate step: distinguishing between changes of different kinds

To search for a suitable set of dynamic models, species and parameters, it is necessary to identify the types of change that might occur between two different states of the climate/vegetation system. We have attempted to investigate this by modifying the approach used by Smith & Shugart (1993) in several ways: a) instead of the Holdridge climate classification, we have used a plant functional types based model (BIOME 1.1 – Prentice *et al.* 1993b); b) rather than employing the step-wise climate change from pre-industrial to 'post-industrial' conditions, we assume that the climate of a $3 \times CO_2$ scenario (Perlwitz 1992)[2] requires 100 years to be reached from today; and c) we assume that, at any location, ecosystem response begins only when the local climate has changed to conditions outside the current climatic range of the local ecosystem. Because this assessment is made with the intention of exploring the nature of potential vegetation change itself, we have not assigned carbon release or sequestration rates to it – instead, we simply want to document the types of changes that could occur under a given scenario.

To classify the changes that can occur between different biome types, we have looked at the structural characteristics of each pair of biomes that could occur before, during and after the 100 years of change. A preliminary investigation of the change matrix yielded the three basic change types given by Smith & Shugart (1993): immigration, dieback and successional replacement. To capture the dynamics of tree/grass mixed systems, such as savannas, we found it necessary to add a fourth type 'compositional change in mixed structural type systems'.

SOME RESULTS

The results can be mapped and investigated in various ways. Here, only a subset is shown in two summary tables. Table 1 illustrates the possible fate of all areas that, under current climate, can support temperate deciduous forests. It shows that about half of these forests (*ca.* 3,041,000 km^2) stays within the same climatic range, i.e. no

---

[2] This scenario was used by calculation relative anomalies of temperature and rainfall between a control run and a $3 \times CO_2$ simulation, interpolating these anomalies to the high resolution grid of our climate data base (Cramer *et al.* in prep.), and then applying the anomalies to the values found in that data base.

significant changes in ecosystem structure and composition are expected in these areas. Of the remainder, 26% are predicted to change into broad-leaved evergreen forests by successional replacement. On average, the time required for this process to begin is 52 years, but the range is broad from only 3 years for some locations to 98 years for others, depending on whether a location is currently already near its climatic limit towards warmer forest or not. We stress that this time represents the first point at which a functional type within the temperate deciduous forest biome is no longer within its climatic limits. The FT may, however, persist for some time after this point if its limitation is due to the inability to produce offspring under the changing climate. Thus, the true 'lag' time may be considerably more than the average 52 years and the changes to the composition of the forest may not become apparent for 100 - 200 yr.

Table1: Areas changing from temperate deciduous forest into other biomes

		stable	broad-leaved everggreen	warm grass/shrub	xerophyllous woods	cool mixed forest	cool grass/shrub	sum
total area (10³km²)		3041	1577	928	249	184	83	6062
percentage		50.2	26.0	15.3	4.1	3.0	1.4	100
change type		-	succ.	dieb.	dieb.	succ.	dieb.	-
initialisation	mean	-	52	41	18	46	36	-
time	min	-	3	1	18	2	1	-
(yrs)	max	-	98	99	18	99	98	-

The changes shown are those due to a 3×CO$_2$ climate change scenario. Estimated times for the initialisation of these changes are also included. (succ. – successional change; dieb – dieback)

The third largest portion of the area is subject to significantly drier conditions (corresponding to warm grass/shrub or xerophytic woodland biomes), indicating that more dramatic changes could occur to ecosystem structure, such as dieback due to drought accompanied by wildfires of increased frequency and extent. In this case, the time predicted for this process to begin could be much closer to the time of actual change in ecosystem structure. In fact, in some areas the change in structure could actually precede the time at which the previous vegetation is no longer in equilibrium with climate. For example, drying forests, as they approach their climatic limits, could be subject to a sequence of wildfires that hastens their conversion to grasslands or open woodlands. In this change type, disturbance (predominantly fire) will play a major role in controlling the dynamics.

Table 2 illustrates another aspect of possible change in the temperate deciduous forest biome. By coincidence, also approximately half of the area available to this biome under the new climate is already covered by temperate deciduous forests. Inspection of the five types that could change into temperate deciduous forest indicates that the pattern is quite complex, showing successional changes from colder forest types, and invasion events into forest-free areas at both the cold and the warm end of the climatic range. Interestingly, also changes through more than one biome type occur, here indicated by the indirect changes in the table. Again, it should be emphasized that these results give an indication of the biome type that can be expected for a given climate if the system is allowed to come into equilibrium; they do not attempt to represent the pathway by which change in ecosystem structure will actually occur.

Table 2: Areas changing from other biomes into temperate deciduous forest.

		stable	cold mixed forest	cool conifer forest	cool mixed forest	warm grass/shrub	cool grass/shrub	S. taiga	S. cold decid. forest	N. taiga	sum
total area (10³km²)		3041	1917	434	321	168	132	179	139	2	6062
percentage		48.0	30.3	6.9	5.1	2.7	2.1	2.8	2.2	0.0	100
change type		-	succ.	succ.	succ.	inv.	inv.	succ.	succ.	succ.	-
initialisation	mean	-	47	36	42	43	33	-	-	-	-
time	min	-	1	20	1	1	2	-	-	-	-
(yrs)	max	-	99	45	98	99	86	-	-	-	-

The changes shown are due to the same climate change scenario as those in Table 1. Estimated times for initialisation are also included. Note that some changes are indirect, i.e. they would occur with an intermediate step (no time estimates are given in these cases). (succ. – successional change; inv. – invasion)

CONCLUSIONS

This preliminary assessment does not claim comprehensiveness concerning the range of dynamics that might affect the broad-scale changes in vegetation distribution. Rather, it was made to indicate that a range of successional processes will likely be involved in any major change in the structure of the biosphere that occurs over time spans of decades to centuries. We do not know the magnitude and direction of climatic change in every location well enough, and there is considerable uncertainty regarding the resilience of particular ecosystems against such changes. Most importantly, human land use is ignored in this preliminary assessment, although it should be included as a further forcing function at a later stage of model development. The estimations of lag times, however simplistic they may appear in this study, are a crucial aspect of the overall investigation of a potential carbon pulse to the atmosphere. Gap models, once they are developed far enough to capture the competitive behaviour of functional types in a wide range of climates, are an important example of ecosystem dynamics models that can be used to describe the pathways from one vegetation type into another with adequate detail and temporal resolution.

ACKNOWLEDGEMENTS

We thank Harald Bugmann and Alberte Fischer for valuable contributions to the discussions about these topics, and Brian Huntley and Martin Sykes for critical reviews of the first draft of the paper. This research contributes to IGBP-GCTE (Core Research Project No. 2305 'The Potsdam Land-Atmosphere-Interaction Model (PLAI)').

REFERENCES

Bonan GB (1995) Land-Atmosphere interactions for Climate System Models: Coupling Biophysical, Biogeochemical, and Ecosystem Dynamical Processes. Remote Sens Environ 51(1):57-73

Box EO (1981) Macroclimate and Plant Forms: An Introduction to Predictive Modeling in Phytogeography. Dr. W. Junk Publishers, The Hague

Bugmann H (in press) Functional types of trees in temperate and boreal forests: Classification and testing. J Veg Sci

Bugmann H, Fischlin A (1994) Comparing the behaviour of mountainous forest succession models in a changing climate. *in* Beniston M (ed) Mountain Environment in Changing Climates, 204-219. Routledge, London

Claußen M (1994) On coupling global biome models with climate models. Clim Res 4(3):203-221

Cramer W, Leemans R, Hutchinson MF Huntley B (in prep) A new climate data base for terrestrial ecosystem modelling with variable spatial resolution.

Gyalistras D, Von Storch H, Fischlin A Beniston M (1994) Linking GCM generated climate scenarios to ecosystems: case studies of statistical downscaling in the Alps. Clim Res 4(3):167-189

Henderson-Sellers A. McGuffie K (1995) Global climate models and 'dynamic' vegetation changes. Gl Change Biol 1(1):63-75

Lüdeke MKB, Badeck F-W, Otto RD, Häger C, Dönges S, Kindermann J, Würth G, Lang T, Jäkel U, Klaudius A, Ramge P, Habermehl S Kohlmaier GH (1994) The Frankfurt Biosphere Model. A global process oriented model for the seasonal and long-term CO_2 exchange between terrestrial ecosystems and the atmosphere. I. Model description and illustrative results for cold deciduous and boreal forests. Clim Res 4(2):143-166

Neilson RP (1993) Vegetation redistribution: a possible biosphere source of CO_2 during climatic change. Wat Air and Soil Poll 70(1-4):659-673

Noblet, N de (1996) Modelling late-Quaternary paleoclimates and paleobiomes. in Huntley B, Cramer W, Morgan AV, Prentice HC, Allen JRM (eds) Past and future rapid environmental changes: The spatial and evolutionary responses of terrestrial biota, 31-52. Springer-Verlag, Berlin

Parton WJ, Ojima DS, Schimel DS Kittel TGF (1993) Development of simplified ecosystems models for applications in Earth system studies: The Century experience. in Ojima DS (ed) Modeling the Earth System, 281-302. UCAR/Office for Interdisciplinary Earth Studies, Boulder, CO

Perlwitz J (1992) Preliminary results of a global SST anomaly experiment with a T42 GCM. in VII General Assembly of the European Geophysical Society Edinburgh, UK, April 6-10, 1992

Pinter PJ, Kimball BA, Mauney JR, Hendrey GR, Lewin KF, Nagy J (1994) Effects of Free-Air carbon dioxide enrichment on PAR absorption and conversion efficieny by cotton. Agric For Met 70(1-4):209-230

Plöchl M, Cramer W (1995) Coupling global models of vegetation structure and ecosystem processes - An example from Arctic and Boreal ecosystems. Tellus B - Chem Phys Meteorol 47(1/2):240-250

Prentice IC (1993) Climate change - process and production. Nature 363(6426):209-210

Prentice IC, Cramer W, Harrison SP, Leemans R, Monserud RA Solomon AM (1992) A global biome model based on plant physiology and dominance, soil properties and climate. J Biogeogr 19(2):117-134

Prentice IC, Sykes MT Cramer W (1993a) A simulation model for the transient effects of climate change on forest landscapes. Ecol Modelling 65(1-2):51-70

Prentice IC, Sykes MT, Lautenschlager M, Harrison SP, Denissenko O, Bartlein PJ (1993b) Modelling global vegetation patterns and terrestrial carbon storage at the last glacial maximum. Gl Ecol Biogeogr Lett 3(3):67-76

Smith TM, Shugart HH (1993) The transient response of terrestrial carbon storage to a perturbed climate. Nature 361:523-526

Solomon AM, Cramer W (1993) Biospheric implications of global environmental change. in Solomon AM, Shugart HH (ed) Vegetation Dynamics and Global Change, 25-52. Chapman and Hall, New York

Steffen WL, Walker BH, Ingram JS, Koch GW (1992) Global Change and Terrestrial Ecosystems: The Operational Plan. International Geosphere-Biosphere Programme

Woodward FI (1987) Climate and Plant Distribution. Cambridge University Press, Cambridge, UK

The biogeographic consequences of forecast changes in the global environment: Individual species' potential range changes

Martin T. Sykes
Global Systems Group
Department of Ecology
Ekologihuset
S-223 62 Lund
Sweden

INTRODUCTION

Present predictions of the impact of increasing greenhouse gases on climate suggest large changes in high latitudes, in particular winters are likely to become warmer. If these forecast changes do occur as rapidly as predicted then the vegetation of these regions is likely to be in severe disequilibrium with the changing climate. In the past species have responded individualistically (Huntley 1991) to climate change and there is no reason to suppose they would not continue to do so in the future. Many species have migrated continually in response to continuous climatic changes (Huntley & Webb 1989). Different species together in present-day plant assemblages could respond therefore in very different ways to a rapidly changing climate as a particular species-specific climatic limit is either exceeded or is no longer reached. Some species may die out quickly, others may remain longer, opportunities may arise for new species to migrate in, transient groups of species may exist together for short or long periods of time as the climate continues to change. Landscapes that seem to be in some sort of dynamic equilibrium at present may as a result become unstable. Too rapid a climate change coupled with wide scale human-induced changes in land use could lead to delays in new species arriving into suitable areas implying a decline both in diversity and biomass in vulnerable landscapes.

Predicting how a species range may change under a future climate is therefore important in trying to understand how a landscape as a whole may respond. To be

NATO ASI Series, Vol. I 47
Past and Future Rapid Environmental Changes:
The Spatial and Evolutionary Responses of
Terrestrial Biota
Edited by Brian Huntley et al.
© Springer-Verlag Berlin Heidelberg 1997

able to do this in any sensible way it is first important to have tools that accurately describe present-day species distributions. STASH is a simple equilibrium model (Sykes *et al.* 1996) which is used as a tool to simulate such distributions. It describes plant species' ranges in terms of a number of bioclimatic variables which represent distinct physiological limiting mechanisms rather than using simple correlations with some climate variable. Three variables are used to define distributions; these are winter cold tolerance, growing degree days including a chilling requirement recognised as relevant for some species (Murray *et al.* 1989) and a drought index (Sykes *et al.* 1996).

The model is used in this paper to examine some of the general biogeographic consequences of a rapid change in climate focusing, as an example, on the consequences for the landscapes of southern Sweden. A number of modelling exercises in this region point to the area being one where some of the biological effects of climate change may be significant. Sykes and Prentice (1996a) using BIOME 1.1 (Prentice *et al.* 1992; Prentice *et al.* 1993a) examined climate change effects on carbon storage in Sweden by simulating present-day biomes and future biomes. These simulations show biomes tracking northwards under an illustrative future climate and suggest the potential in south and west Sweden for an evergreen warm mixed forest biome. Sykes and Prentice (1995) used the forest gap model FORSKA 2 (Prentice *et al.* 1993b) to suggest the possibility of large reductions in biomass and diversity in western Sweden as susceptible species decline under a climate warming. Such a decline can, however, be influenced by changes in the disturbance regime and availability of new species. Sykes and Prentice (1996b) simulating drier forests in eastern Sweden suggest that reduced disturbance could maintain present-day forest mixes for substantial periods, mainly as old-growth forest. A similar view was expressed by Davis (1989) who suggested that forests under frequent logging regimes will adjust to climatic change more rapidly than those undisturbed in reserves. Sykes and Prentice (1995) also suggest that immigration into drier east coast forests, while important for forest diversity, may not always affect forest biomass if a future climate change also means increased precipitation. In that case some species which are already on-site, and which have wide climatic tolerances remain in the landscape and respond positively to the change by increasing their presence; thus site biomass is maintained at pre-change levels.

STASH predicts however that many common tree species of northern Europe may undergo substantial range shifts under climate change (Sykes & Prentice 1995), and this does have implications for diversity in some landscapes. Diversity decline, however, may be offset by immigration from elsewhere in Europe. Some species from southern or central Europe may potentially be new occupiers of areas well outside their present natural ranges.

In this paper the potential of tree species from outside northern Europe to extend their ranges northwards is examined. Such changes could in the longer term lead to forest compositions that are radically different from present day forests.

Shrubs and herbs also form part of the natural landscape and are likely to be also influenced by changes in climate. STASH can also be used to simulate present and future distributions of such species. As an example, a small number of herbaceous species that have their northern boundaries in southern Sweden (Bengtsson *et al.* 1988) are modelled under both present and future climates. Such boundaries are likely to respond early, expanding northwards as the climate warms.

Results from some of these exercises indicate that future natural groupings in a landscape may well comprise a mix of species that have some different members from present day groupings at all levels, including the tree, shrub and herb layers.

METHODS

Model parameters

The detailed structure of the STASH model, and comparisons of simulated output with natural distributions, have been described elsewhere (Sykes *et al.* 1996). In this paper the model is used to simulate distributions of southern and central European tree species as well as a number of herbaceous species from grasslands and forest herb layers at their northern limit in southern Sweden.

Three climatic variables, winter cold tolerance (mean temperature of the coldest month), growing degree days plus a chilling requirement, and a drought index are used. Minimum winter temperature is critical for many woody species and very low temperatures cause death or damage (Woodward 1987). Absolute minimum temperature is well correlated with mean temperature of the coldest month (Prentice *et al.* 1992).

The number of growing degree days (GDD) as a measure of energetic requirements to complete the life cycle is well known to limit northern distributions of many species (Woodward 1987). However, some species seem to have a low tolerance to late spring frosts, and have been shown to have mechanisms which avoid the possibility of damage by requiring a period of chilling (temperatures $< 5°C$) before budburst can occur. Murray *et al.* (1989) have experimentally measured this requirement for some tree species; among them, *Fagus sylvatica* L. has been shown to have a high requirement for chilling. The north-western limit of *Fagus sylvatica*, for example, does not follow the usual GDD limits, but rather a limit that is influenced by the need to fulfil its chilling requirements.

The Priestley-Taylor coefficient as the ratio of actual to equilibrium evapotranspiration is used as an index of plant-available moisture. In STASH the coefficient is evaluated over the total assimilation period for evergreen trees (temperatures $> -4°C$) and over the growing season for deciduous trees, shrubs and herbs (temperatures $> 5°C$) to give the drought index (Sykes *et al.* 1996).

Climate data

STASH uses long-term mean monthly temperature, precipitation and sunshine (as a proportion of possible sunshine hours) data based on weather station records. These records form an updated version (W. Cramer, unpublished) of the Leemans and Cramer (1991) climate data set which has been interpolated with a surface fitting technique (Hutchinson 1989) to a 10′ European grid ($25° - 70°E$, $35° - 75°N$).

A future 'illustrative' climate scenario ($2 \times CO_2$) was calculated using four general circulation models (GCMs): GFDL (Manabe & Wetherald 1987), OSU (Schlesinger & Zhao 1989), UKMO (Mitchell 1983) and GISS (Hansen *et al.* 1988). They all give in general a qualitatively similar though quantitatively different increase in temperature and precipitation over the selected window. Temperature and precipitation anomalies

---

Figure 1. Simulated potential distributions of *Picea abies*, *Fagus sylvatica* and *Tilia cordata* under the present climate and a mean $2 \times CO_2$ climate as predicted by four atmospheric general circulation models.

Picea abies current climate

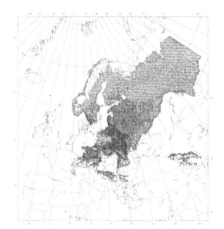

Picea abies mean GCM climate

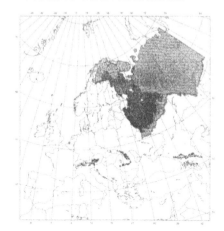

Fagus sylvatica current climate

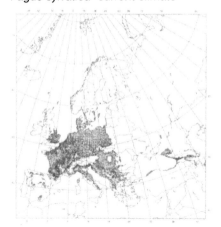

Fagus sylvatica mean GCM climate

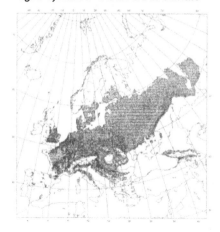

Tilia cordata current climate

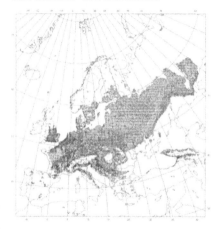

Tilia cordata mean GCM climate

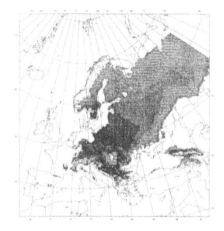

(the difference between simulated $2 \times CO_2$ and control model runs) were calculated and interpolated to each grid cell. The interpolated anomalies were then added to the interpolated current climate data set. Sunshine anomalies were not available so present day sunshine was used. The mean of the four scenarios gave a 'mean GCM' $2 \times CO_2$ scenario for the STASH simulations.

RESULTS

Examples of tree species at present naturally found in southern Sweden

Picea abies (L.) Karsten (Norway spruce): The present boundaries of *P. abies* are simulated well by STASH (Sykes *et al.* 1996), in particular much of the complex north western boundaries are described, as well as its exclusion from parts of the western and northern Norwegian coast and the far south of Sweden. Under the future climate it retreats northwards and eastwards and becomes excluded from most of Sweden except the far north (Figure 1).

Fagus sylvatica L. (beech): The present-day distribution is in general described well by the model (Figure 1). Its north- western limit is controlled by its requirement for chilling and without such a requirement could extend throughout Ireland (M. Sykes, unpublished). A climate change of the magnitude of the mean GCM scenario would extend *Fagus* throughout Sweden well into the boreal zone, although it might be expected to retreat from its present south western Swedish borders in response to an inability to fulfil its chilling requirement.

Tilia cordata Millar (small-leaved lime): In the present climate *Tilia* reaches its northern limit in the southern boreal zone in Sweden as simulated by STASH (Figure 1). Huntley and Birks (1983) suggest that temperature is the critical control on the northern limit of *T. cordata*. Thus, under a warmer climate the northern and eastern expansion predicted by the model would seem possible. At the same time it withdraws from southern and coastal regions of Europe.

Figure 2. Simulated potential distributions of *Abies alba, Acer pseudoplatanus* and *Quercus ilex* under the present climate and a mean $2 \times CO_2$ climate as predicted by four atmospheric general circulation models.

Abies alba current climate

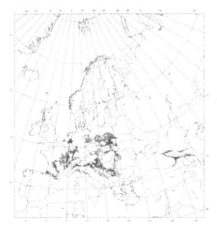

Abies alba mean GCM climate

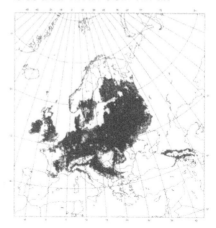

Acer pseudoplatanus current climate

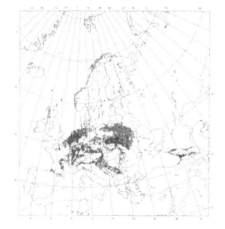

Acer pseudoplatanus mean GCM climate

Quercus ilex current climate

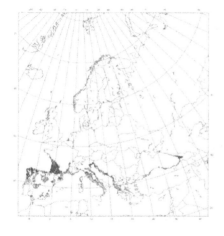

Quercus ilex mean GCM climate

Examples of candidates for establishment in southern Sweden

Abies alba Millar (silver fir): A species of the deciduous–coniferous montane forest zone of central Europe where it co-dominates with *Fagus* (Huntley & Birks 1983). Atlas Florae Europaeae (AFE) describes the present natural distribution of the conifer *Abies* as concentrated in south and central Europe (Jalas & Suominen 1973). It is also planted beyond this range. The model in general simulates the natural distribution well, although it cannot always pick up the distribution in some mountainous areas, e.g. the Alps (Figure 2). This discrepancy may be the result of the model using the grid-cell modal elevation in its calculations whereas the AFE maps record a presence at any elevation in the grid cell (Sykes *et al.* 1996). Under the future climate scenario *A. alba* extends through much of eastern Europe and well into southern and central Sweden (Figure 2).

Acer pseudoplatanus L. (sycamore): The natural range of the deciduous *Acer* is limited to the eastern and central European mainland (Figure 2), it is, however, planted and regenerating beyond this limit. Under a climate warming it could be expected to spread rapidly throughout Sweden (Figure 2).

Quercus ilex L. (evergreen oak): The model describes the present south European natural distribution of this species particularly well (Figure 2). If, as the BIOME 1.1 (Sykes & Prentice 1996a) output suggests, a possible future climate scenario could lead to the potential of an evergreen warm mixed forest, *Quercus ilex* is a possible candidate. It is already planted and naturalising in England (Clapham *et al.* 1987). A climate change of the magnitude described could indeed allow a large range expansion and an extension into eastern and western coastal regions of Sweden.

Potential range changes in some herbaceous species.

Prunella grandiflora (L.) Scholler: A perennial herbaceous species of open grassland which is found at its northern range margin in a few sites in southern Sweden, particularly Öland (Mossberg *et al.* 1992), although it is widely distributed in central and eastern Europe (Hultén & Fries 1986). It is absent from most of north and western Europe including Britain and western parts of France (Figure 3). Under the

---

Figure 3. Simulated potential distributions of *Prunella grandiflora*, *Trifolium striatum* and *Corydalis cava* under the present climate and a mean $2 \times CO_2$ climate as predicted by four atmospheric general circulation models.

Prunella grandiflora current climate

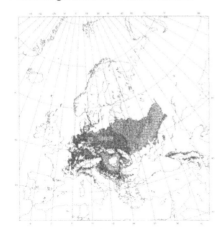

Prunella grandiflora mean GCM climate

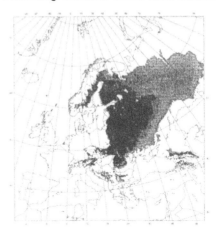

Trifolium striatum current climate

Trifolium striatum mean GCM climate

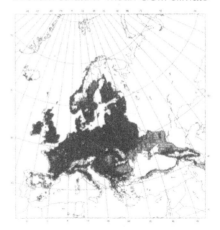

Corydalis cava current climate

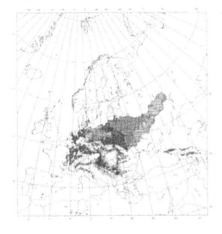

Corydalis cava mean GCM climate

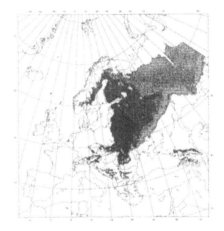

future climate scenario *Prunella* spreads further east and north but retreats from most of western and central Europe. In Sweden it would be restricted to northern and central areas.

Trifolium striatum L.: The simulation of the present distribution of the annual species *T. striatum* corresponds quite well to its natural distribution (Hultén & Fries1986) (Figure 3). It covers most of western Europe and reaches its northern limit in Sweden on Öland where it has a similar distribution to *Prunella*. However under a future climate its boundaries are predicted to be different from *Prunella* (Figure 3). The simulation shows that it could be found throughout western and eastern Europe extending far to the north in Sweden.

Corydalis cava (L.) Koerte: *C. cava* is a species of deciduous forests especially beech forests (Mossberg *et al.* 1992) with a central and eastern distribution in Europe reaching its northern margin in southern Sweden in Skåne (Jalas & Suominen 1991). The simulated distribution in the current climate picks up much of this boundary (Figure 3). Under a future climate *C. cava* could retreat eastwards and into the north of Sweden well beyond its present boundary. Those areas in southern Sweden and northern Germany where it is found at present would become outside its climatic range (Figure 3), even though much of these areas under a future climate would be occupied by beech forests (Figure 1) where it might normally be expected to form part of the herb layer.

DISCUSSION AND CONCLUSIONS

Climate changes continuously and species seem to migrate individualistically in response. Future rapid climate change may have serious consequences for the plant communities of northern Europe. Important tree species of the region such as *Picea abies* are particularly vulnerable to warmer winters. Evidence, for example, from the last 10,000 yr shows *P. abies,* a present-day common co-dominant with *Pinus sylvestris* in many Swedish forests, to have been very mobile in the past, expanding westwards from two eastern populations (Huntley & Webb 1989), while *P. sylvestris* – once common in western Europe – has declined and is now absent from many areas (Bennett 1984; Huntley & Birks 1983). Under a rapid climate change it is likely that both species would retreat northwards and eastwards (Sykes & Prentice 1995).

Such a scenario is dependent both on the rate and degree of climate change, as well as on other factors such as the level of future disturbances, e.g. logging, fire, etc. (Sykes & Prentice 1996b; Davis 1989). In particular present outputs from climate models are by their very nature speculative; not only are the methodologies and simplifications primitive in comparison to the real world, but there is no real indication when and if the present rate of increase in greenhouse gas forcing will be reduced or eliminated. This makes an endpoint difficult to predict and a CO_2 level four times pre-industrial levels may indeed be a possibility. Given our limited knowledge, therefore, on climate–biosphere interactions, the future simulations described here are not intended to be a true prediction of the future, rather they are intended to be illustrative of how species ranges are controlled by climate as well as to show the different sensitivities species have to changing climates. They also provide insights and questions about the mechanisms of range shifts and their effects on regional landscapes with regard to diversity and biomass changes. Issues of the future management of the landscape, particularly in areas where extensive land use has reduced natural vegetation and natural routes for migration, are implicit throughout.

The STASH simulations show that some species already present in an area, e.g. *Fagus sylvatica*, are likely to undergo large range expansions but may themselves eventually also become vulnerable in some parts of their range to a rapid or extended period of climate change (Sykes & Prentice 1995). This is particularly likely to be the case in west Sweden where chilling requirements may become harder to fulfil as winters warm. Other local species such as *Quercus robur* L. and *Q. petraea* (Mattuscka) Liebl. may expand, but dynamic simulations show that at least in some forests there may be a substantial reduction in biomass (Sykes & Prentice 1995). Tree species from outside Scandinavia may be required to complement the remaining species in these forests. STASH shows that some of the tree species of central and southern Europe could grow in these areas under a new climate. They include *Quercus ilex, Abies alba, Castanea sativa* etc. which in the simulations expand their ranges at least to coastal Sweden and in some cases much further. These results have implications for the vegetation of Europe and the migratory rates of species. Davis and Zabinski (1992) quote examples of migration rates ranging from 10 to 200 km per century, but point out that many future scenarios predict climates where these rates of migration are substantially less than what would be required, even if suitable routes were available. Northward natural migration of species from southern and central Europe may be possible, but given the heavily modified landscape and distances in excess of 2000 km required, may not occur fast enough.

Management, which has already changed much of the landscape in Europe is possibly the most reliable way of introducing new, suitable species, in climatic terms, into new areas. Davis and Zabinski (1992) suggest plantations of suitable trees wherever they can grow. The STASH simulations, however, warn against too hasty selections under scenarios of continuous climate change. Many potential tree species are already planted beyond their range as garden specimens. Some of these species could spread rapidly out of cultivation, a prime example is *Acer pseudoplatanus* which seems to be able to rapidly naturalise in the right conditions. Such rapid migration could be a good indicator for climate change.

The simulations suggest that other species such as herbs may undergo extensive range shifts, and the response time of these species may be more rapid than for tree species, given their more frequent generations. Although some of their present limits seem to be similar, the model runs show that future range boundaries could be different as species respond individualistically to changes. So present-day plant assemblages break up and new assemblages form elsewhere with different species mixes. The model shows, for example, that while beech forest may extend in future throughout Sweden, some of the herbs (e.g. *Corydalis cava*) from the beech forest assemblage have different distributional ranges which may not always be congruent with a future beech forest.

STASH is an equilibrium model with quantitative aspects (Sykes *et al.* 1996) and as such has no vegetation dynamics; rather it was designed to simulate regional biogeography, something which is not yet possible in a dynamic way. It is, however, likely that a mix of species that is successful under a new climate will be dependent both on the biogeography of the species involved and on the outcome of competition both among local species but also between them and exotic species. This argues for the development of regional scale dynamic models that could examine such problems. Within the development of such models the questions of migration, management and changing disturbance rates would need to be addressed.

REFERENCES

Bengtsson K, Prentice HC, Rosén E, Moberg R, Sjögren E (1988) The dry alvar grasslands of Öland: ecological amplitudes of plants species in relation to vegetation composition. Acta Phytogeogr Suec 76:21-46

Bennett KD (1984) The post-glacial history of *Pinus sylvestris* in the British Isles. Quat Sci Rev 3:133-155

Clapham AR, Tutin TG, Moore DM (1987) Flora of the British Isles. Cambridge University Press, Cambridge

Davis MB (1989) Lags in vegetation response to greenhouse warming. Clim Change 15:75-82

Davis MB, Zabinski C (1992) Changes in geographical range resulting from greenhouse warming: Effects on biodiversity in forests. *in* Peters RL, Lovejoy TE (eds) Global warming and biological diversity, 297-308. Yale University Press, Yale

Hansen J, Fung I, Lacis A, Rind D, Russell G, Lebedeff S, Ruedy R (1988) Global climate changes as forecast by the GISS-3-D model. J Geophys Res 93:9341-9364

Hultén E, Fries M (1986) Atlas of North European vascular plants north of the Tropic of Cancer. Three volumes. Koelyz, Kvnigstein, Germany

Huntley B (1991) How plants respond to climate change: migration rates, individualism and the consequences for plant communities. Ann Bot 67:15-22

Huntley B, Birks HJB (1983) An atlas of past and present pollen maps for Europe: 0-13 000 years ago. Cambridge University Press, Cambridge

Huntley B, Webb T III (1989) Migration: species responses to climatic variations caused by changes in the earth's orbit. J Biogeog 16:5-19

Hutchinson MF (1989) A new objective method for spatial interpolation of meteorological variables from irregular networks applied to the estimation of monthly mean solar radiation, temperature, precipitation and windrun. Need for climatic and hydrologic data in agriculture in southeast Asia 89/5:95-104. CSIRO, Canberra, Australia

Jalas J, Suominen J (eds) (1973) Atlas Florae Europaeae: Volume 2. Gymnospermae (Pinaceae to Ephedraceae). Societas Biologica Fennica Vanamo, Helsinki.

Jalas J, Suominen J (eds) (1991) Atlas Florae Europaeae: Volume 9. Paeoniaceae to Capparaceae, Societas Biologica Fennica Vanamo, Helsinki.

Leemans R, Cramer W (1991) The IIASA Climate Database for Monthly Mean Values of Temperature, Precipitation and Cloudiness on a Terrestrial Grid. RR-91-18 Laxenburg IIASA

Manabe S, Wetherald RT (1987) Large scale changes in soil wetness induced by an increase in carbon dioxide. J Atmos Sci 44:1211-1235

Mitchell JFB (1983) The seasonal response of a general circulation model to changes in CO_2 and sea temperature. Quart J Roy Met Soc B 109:113-152

Mossberg B, Stenberg L, Ericsson S (1992) Den Nordiska Floran. Wahlström and Widstrand, Stockholm

Murray MB, Cannell MGR, Smith I (1989) Date of budburst of fifteen tree species in Britain following climatic warming. J Appl Ecol 26:693-700

Prentice IC, Cramer W, Harrison SP, Leemans R, Monserud R, Solomon AM (1992) A global biome model based on plant physiology and dominance, soil properties and climate. J Biogeog 19:117-134

Prentice IC, Sykes MT, Lautenschlager M, Harrison SP, Denissenko O, Bartlein PJ (1993a) Modelling global vegetation patterns and terrestrial carbon storage at the last glacial maximum. Global Ecol Biogeogr Lett 3:67-76

Prentice IC, Sykes MT, Cramer W (1993b) A simulation model for the transient effects of climatic change on forest landscapes. Ecol Modelling 65:51-70

Schlesinger ME, Zhao ZC (1989) Seasonal climatic changes induced by doubled CO_2 as simulated by the OSU atmospheric GCM/mixed-layer ocean model. J Clim 2:459-495

Sykes MT, Prentice IC (1995) Boreal forest futures: Modelling the controls on tree species range limits and transient responses to climate change. Water, Air, Soil Pollution 82:415-428

Sykes MT, Prentice IC (1996a) Carbon storage and climate change in Swedish forests: A comparison of static and dynamic modelling approaches. in Apps MJ, Price DT (eds) The role of global forests ecosystems and forest resource management in the global carbon cycle, 69-78. NATO ASI Series vol I 40 Springer-Verlag, Berlin

Sykes MT, Prentice IC (1996b) Climate change, tree species distributions and forest dynamics: a case study in the mixed conifers/northern hardwoods zone of northern Europe. Clim Change (in press)

Sykes MT, Prentice IC, Cramer W (1996). A bioclimatic model for the potential distribution of northern European tree species under present and future climates. J Biogeogr (in press)

Woodward FI (1987) Climate and plant distribution. Cambridge University Press, Cambridge

Gap models, forest dynamics and the response of vegetation to climate change

Harald Bugmann[1]
Systems Ecology
Institute of Terrestrial Ecology
Swiss Federal Institute of Technology Zürich (ETHZ)
Grabenstrasse 3
CH-8952 Schlieren
Switzerland

INTRODUCTION

Mathematical models of successional processes in forests that are based on the concept of gap dynamics (Watt 1947) are among the most prominent tools to test ecological hypotheses on long-term forest dynamics (Botkin *et al.* 1972; Shugart 1984). Moreover, these models have also gained an important role for studying a wide range of applied environmental problems such as growth enhancement through increasing atmospheric CO_2 content (e.g. Shugart & Emanuel 1985), air pollution (e.g. Kercher & Axelrod 1984) and climatic change (e.g. Solomon 1986).

A large number of studies using gap models have been conducted to evaluate the consequences of future climatic change on the carbon storage and species composition of forests (for a review, see Bugmann 1994). Most of these studies were based on two fundamental assumptions:

1. It was assumed that there is an equilibrium between the present climate and the present vegetation structure. Technically speaking, this means that the models were run to their equilibrium species composition under current climate, starting

---

[1] Present address: Department of Global Change and Natural Systems, Potsdam Institute for Climate Impact Research, P.O. Box 60 12 03, D-14412 Potsdam, Germany

NATO ASI Series, Vol. I 47
Past and Future Rapid Environmental Changes:
The Spatial and Evolutionary Responses of
Terrestrial Biota
Edited by Brian Huntley et al.
© Springer-Verlag Berlin Heidelberg 1997

from bare soil. A number of palaeoecological modelling studies of the distant past have been conducted (e.g. Solomon *et al.* 1980; Lotter & Kienast 1992), but there is only one study that dealt with forest dynamics as influenced by the climatic variations of the past few centuries (Campbell & McAndrews 1993). These authors assumed a linear decrease of the annual mean temperature from the year 1200 to 1850 AD to mimic Little Ice Age cooling and to assess the impact of this climate signal on the simulated species composition of a forest in southern Ontario. Apparently, a more accurate reconstruction of the climatic input data required by their model was not possible. In a unique effort, Pfister (1988) reconstructed the time series of monthly temperature and precipitation data for the last 450 years in Switzerland. The first aim of the present study is to use this set of data to evaluate the impact of recent climatic variations on forest dynamics in Europe as compared to projected future impacts of climate.

2. It was assumed that climate would change linearly between the current and a hypothesized future constant climate. Although this assumption is unrealistic and one has to expect that the exact time course of climatic change influences the simulated forest dynamics, this issue has never been evaluated quantitatively. Thus, the second aim of the present study is to compare the influence of various assumptions on *how* climate could change on the simulated species composition of forests.

By addressing these two issues using the forest gap model ForClim (Bugmann 1994), it will be possible to evaluate the sensitivity to climatic parameters of successional processes in forests, thus complementing the theoretical study by Davis and Botkin (1985) and the assessment for southern Ontario by Campbell and McAndrews (1993). Conclusions will be drawn concerning the importance of migration vs. evolution as responses to future climatic change, and it will be defined more precisely how important it is to obtain detailed scenarios on the transient behaviour of climate.

MATERIALS AND METHODS

The forest model FORCLIM

The FORCLIM model (Bugmann 1994; Fischlin *et al.* 1995) is a so-called "gap" model (Botkin *et al.* 1972; Shugart 1984), i.e. the establishment, growth and mortality of trees are simulated on small patches of land ($100 - 1000$ m^2) as a mixture of deterministic and stochastic processes. FORCLIM was scrutinized to incorporate reliable yet simple formulations of climatic influences on ecological processes so as to produce plausible results when applied along climate gradients. Moreover, the model was designed to include only a minimum number of ecological assumptions.

FORCLIM consists of three modular submodels, each of which can be run independently, or combined: (1) FORCLIM-E, a submodel for the abiotic environment containing a stochastic weather generator and a bucket model of soil moisture balance with a monthly time step; (2) FORCLIM-S, a submodel for soil carbon and nitrogen turnover modified from Pastor and Post (1985); (3) FORCLIM-P, a submodel for tree population dynamics that is based on the gap dynamics hypothesis (Watt 1947, Shugart 1984). The model structure, all equations and parameter values of the model version 2·4, which was used here, were described by Bugmann (1994).

The behaviour of FORCLIM was tested systematically in central Europe (Bugmann 1994) and in eastern North America (Bugmann & Solomon 1995); it was found to reproduce known tree species composition of unmanaged stands for a wide range of environmental conditions on both continents.

Effects of historical climate anomalies on simulated forest dynamics

In a unique effort, Pfister (1988) developed a system of monthly thermic and hydric indices to characterize the temperature and precipitation regime of every month between 1525 and 1979 AD in Switzerland. The indices were based on a wealth of historical data sources, ranging from temperature measurements at a few sites and written records of extreme events (e.g. lake glaciations) to agricultural yield data and tree-rings. The temperature indices refer to the site Basel, while the precipitation indices are an average of the sites Bern, Cottens/Begnins, Rickenbach, Basel, Geneva and Zürich (Pfister 1988).

Based on these indices and the regression equations developed by Pfister (1988), the time series of monthly temperature and precipitation data from 1525 to 1979 AD were reconstructed for the present study, thus defining a virtual test site that was named "CLIMINDEX" after the name of the database.

The following simulation experiment was designed based on the CLIMINDEX data. First, the model was run to steady-state under current climatic conditions from 525-1525 AD by simulating 200 patches, each starting from bare ground (Bugmann & Fischlin 1992; Bugmann *et al.* 1996). For this part of the experiment, the weather data were sampled stochastically from the long-term distributions of the CLIMINDEX variables of the years 1901 – 1960, which are more accurate than an earlier period (such as 1525 – 1585) and where no values are missing. Then the simulation was continued through the years 1525 – 1979 using the reconstructed series of weather data from Pfister (1988) to calculate the bioclimatic variables in FORCLIM-E deterministically. The temperature and precipitation of months with missing values in CLIMINDEX were assumed not to deviate from the long-term statistics. The site-specific value of the soil water holding capacity was set to 30 cm, and nitrogen availability to 100 kg ha^{-1} (cf. Bugmann 1994).

Sensitivity of *FORCLIM* to assumptions on the course of climatic change

Most of the previous impact assessments using forest gap models adopted a linear change of climatic parameters over time (e.g. Solomon 1986; Pastor & Post 1988; Kienast 1991). On the other hand, it is a common practice in systems theory to explore the response of a system to a step change in the input data, and this approach has been used as well (e.g. Bugmann & Fischlin 1994; Fischlin *et al.* 1995). In reality, however, climatic change will follow neither of these assumptions, and a more gradual, e.g. sigmoid, change would be more likely to occur. Thus, these three types of climatic changes were used to evaluate the sensitivity of gap models to changing mean climatic parameters. The distributions of the variables were assumed not to deviate from the current values.

To define a future constant climate, the regionalized scenarios of climatic change from Gyalistras *et al.* (1994) for the European test sites Bever, Davos and Bern were selected. The three sites encompass widely differing climatic conditions from cold-adapted coniferous forests close to the alpine timberline to low-elevation mixed deciduous forests on the Swiss Plateau. The methodology developed by Gyalistras *et al.* (1994) relates continental-scale temperature and pressure anomalies to local weather anomalies by means of principal component analysis and canonical correlation analysis (so-called statistical downscaling). The regionalized climate data for each site were used to define an equilibrium climate scenario for the year 2100 (Table 1). For details of this derivation see Bugmann (1994).

Table 1: Current climate and regionalized projections of climatic change for the year 2100.

Site	Elevation [m]	T_{annual} [°C]	P_{annual} [mm/yr]	ΔT_{Summer} [°C]	ΔP_{Summer} [mm/month]	ΔT_{Winter} [°C]	ΔP_{Winter} [mm/month]
Bern	570	8.4	1006	+2.64	+39.8	+3.76	+31.3
Davos	1590	3.0	1007	+3.28	+9.1	+3.00	+21.4
Bever	1712	1.5	841	+4.16	+38.2	+1.48	+25.4

Projected climatic change values obtained from Gyalistras *et al.* (1994) by extrapolating the linear downscaled trends from an uncorrected 100-year (1986-2085) transient run of the ECHAM GCM for the IPCC "Business-As-Usual" scenario A (Cubasch *et al.* 1992). (T – mean temperature, P – precipitation sum)

At each site, FORCLIM was allowed to reach the steady-state species composition under current climate during the first 800 simulation years, starting from bare soil and calculating the average dynamics on 200 patches (Bugmann & Fischlin 1992; Bugmann *et al.* 1996). Then the transient climatic change was applied for 100 years, and after the year 900 the future climate was assumed to be constant until the end of the simulation in the year 1500. For the step scenario, the climatic change was assumed to take place in the simulation year 800.

RESULTS AND DISCUSSION

Effects of historical climate anomalies on simulated forest dynamics

The simulation results from the first phase of the experiment (Figure 1, *top*) are typical of low-elevation mixed deciduous forests of the Swiss Plateau and large areas of central Europe. The simulated forest is characterized by a strong dominance of beech *Fagus silvatica* accompanied by silver fir *Abies alba* and oak *Quercus* spp. (plant nomenclature follows Hess *et al.* 1980). Due to the comparably high temperature, spruce *Picea excelsa* is outcompeted by those species. These results appear to be plausible (Ellenberg & Klötzli 1972).

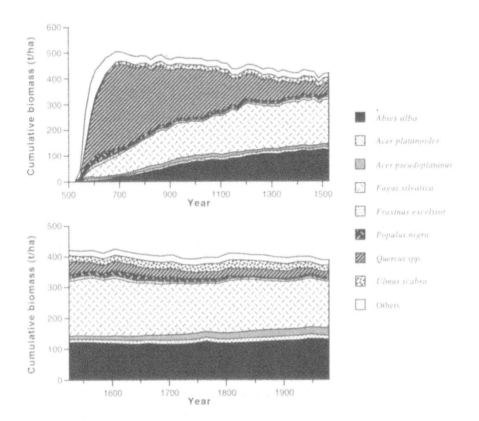

Figure 1: Simulation results at the site "CLIMINDEX". *Top*: Years 525-1525 AD, sampling the weather stochastically from the long-term statistics. The simulation starts with a bare patch in the year 525. *Bottom*: Years 1525 to 1979 AD when using the reconstructed weather data from Pfister (1988) to drive the FORCLIM-E model.

Of the three bioclimatic indices used in FORCLIM, the simulated winter temperature during the period from 1525 to 1979 was always above –5 °C; thus its ecological significance in the model is negligible. Similarly, the simulated evapotranspiration deficit is mostly below 10%, with a few exceptions where stronger drought occurred, the strongest being almost 30% in the "mediterranean" year 1540 (cf. Pfister 1988). The third variable, the annual sum of degree-days (Figure 2), thus could have the largest effect on the simulated forest dynamics. Most of the periods outlined by

Pfister (1988) as mid-term climatic variations are evident from Figure 2, such as the warm period from 1530 – 1564, the maximum of the Little Ice Age from 1688 – 1701, the rapid warming from 1702 – 1730, and the cool phase from 1812 – 1860. These variations had strong effects on agricultural yield (Pfister 1988) – what was their effect on the characteristics of near-natural forests, such as species composition and aboveground biomass?

The simulated forest dynamics from 1525 – 1979 (Figure 1, *bottom*) do not show any relationship to the climatic variations of Figure 2. There is almost no variability of the simulated species composition, and the variability of above ground biomass is due to the stochastic formulation of tree establishment and mortality in FORCLIM, not to the changing abiotic environment.

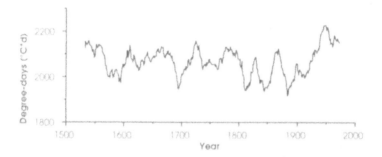

Figure 2: Simulated annual sum of degree-days (1525-1979 AD) based on the monthly temperature indices from Pfister (1988). The graph shows moving averages over 15 years.

We may conclude that the species composition simulated at the CLIMINDEX site is well buffered against climatic variations of the duration and magnitude that occurred during the last 450 years, corroborating the findings by Davis and Botkin (1985). From an evolutionary point of view, these results are plausible. Trees typically have lifespans of several centuries; given the fact that climatic variations like the ones reconstructed by Pfister (1988) occur on the timescale of decades, trees must be capable of surviving such anomalies, otherwise they could not grow to adult size and would not be able to reproduce.

Campbell and McAndrews (1993) found that Little Ice Age cooling had a strong impact on the species composition of mixed forests in southern Ontario, which appears to contradict the findings from the present FoRClIM experiment. Two major factors are responsible for the diverging results. First, Campbell and McAndrews (1993) assumed a linear cooling of 2°C over 650 years, which corresponds to a decrease of the annual sum of degree-days of *ca.* 400°C·d in southern Ontario. The ClIMINDEX database, however, does not suggest that climatic changes of this magnitude or length have occurred in central Europe (cf. Figure 2). Second, the forest studied by Campbell and MacAndrews (1993, p. 338) is of "ecotonal nature", whereas the low-elevation site "ClIMINDEX" is typical of locations like Bern (cf. Table 1) where forests are neither "ecotonal" nor strongly limited by temperature (Bugmann 1994).

Hence, the inertia to climatic variations of the simulated species composition probably is characteristic of real forests of the Swiss Plateau, but the study by Campbell and McAndrews (1993) suggests that these findings can not readily be generalized to other areas. For example, under conditions of strong environmental stress, such as close to the alpine or the dry timberline, it is conceivable that climatic variations on the timescale of decades may influence competition, e.g. by inducing decline phenomena. Further studies would be required to address this issue.

Sensitivity of FoRClIM to assumptions on the course of transient climatic change

The results of the transient simulation based on scenarios of step, linear, and sigmoid climatic change show only minor differences at any site. The results from Davos (Figure 3) are used here as a typical example. At Bever, slightly larger differences occur, whereas at Bern the behaviour of FoRClIM across the three scenarios is virtually identical.

At the site Davos, a larch *Larix decidua* – Norway spruce *Picea excelsa* forest is simulated under current climate (years 0 – 800) where spruce is dominating in the late successional phase (Figure 3, *top left*). This corresponds favourably to the description of the Larici-Piceetum according to Ellenberg and Klötzli (1972), and this community is also typical of the Davos area. Under the new climate, *Fagus silvatica* and *Abies alba* invade at the expense of *P. excelsa*, i.e. a montane beech-spruce-fir forest is formed (Figure 3, *top left*), as it is typical of lower elevations under current climate.

When we focus on the time window from the years 700 – 1300 (Figure 3, other panels), it becomes evident that the time course of climatic change over 100 years has an almost negligible influence on the simulated above-ground biomass and

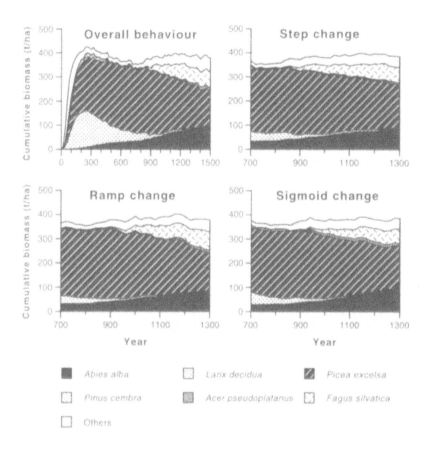

Figure 3: Effect of various assumptions about the nature of transient climatic change between the simulation years 800 and 900 on the transient behaviour of FORCLIM-E/P at the site Davos. *Top left*: Behaviour of the model under the step change scenario for the full simulation experiment. Excerpts from the simulation results between the years 700 and 1300 are shown for the step change scenario (*top right*), the ramp change scenario (*bottom left*) and the sigmoid change scenario (*bottom right*).

species composition. Thus, for a climatic change of the magnitude projected by Gyalistras *et al.* (1994) taking place during the relatively short time of one century, assumptions about how the climate changes are not crucial for gap models because such a climatic change proceeds very fast compared to the successional dynamics. We may conclude that the assumptions about the level of a hypothesized future *constant* climate are much more important than how the mean climate changes towards these new values (Figure 3).

CONCLUSIONS

In the present study, the gap model FORCLIM was used to reconstruct the dynamics of a mixed deciduous forest in central Europe through the last 450 years, and the sensitivity of the model to various assumptions on how the climate might change towards a new equilibrium was tested.

The reconstruction of past forest dynamics showed that the climatic variations as compiled by Pfister (1988) did not have an effect on the simulated above-ground biomass nor on the species composition. This result is plausible from an evolutionary point of view because it is vital for long-lived trees to be buffered against climatic variations on the time scale of decades if they are to reach sexual maturity and reproduce. These findings have a number of implications:

1. Simulation studies suggest that the species composition of forests equilibriates within 600 – 800 years (e.g. Bugmann & Fischlin 1994), starting from bare ground. Although we are still lacking an exact reconstruction of the pre-1500 climate, there is no evidence that large climatic changes took place between 1200 and 1500 AD in Europe. Hence we may surmise that today's near-natural forests of central Europe are in equilibrium with climate.

2. Many forest ecosystems appear to be buffered well against short-term climatic variations. This and the long generation time of most forest trees makes it unlikely that the population dynamics of trees will be affected strongly by evolutionary changes within the next 100 – 200 years. Rather, it is likely that the main response to climatic change of trees will occur through migration.

The comparison of linear, step and sigmoid climatic changes during 100 years showed that the choice of the transient scenario does not affect the simulated forest characteristics strongly because the change of the abiotic conditions in all three cases proceeds fast compared to the successional dynamics. The following are worth while noting:

1. The inertia of the simulated forests implies long lags until the effects of climatic change may become visible. For example, at the site Davos the simulated species composition changed only gradually, and the first deviations were not detectable until about 100 – 150 years after the onset of climatic change. Hence, current forests may appear to be healthy but, for example, may lose their regenerative capability as a consequence of climatic change.

2. Some of the simulated forests show a strong response to climatic change, and most species currently growing there could not grow or reproduce any more under the changed climate. Given that these changes may occur within less than a century, it is unlikely that the speed of migration of many tree species would be sufficient to reach favourable habitats (e.g. Davis 1989).

3. If climatic change should continue for several centuries, i.e. when the time scale of climatic change approaches the time scale of forest succession (Bugmann & Fischlin 1994), the impact of the various scenarios of transient climatic change certainly would differ increasingly. It should be kept in mind that there is no evidence that climatic change would come to a halt by the end of the next century, as assumed in the present study. Moreover, the findings of this study may hold neither for changes of the variance of climatic parameters, i.e. the distribution of extreme events such as frosts, which may have catastrophic consequences for forests, nor for very strong short-term fluctuations of climatic parameters as they occured e.g. at the end of the last glaciation (Mayewski *et al.* 1993; Alley *et al.* 1993), or during the last interglacial period (GRIP Members 1993).

ACKNOWLEDGEMENTS

This study is based on the author's doctoral dissertation submitted to the Swiss Federal Institute of Technology Zürich (ETHZ). The advice by Andreas Fischlin and Hannes Flühler of the Institute of Terrestrial Ecology ETHZ as well as the financial support of ETHZ are gratefully acknowledged. Warm thanks are due to Christian

Pfister (University of Bern, Switzerland) for the access to the CLIMINDEX database and fruitful discussions about its applicability in ecological research Two reviewers provided constructive comments on the article.

REFERENCES

Alley RB, Meese DA, Shuman CA, Gow AJ, Taylor KC, Grootes PM, White JWC, Ram M, Waddington ED, Mayewski PA, Zielinski GA (1993) Abrupt increase in Greenland snow accumulation at the end of the Younger Dryas event. Nature 362:527-529

Botkin DB, Janak JF, Wallis JR (1972) Some ecological consequences of a computer model of forest growth. J Ecol 60:849-872

Bugmann H (1994) On the ecology of mountainous forests in a changing climate: A simulation study. Ph.D. thesis no. 10638, Swiss Federal Institute of Technology Zurich, Switzerland

Bugmann H, Fischlin A (1992) Ecological processes in forest gap models – analysis and improvement. in Teller A Mathy P Jeffers JNR (eds) Responses of forest ecosystems to environmental changes, 953-954. Elsevier Applied Science, London & New York

Bugmann H, Fischlin A (1994) Comparing the behaviour of mountainous forest succession models in a changing climate. In: Beniston, M. (ed.), Mountain environments in changing climates. Routledge, London, 204-219.

Bugmann HKM, Solomon AM (1995) The use of a European forest model in North America: A study of ecosystem response to climate gradients. J Biogeogr 22:477-484

Bugmann H, Fischlin A, Kienast F (1996) Model convergence and state variable update in forest gap models. Ecol Modelling (in press)

Campbell ID, McAndrews JH (1993) Forest disequilibrium caused by rapid Little Ice Age cooling. Nature 366:336-338

Cubasch U, Hasselmann K, Höck H, Maier-Reimer E, Mikolajewicz U, Santer B Sausen R (1992) Time-dependent greenhouse warming computations with a coupled ocean-atmosphere model. Climate Dynamics 8:55-69

Davis MB (1989) Lags in vegetation response to greenhouse warming. Clim Change 15:75-82

Davis MB, Botkin DB (1985) Sensitivity of cool-temperate forests and their fossil pollen record to rapid temperature change. Quat Res 23:327-340

Ellenberg H, Klötzli F (1972) Waldgesellschaften und Waldstandorte der Schweiz. Eidg Anst Forstl Versuchswes Mitt 48:587-930

Fischlin A, Bugmann H, Gyalistras D (1995) Sensitivity of a forest ecosystem model to climate parametrization schemes. Environ Pollut 87:267-282

GRIP (Greenland Ice-core Project) Members (1993) Climate instability during the last interglacial period recorded in the GRIP ice core. Nature 364:203-207

Gyalistras D, Storch H von, Fischlin A, Beniston M (1994) Linking GCM generated climate scenarios to ecosystems: Case studies of statistical downscaling in the Alps. Clim Res 4:167-189

Hess HE, Landolt E, Hirzel R (1980) Flora der Schweiz. Birkhäuser, Basel & Stuttgart, 4 Vol., 2nd ed

Kercher JR, Axelrod MC (1984) Analysis of SILVA: a model for forecasting the effects of SO_2 pollution and fire on western coniferous forests. Ecol Modelling 23:165-184

Kienast F (1991) Simulated effects of increasing CO_2 on the successional characteristics of Alpine forest ecosystems. Landscape Ecology 5:225-238

Lotter A, Kienast F (1992) Validation of a forest succession model by means of annually laminated sediments. in Saarnisto M. & Kahra A (eds) Proceedings of the INQUA workshop on laminated sediments, June 4-6, 1990, Lammi, Finland. Geological Survey of Finland, Special paper series 14:25-31

Mayewski PA, Meeker LD, Whitlow S, Twickler MS, Morrison MC, Alley RB, Bloomfield P, Taylor K (1993) The atmosphere during the Younger Dryas. Science 261:195-197

Pastor J, Post WM (1985) Development of a linked forest productivity-soil process model. U.S. Dept of Energy ORNL/TM-9519

Pastor J, Post WM (1988) Response of northern forests to CO_2-induced climate change. Nature 334:55-58

Pfister C (1988) Klimageschichte der Schweiz 1525-1860: Das Klima der Schweiz von 1525-1860 und seine Bedeutung in der Geschichte von Bevölkerung und Landwirtschaft. Haupt, Bern

Shugart HH (1984) A theory of forest dynamics. The ecological implications of forest succession models. Springer, New York

Shugart HH, Emanuel WR (1985) Carbon dioxide increase: the implications at the ecosystem level. Plant, Cell and Environment, 8:381-386

Solomon AM (1986) Transient response of forests to CO_2-induced climate change: simulation modeling experiments in eastern North America. Oecologia 68:567-579

Solomon AM, Delcourt HR, West DC, Blasing TJ (1980) Testing a simulation model for reconstruction of prehistoric forest-stand dynamics. Quat Res 14:275-293

Watt AS (1947) Pattern and process in the plant community. J Ecol 35:1-22

Natural migration rates of trees: Global terrestrial carbon cycle implications

Allen M. Solomon

National Health and Environmental Effects Research Laboratory

Environmental Protection Agency

200 SW 35th Street

Corvallis, OR 97333

U.S.A.

INTRODUCTION

Migration of populations or species of trees ('tree migration') in response to climate change is of interest both to palaeoecologists who assess past vegetational responses to climate change, and to global ecologists concerned with future climate change induced by increasing greenhouse gases (GHGs). A major difference between climate-driven tree migrations in prehistory and those expected in the future is the high speed of the latter climate change. The 4–6 km which temperate-zone July isotherms are predicted to move northward annually (Solomon et al. 1984) are about an order of magnitude more rapid than prehistoric rates deduced from palaeoecological evidence. Assuming prehistoric rates of warming matched the rate of tree migration (T Webb 1986; Prentice et al. 1991), fossil pollen data allow inference of 400 m yr^{-1} (Davis 1983) to 800 m yr^{-1} (Gear & Huntley 1991) of climate change and tree migration at most. The rate may be even slower if tree migration includes the establishment and maturity of the tree population (Bennett 1986) as well as the processes of seed transport, establishment, growth and seed production, normally defined as migration (e.g. Davis 1989; MacDonald et al. 1993).

The difference in definition is important for predicting the amount of carbon (CO_2 is the most important of the GHGs) that will reside in the atmosphere in the future. The oceans provide the ultimate long-term control on atmospheric carbon concentrations (e.g. Sundquist 1985; Prentice et al. 1993). However, the terrestrial biosphere modulates the shorter-term changes in carbon content, measured over a few

NATO ASI Series, Vol. I 47
Past and Future Rapid Environmental Changes:
The Spatial and Evolutionary Responses of
Terrestrial Biota
Edited by Brian Huntley et al.
© Springer-Verlag Berlin Heidelberg 1997

decades or centuries (Gammon *et al.* 1985; Keeling *et al.* 1995; Denning *et al.* 1995). Forests store about 2/3 of above-ground terrestrial organic carbon and over half of the carbon present in the world's soils (Dixon *et al.* 1994). The presence of a few trees on the landscape (e.g. MacDonald *et al.* 1993), indicated by establishment and reproductive maturity of seed trees, contributes little carbon to terrestrial stocks. Instead, closed-canopy stands of mature mixed or pure species provide the dense carbon stocks of interest. These are associated with mature, stable populations.

Projections of global terrestrial carbon cycle dynamics under warmer climates of a doubled GHG concentration have used static vegetation models (Prentice & Solomon 1990). These projections hinge on the critical assumption that the migration of trees and the formation of mature, stable populations at new locations proceeds at the same rate as the climate change to which it is responding (Sedjo & Solomon 1989; Leemans 1989; Prentice & Fung 1990; Smith *et al.* 1992a, b; Smith & Shugart 1993a, b; Solomon *et al.* 1993). To date, these static model exercises have projected increased global terrestrial carbon storage under future warming, because large new land areas suitable for forest growth are created either by warming of high latitude treeless tundra, or by increased hydrologic cycle intensity in treeless steppe.

Yet, unchanged or decreased rather than increased carbon storage may result if forests cannot migrate and establish in the time required to attain the doubled GHG benchmark. The objective of the current paper is to estimate the time required for forests to develop in regions new to them, to estimate the time required for forests to die out where they become climatically obsolete, then to calculate the impacts of those times on future terrestrial carbon stocks.

MIGRATION RATE LIMITS

For our purposes, three time-variable steps may be distinguished in the migration process: Seed transport, coupled with establishment of seedlings; reproductive maturation of individuals; and, forest maturation i.e. maturation of populations in closed canopy forests.

Seed Transport

Tree seed transport, whether by wind, animals or running water, requires little more than a day. Establishment of tree seeds (seed germination and growth of a taproot into mineral soil) requires one to two growing seasons. Taken together, transport and establishment are instantaneous compared with the multiple decades required for warming. However, transport and establisment is not a singular process but rather, must be repeated multiple times, each consisting of several time-variable processes.

Some shifting populations possess sharp boundaries which resemble a slowly moving wave (Davis 1987). These depend upon the regular transport of seeds 10 - 100 m, followed by a few decades to centuries of tree maturity before another 10 - 100 m 'step' is taken. Other population boundaries consist of 'infection sites,' located well beyond the margin of the main population, surrounded by population voids (Davis 1987, 1990), and derived from irregular transport events. The latter pattern results from migration comprised of rare long-distance transport and establishment events, and is followed by local population growth via transport and establishment between founder seed sources (Bennett 1984; Davis 1987). This migration form probably produces the most rapid migration rates (Leishman *et al.* 1992; Collingham *et al.* 1996). Multiple rare events, by definition, form a (long) time-ordered process. Jumps of 100 – 200 km have been detected in the Holocene at about 1000 year intervals (SL Webb 1986; Davis *et al.* 1986).

Tree Maturation

The time required to complete tree life cycles varies considerably. Most evergreen and deciduous conifers can reproduce within 10 years of seed germination and deciduous hardwoods within 15 years (Harlow *et al.* 1979). Yet, these times apply to trees growing in full sunlight (i.e. without other trees nearby), and to reproduction by trees still the size of saplings (i.e. 2 – 5 m tall). Annual seed production in this case is very small (1000s instead of 100,000s of seeds per tree; Burns & Honkala 1990), and transport distances may be quite small because of the low stature of the seed sources (e.g. < 50 m for trees < 5 m tall, according to experiments by Greene & Johnson 1989). The short life cycling times occur when tree populations invade treeless areas and form clumped seed sources (with low carbon densities), rather than when populations expand to form areas of continuous forest between seed sources, or grow rapidly within stands already populated by mature forests.

Forest Maturation

The development of scattered individuals and stands of trees into closed canopy forests containing maximum carbon densities takes considerably longer than maturation of isolated trees.

- First, the areas between trees and stands must be occupied. If 10 – 15 year-old trees in each generation produce seedlings 50 – 100 m away, even a founder population of one tree km^{-2} (not a rare-event distribution but one calculated by Greene & Johnson (1995) as rare enough to constrain metapopulation expansion) requires 5 – 10 generations to plant seeds in all parts of the square kilometre. The sequential completion of 5 or 10 generations could require 50 – 100 years in conifers and 75 – 150 years in deciduous hardwoods from dense populations, and a millennium from sparse tree densities (*ibid.*). The minimum number of life cycles obviously depends on the distances between seed sources, but would usually exceed the 5 – 10 generations exemplified at distances beyond about 2 – 10 km from the original forests (*ibid.*). This is consistent with estimates of mid-Holocene hemlock occupation of northern lower Michigan in about 500 years (Davis *et al.* 1986) and documentation of periods exceeding 1000 years for local forest development during the forestation of the British Isles (Bennett 1986).

- Second, closed forests must develop. Tree growth rates are much slower in the shade than in the open. In closed forests, trees must grow to reach the canopy surface before producing significant numbers of seeds (Daubenmire, 1959; Waring & Schlesinger, 1985). This requires a minimum of 25-50 years in most temperate and boreal regions in which closed canopy forests exist (Harlow *et al.* 1979). Certain shade-tolerant species may require 150 (*Acer saccharum*, Canham 1988) to 450 years (*Tsuga canadensis*, Godman & Lancaster 1990) to reach the canopy and complete the cycle. Even the 25 – 50 years for life cycle completion in closed canopy forests presumes optimum rates of height and diameter growth. Yet, warming is likely to slow growth rates, as it has in the past (e.g. Fritts 1976; Briffa *et al.* 1995).

In sum, the minimum time required for development of mature forests by tree species that originate elsewhere is between 100 years (conifers that spread from initial seed sources in an arbitrary five 10-year generations and grow to canopy height in 25 years) and 200 years (deciduous hardwoods that spread from initial seed sources in ten 15-year generations and grow to canopy height in 50 years). This 'transient response' to climate changes is consistent with forest maturity rates measured in the past and simulated by mechanistic gap models (e.g. Solomon 1986; Bugmann 1993). The latter models exclude the time needed for development of isolated tree seed sources but include crude effects of chronic climate change on tree growth. In reality, the rate will probably be much slower, perhaps approaching the > 1000 yr period inferred by Bennett (1986, 1988) in prehistoric data describing forest initiation and maturation.

Forest Tree Mortality

The other half of the migration question for calculating carbon stocks is the concomitant mortality of trees which have become climatically and fatally 'obsolete' (i.e. after climate variables exceed their climate tolerances). Stress induced by warming and drought may directly slow growth of individual trees, and thereby induce mortality, usually in less than a decade (Nichols 1968; Franklin *et al.* 1987; Auclair 1992), especially among seedlings and saplings (Peet & Christiansen 1987). However, climate stresses more commonly predispose trees to succumb to other mortality agents (Waring 1987), such as air pollutants (Hinrichsen 1987), wildfire (Payette 1992), insect infestations (Holling 1992) and windthrow (Webb 1989). Such direct and indirect stress-induced mortality can be viewed as chronic decline over several decades if it is measured at the scale of a large region, where new individuals, populations and stands are dying each year (Mueller-Dombois 1992).

The amount of time required to extirpate populations from regions in which they have become climatically obsolete is exceedingly complex to predict. The fatal obsolescence itself is a non-linear time-transgressive property of the rate and spatial distribution of climate change. Trees growing in areas undergoing the most rapid climate changes, growing near limits of their distribution and growing in stressed habitats may die in only a few years, while others may just begin to sense stress by the time GHGs have doubled. Following mortality, release of carbon from dead trees may require additional decades. Harmon *et al.* (1986) cite 50% volume loss times of 14 – 172 years for log mineralization of softwoods under temperate climates inducing slow decomposition, and 2 – 24 years for hardwoods.

Although seedlings and saplings should disappear quickly, the loss of mature trees which store most of the carbon is more relevant. Increased, climate-induced mortality of mature trees, and carbon losses to mortality greatly exceeding carbon gains from new growth (Kirschbaum & Fischlin 1996), is expected to generate a future pulse of increased atmospheric CO_2 (Solomon 1986; King & Neilson 1992; Smith & Shugart 1993a,b). Epidemics of pathogens and insects, mortality agents through which climate change may act, have required about 50 years in eastern North America to kill most individuals of American chestnut (Odum 1969) and American elm (Gibbs 1978) in the 20th century and eastern hemlock in mid-Holocene time (Davis 1981). Based on the foregoing, I assumed that trees would be extirpated from areas of fatal obsolescence in the 60-70 years required to reach climate of doubled GHGs.

MIGRATION RATES APPLIED

The minimum of 100 to 200 years required to develop forests composed of new species in a given region following the imposition of a changed climate can be compared to the expected time climate will take to change. Frequently, the time needed to impose the climate of doubled concentrations of CO_2 or of all GHGs (e.g. Houghton *et al.* 1995) is used, currently expected to occur in 60 to 70 years or by about the year 2050 (Greco *et al.* 1994). At that rate, forest migration would be very incomplete at best, even if the forest development began when climate change began, rather than following sometime after initiation of the climate change.

This lag is incompatible with the assumption of instant migration and forest maturity incorporated by the static models discussed in the introduction, above. Consider that the static vegetation models utilize a half-degree latitude and longitude grid (one degree for Prentice & Fung 1990), with a latitudinal resolution of about 55 km at the equator. Measurement of any tree migration which sequesters carbon by the 70 year doubling of GHGs is highly unlikely, if not impossible, even at the normal 150-400 m yr^{-1} migration rate Davis (1983) estimated from fossil pollen data (producing tree migration of only 11 – 28 km during the 70 yr). A more realistic assumption is that no forest migration occurs in the 70-year time span needed to reach a doubling of CO_2 or GHGs, although migration would occur eventually. Belotelov *et al.* (1996) applied this model condition to the Holdridge Life Zone System to define the absolute range of carbon values possible for specified future climate scenarios in the former Soviet Union.

Solomon and Kirilenko (1996) modified the Biome 1.1 model (Prentice *et al.* 1992, 1993) to reflect this 'delayed migration' condition. They assumed that tree functional types (TFTs) which were incompatible with doubled GHG climates would disappear during the 60 or 70 years from areas acceptable under the initial climate, but that TFTs could not appear in areas in which they were absent under the initial climate. Non-arboreal plant functional types (shrubs, grasses; NAFTs) were modeled as migrating instantly, on the assumption that they are able to produce seed as quickly as their first or second growing seasons, greatly reducing lags in response to rapid climate change.

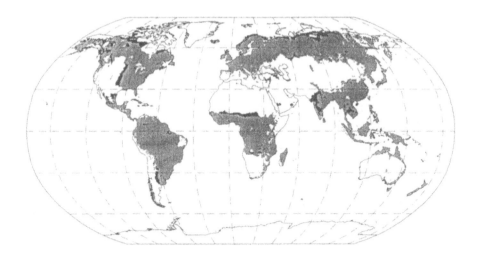

Figure 1. Global distribution of land occupied by forest, projected by the BIOME 1.1 model (Prentice *et al.* 1993) under the UKTR climate scenario (Murphy & Mitchell 1996). Areas in black show differences between assuming instant migration and absent migration and areas in grey are forested under either instant or absent migration.

Figure 1 illustrates a Biome 1.1 model run using climate output from the UKTR coupled ocean–atmosphere GCM for climate of current and of doubled GHG concentration (Solomon & Kirilenko 1996). Note that differences between instant migration and no migration are present at all latitudes. Reduced temperature constraints on tree growth in high latitudes permit occupation of latitudinal bands of non-arboreal tundra by boreal forests under instant migration assumptions. Similarly, increased soil moisture and instant tree migration produces forest occupation of longitudinally-oriented steppe and savanna areas by dry temperate forests in North and South America and by tropical forests in Africa, India and Asia. In part the patterns reflect the great difference in sensitivity of temperature to GHGs at high latitudes compared to that at low latitudes which undergo little warming (Greco *et al* 1995).

RESULTS AND DISCUSSION

The paper is written to estimate the time required to form closed canopy, mature forests by immigrating species and the time required to eliminate trees which have become fatally obsolete. The objective is to define the implications of this lagged climate response to the global carbon cycle. Prentice *et al.* (1993) have transformed biome areas into carbon stocks by appying carbon density estimates for above-ground and below-ground carbon.

Table 1 compares terrestrial carbon storage under future climate with migration versus no-migration assumptions. Like other static models, this exercise with instant migration generated more terrestrial carbon under doubled GHG climate than under modern climate. Simulated tropical forests invade large areas of grassland and savanna due to the positive moisture balance this particular GCM projects for the tropics. The difference between modern and future carbon with instant migration is considerably less than others have projected (Prentice & Fung 1990; Leemans 1989; Smith *et al.* 1992a, b; Smith & Shugart 1993a, b; Solomon *et al.* 1993). The UKTR coupled ocean–atmosphere GCM is considerably less GHG-sensitive than those used in the past. In contrast to carbon increases associated with instant migration, the amount of carbon stored in terrestrial vegetation and soils declines under GHG-induced climate when one assumes that trees do not migrate (Table 1). All forests decline in biomass and do so both above (40 Pg) and below ground (40 Pg). Non-arboreal biomass that replaced forest biomass (9 Pg above ground, 30 Pg below

Table 1. Above-ground (AG)[1], below-ground (BG)[2] and total (TOTAL) biomass in Petagrams (Pg) under modern and doubled GHG climate, with and without tree migration

| | MODERN CLIMATE[3] | | | DOUBLED GHG CLIMATE[4] | | | | | |
| | | | | WITH MIGRATION | | | NO MIGRATION | | |
BIOMES	AG	BG	TOTAL	AG	BG	TOTAL	AG	BG	TOTAL
Boreal Forests	170	300	470	179	298	477	165	292	457
Temperate Forests	169	360	529	174	343	517	147	336	483
Tropical Forests	300	289	589	346	330	676	287	281	568
Boreal Non-forest	11	177	188	8	127	134	12	151	163
Temperate Non-forest	22	189	211	25	191	216	33	261	294
Tropical Non-forest	52	144	196	51	132	183	49	128	177
Forest Biomass	639	949	1588	699	971	1670	599	909	1508
Total Biomass	723	1459	2177	783	1420	2203	693	1449	2142

[1] from table converting area to biomass in Prentice *et al.* (1993), based on Olson *et al.* (1983).
[2] from table converting area to biomass in Prentice *et al.* (1993), based on Zinke *et al.* (1984).
[3] from data of Leemans and Cramer (1991).
[4] from temperature and precipitation differences between $2 \times CO_2$ and $1 \times CO_2$ of Murphy and Mitchell (1996) applied to data of Leemans and Cramer (1991).

ground) does not make up for the forest carbon losses. Despite the moderate future climate scenario used, and hence the moderate amount of forest response projected, the difference in forest biomass between the two assumptions is significant. The above-ground forest biomass difference is 100 Pg or about 16% of the initial 639 Pg, and the total forest biomass difference is 162 Pg, or 10% of the initial 1583 Pg.

The foregoing values, for vegetation composition responses to climate change alone, exceed the calculated amount of carbon sequestered in terrestrial vegetation from carbon fertilization, as hypothesized from laboratory experiments (Melillo *et al.* 1993; Schimel *et al.* 1995), or calculated by difference from ocean uptake (Denning *et al.*

1995). Although the global difference for all biomes is considerably less (61 Pg), it is the forests, especially those at high latitudes, which are suspected of increasing storage of the carbon not accounted for in ocean–atmosphere models. Clearly, where this view depends on the future distribution of forests (e.g. Melillo *et al.* 1993; VEMAP Participants 1995), it needs to be re-examined. In any case, the amount of carbon tabulated under the traditional instant-migration assumption should be recalculated based on a concept of imperceptibly slow forest immigration response to climate change in the 21st century.

ACKNOWLEDGEMENTS

I thank Joe Alcamo who inadvertently convinced me to examine the ecological implications of the concept, and Andrew Kirilenko, who first suggested to me the idea of non-migration of trees under changed climate, and who modified the Biome 1.1 model to run without tree migration. Harold Mueller provided the figure. Critical reviews by Wolfgang Cramer, David Hollinger, George King, Andrew Kirilenko, William Leak, Dale Solomon and an anonymous reviewer improved the paper considerably. This paper was subjected to peer and administrative review by the U.S. Environmental Protection Agency and was approved by the Agency for publication.

REFERENCES

Auclair AND (1992) Extreme climatic fluctuations as a cause of forest dieback in the Pacific Rim. Air Water and Soil Pollut 66:207-229

Bennett KD (1984) The post-glacial history of *Pinus sylvestris* in the British Isles. Quat Sci Rev 3:133-155

Bennett KD (1986) The rate of spread and population increase of forest trees during the postglacial. Phil Trans R Soc London B134:523-531

Bennett KD (1988) Modeling changes in beech populations: A reply to Dexter *et al.* (1987). Rev Palaeobot Palynol 56:361-364

Belotelov NV, Bogatyrev BG, Kirilenko AP, Venevsky SV (1996) Modelling time-dependent biome shifts under global climate changes. Ecological Modelling (in press)

Briffa KR, Jones PD, Schweingruber FH, Shiyatov SG, Cook ER (1995) Unusual twentieth-century summer warmth in a 1,000-year temperature record from Siberia. Nature 376:156-159

Bugmann HKM (1993) On the ecology of mountainous forests in a changing climate: A simulation study. Ph.D. Dissertation, Swiss Federal Institute of Technology, Zürich

Burns RM, Honkala BH (eds) (1990) Silvics of North America. Volume 1, Conifers; Volume 2, Hardwoods. Agr Hndbk 654. USDA Forest Service, Washington DC

Canham CD (1988) Growth and canopy architecture of shade-tolerant trees: response to canopy gaps. Ecology 69:786-795

Collingham YC, Hill MO, Huntley B (1996) The migration of sessile organisms: a simulation model with measureable parameters. J Veg Sci (in press)

Daubenmire RF (1959) Plants and environment. Wiley, New York

Davis MB (1981) Outbreaks of forest pathogens in Quaternary history. Proc IV Int Palynol Conf, Lucknow (1976-1977) 3:216-227

Davis MB (1983) Quaternary history of deciduous forests of eastern North America and Europe. Ann Mo Bot Gard 70:550-563

Davis MB (1987) Invasions of forest communities during the Holocene: Beech and hemlock in the Great Lakes Region. in Gran AJ, Crawley MJ, Edward PH (eds) Colonization and stability, 373-393. Blackwell Scientific Publications, Oxford

Davis MB (1989) Lags in vegetation response to greenhouse warming. Clim Change 15:75-82

Davis MB (1990) Climatic change and the survival of forest species. in Woodwell GM (ed) The earth in transition: Patterns and processes of biotic impoverishment, 99-110. Cambridge Univ Press, Cambridge

Davis MB, Woods KD, Webb SL, Futyma R (1986) Dispersal versus climate: Expansion of Fagus and Tsuga into the Upper Great Lakes region. Vegetatio 67:93-103

Denning AS, Fung IY, Randall D (1995). Latitudinal gradient of atmospheric CO_2 due to seasonal exchange with land biota. Nature 376:240-243

Dixon RK, Brown S, Houghton RA, Solomon AM, Trexler MC, Wisneiwski J (1994) Carbon pools and flux of global forest ecosystems. Science 263:185-190

Franklin JF, Shugart HH, Harmon ME (1987) Tree death as an ecological process. BioScience 37:550-556

Fritts HA (1976) Tree rings and climate. Academic Press, New York

Gammon RH, Sundquist ET, Fraser PJ (1985) History of carbon dioxide in the atmosphere in Trabalka JR (ed) Atmospheric carbon dioxide and the global carbon cycle, 25-62. DOE/ER-0239, US Dept. of Energy, Washington DC

Gear AJ, Huntley B (1991) Rapid changes in the range limits of scots pine 4000 years ago. Science 251:544-547

Godman RM, Lancaster K (1990) Tsuga canadensis (L) Carr., Eastern Hemlock. in Burns RM, Honkala BH (eds) (1990a) Silvics of North America. Volume 1, Conifers, 604-612. Agr Hndbk 654, USDA Forest Service, Washington D.C.

Greco S, Moss RH, Viner D, Jenne R (1995) Climate Scenarios and Socioeconomic Projections for IPCC WG II Assessment. Material assembled for Lead Authors by IPCC WG II TSU, Washington, DC

Greene DF, Johnson EA (1989) A model of wind dispersal of winged or plumed seeds. Ecology 70:339-347

Greene DF, Johnson EA (1995) Long-distance wind dispersal of tree seeds. Can J Bot 73:1036-1045

Gibbs JN (1978) Intercontinental epidemiology of Dutch elm disease. Ann Rev Phytopath 16:287-307

Harlow WM, Harrar ES, White FM (1979) Textbook of dendrology, 6th edn. McGraw-Hill, New York

Harmon ME, Franklin JF, Swanson FJ, Sollins P, Gregory SV, Lattin JD, Anderson NH, Cline SO, Aumen NG, Sedell JR, Lienkaemper GW, Cromack K, and Cummins KW (1986) Ecology of coarse woody debris in temperate ecosystems. Adv Ecol Res15:133-302

Hinrichsen D (1987) The forest decline enigma. BioScience 37:542-546

Holling CS (1992) The role of forest insects in structuring the boreal landscape. *in* Shugart HH, Leemans R, Bonan GB (eds) A Systems Analysis of the Global Boreal Forest, 170-191. Cambridge University Press, Cambridge

Houghton JT, Meira-Filho LG, Bruce J, Lee H-S, Callander BA, Haites E, Harris N, Maskell K (eds) (1995) Climate change 1994: Radiative forcing of climate change and an evaluation of the IPCC IS92 emission scenarios. Cambridge Univ. Press, Cambridge

Keeling CD, Whorf TP, Wahlen M, van der Plicht J (1995) Interannual extremes in the rate of rise of atmospheric carbon dioxide since 1980. Nature 375:666-670

King GA, Neilson RP (1992) The transient response of vegetation to climate change: A potential source of CO_2 to the atmosphere. Water, Air and Soil Pollut 64:365-383

Kirschbaum M, Fischlin A, Cannell MGR, Cruz RVO, Galinski W, Cramer WP *et al.* (1996) The impacts of climate change on forest ecosystems. *in* Climate Change 1995. Impacts, adaptations and mitigation of climate change: Scientific – Technical analyses. IPCC WG II Second Assessment Report, Chapter 1. Cambridge University Press, Cambridge

Leemans R (1989) Possible changes in natural vegetation patterns due to a global warming. *in* Hackl A (ed) Der Treibhauseffekt: Das Problem-Mögliche Folgen-Erforderliche Massnahmen, 105-122. Akadamie für Umwelt und Energie, Laxenburg, Austria

Leemans R, Cramer WP (1991) The IIASA database for mean monthly values of temperature, precipitation and cloudiness on a global terrestrial grid. Research Report RR91-18. Internat Inst for Applied Sys Anal, Laxenburg, Austria

Leishman MR, Hughes L, French K, Armstrong D, Westoby M (1992) Seed and seedling biology in relation to modeling vegetation dynamics under global climate change. Aust J Bot 40:599-613

MacDonald GM, Edwards TWD, Moser KA, Pienitz R, Smol JP (1993) Rapid response of treeline vegetation and lakes to past climate warming. Nature 361:243-246

Melillo JR, McGuire AD, Kicklighter DW, Moore B, Vorosmarty CJ, Schloss AL (1993) Global climate change and terrestrial net primary production. Nature 363:234-239

Mitchell JFB, Johns TC, Gregory JM, Tett SFB (1995) Climate response to increasing levels of greenhouse gases and sulphate aerosols. Nature 376:501-504

Mueller-Dombois D (1992) Potential effects of the increase in carbon dioxide and climate change on the dynamics of vegetation. Water Air and Soil Pollut 64:61-79

Murphy JM, Mitchell JFB (1996) Transient response of the Hadley Centre coupled ocean–atmosphere model to increasing carbon dioxide. Part II: Spatial and temporal structure of the response. Jour of Climate (in press)

Nichols JO (1968) Oak mortality in Pennsylvania: A ten-year study. J For 66:681-694

Odum EP (1969) Fundamentals of Ecology. WB Saunders Co, Philadelphia

Olson JS, Watts JA, Allison LJ (1983) Carbon in Live Vegetation of Major World Ecosystems. ORNL/TM-5862, Oak Ridge National Laboratory, Oak Ridge

Payette S (1992) Fire as a controlling process in the North American boreal forest. *in* Shugart HH, Leemans R, Bonan GB (eds) A Systems Analysis of the Global Boreal Forest, 144-169. Cambridge Univ Press, Cambridge

Peet RK, Christiansen NL (1987) Competition and tree death. BioScience 37:586-595

Prentice K, Fung IY (1990) The sensitivity of terrestrial carbon storage to climate change. Nature 346:48-50

Prentice IC, Solomon AM (1990) Vegetation models and global change. *in* Bradley RS (ed) Global Changes of the Past, 365-383. OIES, UCAR, Boulder

Prentice IC, Bartlein PJ, Webb T III (1991) Vegetation change in eastern North America since the last glacial maximum: A response to continuous climatic forcing. Ecology 72:2038-2056

Prentice IC, Cramer WP, Harrison SP, Leemans R, Monserud RA, Solomon AM (1992) A global biome model based on plant physiology and dominance, soil properties and climate. J Biogeogr 19:117-134

Prentice IC, Sykes MT, Lautenschlager M, Harrison SP, Denissenko O, Bartlein PJ (1993) Modelling global vegetation patterns and terrestrial carbon storage at the last glacial maximum. Global Ecol and Biogeogr Let 3:67-76

Schimel D, Enting IG, Heimann M, Wigley TML, Raynaud D, Alves D, Siegenthaler U (1995) CO_2 and the Carbon Cycle. *in* Houghton JT, Meira-Filho LG, Bruce J, Lee H-S, Callander BA, Haites E, Harris N, Maskell K (eds) Climate change 1994: Radiative Forcing of Climate Change and an Evaluation of the IPCC IS92 Emission Scenarios, 39-71. Cambridge Univ Press, Cambridge

Sedjo RA, Solomon AM (1989) Climate and forests. *in* Rosenberg NJ, Easterling WE, Crosson PR, Darmstadter J (eds) Greenhouse Warming: Abatement and Adaptation, 105-119. Resources for the Future, Washington

Smith TM, Leemans R, Shugart HH (1992a) Sensitivity of terrestrial carbon storage to CO_2-induced climate change: Comparison of four scenarios based on general circulation models. Clim Change 21:367-384

Smith TM, Shugart HH, Bonan GB, Smith JB (1992b) Modeling potential response of vegetation to global climate change. Adv Ecol Res 22:93-116

Smith TM, Shugart HH (1993a) The transient response of terrestrial carbon storage to a perturbed climate. Nature 361:523-526

Smith TM, Shugart HH (1993b) The potential response of global terrestrial carbon storage to a climate change. Water Air and Soil Pollut 70:629-642

Solomon AM (1986) Transient response of forests to CO_2-induced climate change: Simulation experiments in eastern North America. Oecologia 68:567-79

Solomon AM, Kirilenko AP (1996) Simplifying assumptions in modeling terrestrial carbon stocks under changing climate: What if trees do not migrate? (ms. submitted)

Solomon AM, Tharp ML, West DC, Taylor GE, Webb JW, Trimble JL (1984) Response of Unmanaged Forests to Carbon Dioxide-Induced Climate Change: Available Information, Initial Tests, and Data Requirements. TR-009, United States Department of Energy Washington

Solomon AM, Prentice IC, Leemans R, Cramer WP (1993) The interaction of climate and land use in future terrestrial carbon storage and release. Water Air and Soil Pollut 70:595-614

Sundquist ET (1985) Geologic analogs: Their value and limitations in carbon dioxide research. *in* Trabalka JR, Reichle DE (eds) The Changing Carbon Cycle: A Global Analysis, 371-402. Springer-Verlag, New York

VEMAP Participants (1995) Comparing biogeography and biogeochemistry models in a continental-scale study of terrestrial ecosystem responses to climate change and CO_2 doubling. Global Biogeochem Cyc 9:407-437

Waring RH (1987) Characteristics of trees predisposed to die. BioScience 37:569-574

Waring RH, Schlesinger WH (1985) Forest Ecosystems: Concepts and Management. Academic Press New York

Webb SL (1986) Potential role of passenger pigeons and other vertebrates in the rapid Holocene migrations of nut trees. Quat Res 26:367-375

Webb SL (1989) Contrasting windstorm consequences in two forests, Itasca State Park, Minnesota. Ecology 70:1167-1180

Webb T III (1986) Is vegetation in equilibrium with climate? How to interpret late-Quaternary pollen data. Vegetatio 67:75-91

Zinke PJ, Stangenberger AG, Post WM, Emanuel WR, Olson JS (1984) Worldwide organic soil carbon and nitrogen data. ORNL TM-8857, Oak Ridge National Laboratory, Oak Ridge

Seasonal features of global net primary productivity models for the terrestrial biosphere

Alberte Fischer[1][†]
CESBIO
18 av. E.Belin, bpi 2801
F-31055 Toulouse Cedex
France

INTRODUCTION

The prediction of future environmental changes, in particular the rapid ones, requires the development of coupled global vegetation-climate models (cf. Noblet 1996). Important components of this modelling effort are, for example, the so-called 'soil-vegetation atmosphere transfer schemes' (SVAT) models for the sensitivity of the climate to the vegetation, the 'biome' models for the description of the vegetation structure in relation to (equilibrium) climate, or the global 'net primary productivity' (NPP) models for the biogeochemical fluxes (energy, H_2O, CO_2) between terrestrial vegetation and the atmosphere. The goal of the NPP models is to help understand the photosynthetically active component in the global carbon balance. This is necessary for the assessment of the biospheric responses to climate change, and their feedbacks to climate.

Global NPP models are recent developments, and one of the major difficulties is the lack of direct validation data. Intercomparisons between different models can provide

---

[1] current address: Dept. for Global Change and Natural Systems, Potsdam Institute of Climate Impact Research, P.O.Box 60 12 03, D-14412 Potsdam, Germany

[†] This work is a partial summary of the efforts of the modelling teams that contributed to the Potsdam NPP Model Intercomparison (for names, see acknowledgements section).

NATO ASI Series, Vol. I 47
Past and Future Rapid Environmental Changes:
The Spatial and Evolutionary Responses of
Terrestrial Biota
Edited by Brian Huntley et al.
© Springer-Verlag Berlin Heidelberg 1997

useful indirect information. IGBP-GAIM/DIS[‡] has initiated a series of workshops for the intercomparison of such global NPP models. At the second workshop, in June 1995 in Potsdam (Germany), monthly outputs from 16 models using standardised input data for climate and soil variables were analysed. From the wealth of outputs, we here focus on NPP, and in the third section on some key intermediate variables: the fraction of the photosynthetically active radiation (FPAR) that the vegetation can absorb, and the leaf area index (LAI). The models may be considered to fall into three classes: *statistical* (empirical relationship between NPP and the major driving variables), *diagnostic* (models rely on the satellite-based estimation of some important vegetation characteristics) and *mechanistic* (sometimes called *prognostic* – the relevant mechanisms for the NPP simulation are described using process-based relationships). These various models and their global NPP results are described in the first section of this contribution. *Seasonal* features are very important for the feedback to the climate, and also for the simulation of the competition mechanisms between species. Therefore the seasonal behaviour of NPP is analysed for two regions with different major climatic gradients. The second section deals with the intertropical band in Africa where NPP is mainly dependent on precipitation. The last section analyses the seasonal NPP in relation to radiative (FPAR) and structural (LAI) variables along a temperature gradient in Europe.

GLOBAL MODELS OF NPP

Statistical models

Precipitation and temperature have been recognised as major limiting climatic variables for the absorption of the photosynthetically active radiation (PAR) and its conversion into dry matter (photosynthates), i.e. primary productivity. The pioneer NPP model was the MIAMI model of Lieth (1975), which related the annual NPP to the annual averages of temperature and precipitation through an empirical regression. Because of its simplicity and its empirical basis, this model is still used as a baseline for evaluation while more sophisticated mechanistic models are

---

[‡] International Geosphere-Biosphere Programme, Task Force 'Global Analysis, Interpretation and Modelling' and Core Project 'Data and Information Systems'.

developed. The most up-to-date version of this class of model is the HRBM (Esser *et al.* 1994). It redistributes the estimated annual NPP monthly according to the soil water balance – it also accounts for land use and soil fertility. The version used here uses a potential vegetation map generated by BIOME (Prentice *et al.* 1992), based on climate.

Diagnostic models

The appearance of global data sets from satellites like the NOAA/AVHRR in the late 70's has provided new opportunities for the global monitoring of the temporal variation of terrestrial ecosystems. The FPAR is usually derived from such data. Following its annual course gives an indication of the timing and length of the growing period. The light use efficiency (LUE) relationship from Kumar and Monteith (1981) determines the NPP as the sum over the growing season of the PAR absorbed by the canopy and then converted into dry-matter. Applied globally, the diagnostic models use estimates of LUE for the various biomes linked to the environmental conditions. Because they are driven by satellite observations, they are expected to provide a simple representation of the seasonal dynamics of one component of the biospheric CO_2 exchanges, the other one being the heterotrophic respiration. Heimann and Keeling (1989) first used this approach to estimate biospheric trace gas fluxes in an atmospheric transport model.

The diagnostic models considered here are CASA (Potter *et al.* 1993), GLO-PEM (Prince & Goward in press), SDBM (Knorr & Heimann 1995) and TURC (Ruimy *et al.* in press). The temporal resolution is determined by the satellite data. For GLO-PEM and TURC, the LUE concept is applied to estimate gross primary production, the autotrophic respiration being computed separately. The constant value of LUE for TURC and the maximum one for CASA are empirically determined. In the latter case, the environmental factors are parameterized depending on climate, soil and vegetation characteristics. GLO-PEM mechanistically estimates the maximum LUE and derives the environmental factors from satellite. SDBM uses a simple and uniform equation for NPP, which is calibrated using CO_2 measurements and an inverted transport model providing constraints for the biospheric fluxes. The spatio-temporal CO_2 features determined by such diagnostic models can be used to validate the more mechanistic models.

Mechanistic models

The mechanistic models simulate the physiological processes which determine the fluxes of water, CO_2 and nutrients between the different components of the vegetation, the soil and the atmosphere, as well as the size of the various pools. The major processes are photosynthesis, growth and maintenance respiration, evapotranspiration, uptake and release of nitrogen, allocation of photosynthates to the various parts, litter production and decomposition, and phenological development. Depending on the model, some particular processes are simulated in great detail with short time steps (typically one hour for the photosynthesis, one day for the growth of the various organs). Others remain quite empirical (e.g. the phenological development). We roughly classify the mechanistic NPP models into the following categories:

1. *Mechanistic estimation of the seasonal biogeochemical fluxes for a prescribed vegetation structure.* CENTURY (Parton *et al.* 1993) and TEM (Raich *et al.* 1991) simulate the fluxes between the different C and N pools at weekly or monthly time steps. TEM was the first global ecosystem model, using NPP measurements in different sites for its calibration. Carbon fluxes are parameterized according to soil properties, soil moisture and some phenological indicators, but the canopy structure (for example the LAI) is not considered explicitly. These models use vegetation maps derived from standard compilations, (e.g. Matthews 1983).

2. *Mechanistic simulation of the fluxes for a prescribed vegetation structure, and with a prescribed seasonal behaviour of the canopy.* Such models, BIOME-BGC (Running & Hunt 1993), KGBM (Kergoat manuscript), and SIB2 (Sellers *et al.* manuscript), focus on detailed process-based representation of the biogeochemical fluxes. At the global scale, the phenology can be provided by satellite data using different rules. BIOME-BGC considers water resources for the estimation of the maximum LAI, and a parameterization of processes depending on the vegetation structure related to biome distribution. The optimal LAI is estimated by KGBM depending on hydric constraints, during an active period determined from satellite data for deciduous ecosystems. The ecophysiological processes are parameterized for each class of a simplified Matthews' map. Both work at a daily time step without calibration. SIB2 simulates land surface processes in detail and with short time steps for implementation in GCMs (at

coarse spatial resolution), with critical parameters supplied from satellite data, including roughness length, albedo, FPAR and LAI.

3. *Mechanistic estimation of seasonal canopy behaviour (LAI) and fluxes for a prescribed vegetation structure.* These models, CARAIB (Warnant *et al.* 1994), FBM (Lüdeke *et al.* 1994), PLAI (Plöchl & Cramer 1995) and SILVAN (Kaduk & Heimann 1996), simulate fluxes and allocation (therefore phenology) at hourly/daily time steps. Plant characteristics and key – parameters used in the process representations are ecosystem dependent; an adaptation of the Matthews' vegetation map for FBM, of that of Wilson and Henderson-Sellers (1985) for CARAIB, and the potential biomes estimated by BIOME 1.0 for PLAI and SILVAN. For FBM, PLAI and SILVAN, calibration is performed within each biome so that the components of the carbon balance are fitted to representative values from the literature values. CARAIB uses a calibration of productivity for the 5 vegetation types co-occurring within each grid.

4. *Coupled models of vegetation structure and trace gas, water and energy fluxes.* These models: BIOME3 (Haxeltine & Prentice manuscript) and HYBRID (Friend *et al.* in press), are designed to simultaneously simulate processes (fluxes) and pattern (vegetation type and structure). The determination of the vegetation types follows some rules of process optimisation (maximisation of the NPP according to the soil-climate conditions, or maximisation of the LAI to satisfy the annual moisture and carbon balances), or is the result of the competition for resources. Therefore, these models do not calibrate the biogeochemical fluxes, and the values of the parameters used in process representation are estimated from the literature. Fluxes are simulated at hourly/daily time steps. Such models appear to be the most adequate candidates to be used in a predictive mode for the response of vegetation to climate change, because they will allow a dynamic coupling of the temporal changes of both structure (e.g. LAI) and function (e.g. the fluxes of carbon, water and nutrients). For a full achievement, mortality, dispersal, migration and succession must be accounted for and coupled to fluxes and structure estimates.

DATA SETS USED FOR THE COMPARISON

All models have been run with the same major climatic variables using the monthly mean values for precipitation, temperature and cloudiness from Cramer and Leemans (pers. comm.; based on Leemans & Cramer 1991). The derivation of the solar radiation from cloudiness, as well as the translation from Zobler soil texture (1986) to field capacity and wilting point, is described in Otto *et al.* (manuscript). Diagnostic models used the monthly 1987 ISLSCP data set (Sellers *et al.* 1994), but SDBM used a different data set (Gallo 1992). One difficulty for the intercomparison was the use of different vegetation maps by some models.

MAIN RESULTS

The global annual net primary productivity estimated by the models ranges from 40.2 to 80.5 Gt C, with an average of 58.5 Gt. Figure 1 shows the global map of annual NPP, averaged across all models. The coefficient of variation is less than 15% for most areas – it is therefore possible to consider the broad features of this figure as a comprehensive representation of similar NPP fluxes estimated by different models.

The highest productivity ($> 1200 \ g \ m^{-2} \ y^{-1} \ C$) is found in tropical biomes (Amazon, Central Africa, Southeast Asia), where both temperature and precipitation requirements are fully satisfied for photosynthesis. Temperate regions have an intermediate NPP ($500\text{-}700 \ g \ m^{-2} \ y^{-1} \ C$), and the lowest NPP ($< 200 \ g \ m^{-2} \ y^{-1} \ C$) is found in cold or arid regions, where either temperature or precipitation is a limiting factor. The intertropical band in Africa shows the highest spatial variability, from the most productive biomes (tropical evergreen forest) near the equator to less productive ones (arid shrublands) around latitudes $20 - 25°$.

The annual NPP values are only a reduced description of the biospheric carbon assimilation. Seasonality is another important feature, both for land-surface parameterization in general circulation models, and for the representation of competition processes in ecosystem models. For example, when several plant functional types co-occur within one biome, a warming climate might result in some

Figure 1. Global map of the annual NPP ($g \ m^{-2} \ y^{-1} \ C$) estimated from the average of the models compared during the 'Potsdam 95' workshop.

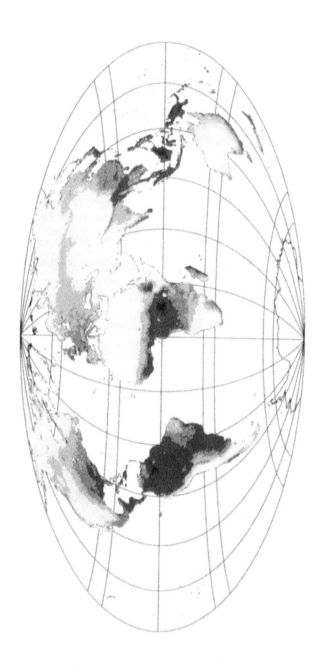

phenological changes in one of them, in turn leading to changing dominance relationships amongst them. The following two sections therefore deal with the *seasonal* NPP fluxes simulated by the models.

SEASONAL NPP IN AFRICA

The intertropical band in Africa is chosen because of its climatic characteristics; temperature remains quite constant throughout the year, as does solar radiation, although the latter decreases towards the equator due to increasing cloudiness. Precipitation is the dominant variable: both its duration and its intensity increase from the tropics to the equator where a short dry season occurs in July. Figure 2 shows the phase diagrams of monthly NPP between 10° and 30° E for all models. From the tropical latitudes towards the equator, the general seasonal behaviour is the following: no productivity (or very low) around the tropics (Sahara, Kalahari), then a short growing period during the rainy season (July in the North, January in the South) that increases in duration as the equator is approached and the more humid savanna ecosystems are reached. In some models, seasonality more and less disappears near the equator, whereas others estimate a reduction in the NPP around July, associated with the short dry season. Figure 2 also shows that models with similar seasonal behaviour can vary in the amplitude of the monthly NPP. A first distinction comes from the fact that only some mechanistic models can simulate a negative monthly NPP. In this part of Africa such negative values might be simulated, for example, in woody savannas during the dry season when photosynthesis is strongly reduced but when the trees still respire. Diagnostic models (SDBM, GLO-PEM, TURC and CASA) generally display smoother seasonal variations than the mechanistic models, although the annual total may not vary significantly. However, the mechanistic models display strong variations in the amplitude of the seasonal dynamics, which may be related to the modelling of LAI. For example, when the models simulate a green active canopy for the whole year up to latitudes as far north as 6 – 7°N (HYBRID, PLAI, FBM), the respiration costs of the dry months (without photosynthesis) result in a negative NPP. Models like PLAI or FBM, which are calibrated at the biome level for representative annual values of NPP, compensate with very high monthly productivity (more than 250 g m^{-2} C) during the active period.

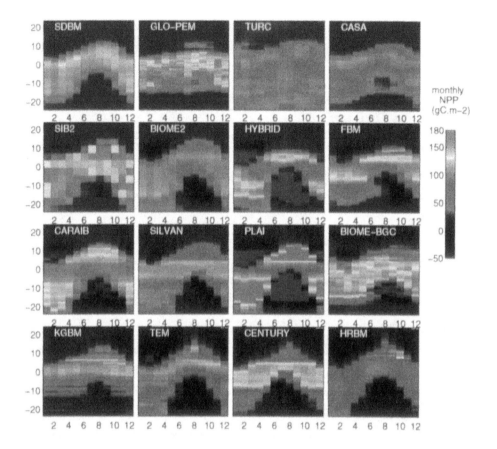

Figure 2. Phase diagrams of the monthly NPP for each model over the intertropical band (23° N–23° S / 10° E–30° E) in Africa. The abscissas are the 12 months, the ordinates are the latitudes. The models are grouped according to the classification made in the text.

SEASONAL NPP/FPAR/LAI INTERCOMPARISON IN EUROPE

In northern latitudes, temperature and radiation are the limiting factors. Models were compared for three points along a temperature gradient from the high Arctic (Novaja Zemlja) to temperate climate (Germany). Table 1 lists the coordinates, the vegetation and the annual NPP (average of all models).

As a key for interpreting the seasonal NPP behaviour, we analysed the intermediate variables LAI and/or FPAR. For the diagnostic models, FPAR is an input variable derived from satellites. LAI is part of the modelling scheme of the mechanistic models, coupling CO_2 assimilation, water exchanges and phenological development. The following computation of FPAR indicates the capacity of such a canopy to absorb the PAR.

Table 1. Sample points considered for Europe

	Coordinates	Vegetation (BIOME 1.0)	Vegetation (Matthews)	annual NPP
P1	52·25° E, 65·25° N	Taiga	Temperate/Subpolar evergreen needle-leaf forest	398 g m^{-2} C
P2	32·25° E, 57·75° N	Cool mixed forest	Cold deciduous forest with evergreens	498 g m^{-2} C
P3	9·75° E, 50·25° N	Temperate deciduous forest	Cold deciduous forest with evergreens	695 g m^{-2} C

Seasonal profiles are shown in Figure 3. The climatic variables show an increase in the length of the 'favourable' growing period from the north-east to the south-west, which can be related to the general increase of the NPP. The figure shows the broad range of variations which can occur for the simulation of LAI between different models, although the NPP is similar. The LAI profiles of the mechanistic models are clearly connected to the biome considered. For example for P1, quasi-constant LAI values occur for models using the Matthews maps, while CARAIB considers some deciduous ecosystems. Extremes are shown by PLAI (LAI > 9 for the whole year) and SILVAN (LAI 2 – 3 for the whole year) although they both work with the same simulated vegetation. However, the FPAR profiles indicate that the sensitivity of the PAR absorption for high LAI saturates nevertheless. Consequently, during the active season, the capacity of absorption simulated by CARAIB, for example, comes close to that of PLAI. Due to the low temperatures in winter, the respiration costs of the evergreen leaves for PLAI (as for FBM or KGBM) are negligible, which may explain that its seasonal NPP behaviour is finally very close to that of CARAIB.

The next point (P2) in the mixed forest displays another feature of the LAI simulations. For BIOME and for the Matthews' map, deciduous trees occur – therefore FBM, KGBM, PLAI and SILVAN attempt to reproduce a seasonal cycle. The LAI derived from satellite data(ISLSCP or CASA) is seasonal, too. Only CARAIB

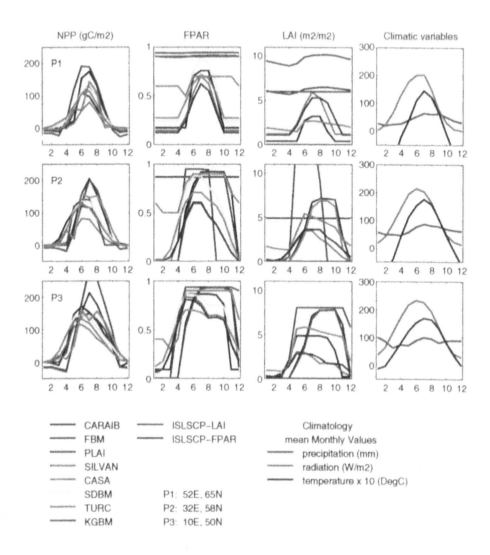

Figure 3. Seasonal variations of NPP, FPAR and LAI as simulated by the models or derived from satellite data, and climatic variables for 3 points along a temperature gradient in Europe.

considers an evergreen ecosystem here but as in the previous case, the NPP profiles are relatively insensitive to these differences, because a green canopy in winter at this latitude does not assimilate so much CO_2 and does not respire so much.

The profiles for the last point (P3) show a slightly better agreement between models with regard to the seasonality of the LAI (all models assume a deciduous forest), but this does not lead to a better agreement between the NPP fluxes. When looking at the FPAR profile for P3, the models seem to agree quite well for the phase of the budburst with the satellite FPAR, although in the FBM and PLAI simulations budburst occurs one month later. This might be due to the fact that they consider only trees, while CARAIB considers a proportion of grass beginning to be active before the trees, and therefore it can simulate a phase similar to the one observed by the satellite (Kohlmaier pers. comm.). Starting later, the NPP fluxes simulated by FBM and PLAI also reach their maximum later. This maximum value is quite high, too – this could be explained by the fact that these models simulate a quite high NPP in summer to compensate its low values in the early spring.

CONCLUSION

This paper has attempted to give an overview of how the seasonal cycle of NPP is simulated by global models. Only the models which have provided monthly NPP outputs for the 'Potsdam 95' workshop have been compared, and it should be recognised that other mechanistic global models for vegetation function do exist (e.g. DEMETER (Foley 1994) and DOLY (Woodward et al. 1995)). Although the seasonal features show large differences between the models, the comparison showed that the effect of two major climatic variables on the seasonality of the NPP is reasonably accounted for. The growing season is linked to the rainy period for the African intertropical band, and it is linked to the warm period for north-east Europe. The differences in the representation of structural variables like LAI appear clearly, however they are not obviously linked to differences in the NPP fluxes.

The diagnostic models driven by satellite observations are expected to provide a simple representation of the seasonal dynamics of the assimilation of the atmospheric CO_2 by the terrestrial biosphere, but it must be kept in mind that the estimated NPP is affected by the way the satellite measurements are processed.

Among the mechanistic models for NPP, we have included those models which do not require calibration and which couple the simulation of vegetation processes and vegetation pattern. These are the most adequate candidates to be used in a predictive mode, when looking for the response of vegetation to climate change. However, the analysis of NPP or LAI behaviours simulated by those models, which are constrained in several ways (prescribed vegetation distribution, prescribed phenology, prescribed productivity at the biome level), is necessary to better understand the relationships between the numerous and complex processes. In fact, even with the assumption of identical biomes, seasonal profiles of both LAI and NPP fluxes displayed different features between the models. A more careful analysis is therefore required. The soil moisture sub-models certainly play a key role in the semi-arid areas, and this must be considered in the comparison. Validation data are still missing, but compilation and organisation of global data bases are currently under way (Prince *et al.* 1995). The use of satellite time-series to evaluate the simulated seasonal behaviour is also a promising strategy, provided that the spatial and temporal scales of the satellite data suitably are adapted, and that the models consider the actual vegetation.

ACKNOWLEDGEMENTS

The author gratefully acknowledges the GAIM/DIS part of IGBP for the global NPP model intercomparison initiative, and thanks all the participants in 'Potsdam 95' for making their data available, as well as for the fruitful discussions on seasonal features during the workshop. Participants in 'Potsdam 95' were: G. Churkina, G. Colinet, W. Cramer, J. Collatz, G. Dedieu, W. Emanuel, G. Esser, C. Field, A. Fischer, L. François, A. Friend, A. Haxeltine, M. Heimann, J. Hoffstadt, J. Kaduk, L. Kergoat, D. Kicklighter, W. Knorr, G. Kohlmaier, B. Lurin, P. Maisongrande, P. Martin, R. McKeown, B. Meeson, B. Moore III, R. Nemani, R. Olson, R. Otto, W. Parton, M. Plöchl, S. Prince, J. Randerson, I. Rasool, B. Rizzo, A. Ruimy, S. Running, D. Sahagian, B. Saugier, A. Schloss, J. Scurlock, W. Steffen, P. Warnant and U. Wittenberg. The global results presented in the first section of this paper were compiled for the GAIM First Science Conference in Garmisch-Partenkirchen, Germany, 25-29 September 1995 by D. Kicklighter, M. Plöchl, and A. Schloss. The paper has been improved thanks to reviews by G.H. Kohlmaier, R. Otto and R. Nemani.

REFERENCES

Esser G, Hoffstadt J, Mack F Wittenberg U (1994) High Resolution Biosphere Model: Documentation Model version 3.00.00. Institut für Pflanzenökologie, Justus-Liebig-Universität Gießen

Foley JA (1994) Net primary productivity in the terrestrial biosphere: The application of a global model. Journal of Geophysical Research 99(D10):20773-20783

Friend AD Stevens AK Knox RG Cannell MGR (in press) A process-based, terrestrial biosphere model of ecosystem dynamics (Hybrid v3.0). Ecological Modelling

Gallo KP (1992) Experimental global vegetation index from AVHRR utilizing pre-launch calibration, cloud and sun-angle screening. Digital data, NOAA, National Geophysical Data Center, Boulder, Colorado, 1992

Haxeltine A Pentice IC (manuscript) A general model for the light-use efficiency of primary production.

Heimann M Keeling CD (1989) A three-dimensional model of atmospheric CO_2 transport based on observed winds: 2. Model description and simulated tracer experiments. Geophysical Monograph 55:237-275

Kaduk J Heimann M (1996) A Prognostic Phenology Scheme for Global Terrestrial Carbon Cycle Models. Climate research 6:1-19

Kergoat L (manuscript) A model of hydrologic equilibrium of leaf area index at the global scale. Journal of Hydrology

Knorr W Heimann M (1995) Impact of drought stress and other factors on seasonal land biosphere CO_2 exchange studied through an atmospheric tracer transport model. Tellus 47B:471-489

Kumar M Monteith JL (1981) Remote sensing of crop growth. in Smith H (ed) Plants and the daylight spectrum, 133-144. Academic Press, New York

Leemans R Cramer W (1991) The IIASA database for mean monthly values of temperature, precipitation and cloudiness of a global terrestrial grid. Research Report, RR-91-18. International Institute for Applied Systems Analysis (IIASA), Laxenburg, Austria

Lieth H (1975) Modelling the primary production of the world. in Lieth H Whittaker RH (eds) Primary productivity of the biosphere, 237-263. Springer-Verlag, Berlin

Lüdeke MKB, Badeck F-W, Otto RD, Häger C, Dönges S, Kindermann J, Würth G, Lang T, Jäkel U, Klaudius A, Ramge P, Habermehl S Kohlmaier GH (1994) The Frankfurt Biosphere Model: a global process-oriented model of seasonal and long-term CO_2 exchange between terrestrial ecosystems and the atmosphere. I. Model description and illustrative results for cold deciduous and boreal forests. Climate Research 4:143-166

Matthews E (1983) Global vegetation and land use: New high-resolution data bases for climate studies. Journal of Climate and Applied Meteorology 22:474-487

Noblet, N de (1996) Modelling late-Quaternary paleoclimates and paleobiomes. in Huntley B, Cramer W, Morgan AV, Prentice HC, Allen JRM (eds) Past and future rapid environmental changes: The spatial and evolutionary responses of terrestrial biota, 31-52. Springer-Verlag, Berlin

Otto RD, Hunt ER Kohlmaier GH (manuscript) Static and dynamic input data of Terrestrial Biogeochemical Models. Global Biogeochemical Cycles

Parton WJ, Scurlock JMO, Ojima DS, Gilmanov TG, Scholes RJ, Schimel DS, Kirchner T, Menaut J-C, Seastedt T, Garcia Moya E, Kamnalrut A Kinyamario JI (1993) Observations and modeling of biomass and soil organic matter dynamics for the grassland biome worldwide. Global Biogeochemical cycles 7(4):785-809

Plöchl M Cramer W (1995) Coupling global models of vegetation structure and ecosystem processes. An example from Arctic and boreal ecosystems. Tellus 47B:240-250

Potter CS, Randerson JT, Field CB, Matson PA, Vitousek PM, Mooney HA Klooster SA (1993) Terrestrial ecosystem production - a process model based on global satellite and surface data. Global Biogeochemical Cycles 7(4):811-841

Prentice IC, Cramer W, Harrison SP, Leemans R, Monserud RA Solomon AM (1992) A global biome model based on plant physiology and dominance, soil properties and climate. Journal of Biogeography 19(2):117-134

Prince SD Goward SN (in press) Global net primary production: a remote sensing approach. Journal of Biogeography

Prince SD, Olson G, Dedieu G, Esser G Cramer W (1995) Global Primary Production Land Data Initiative - Project Description. Working Paper 12, IGBP-DIS

Raich JW, Rastetter EB, Melillo JM, Kicklighter DW, Steudler PA, Peterson BJ, Grace AL, Moore B III Vörösmarty CJ (1991) Potential net primary productivity in South America: application of a global model. Ecological Applications 1(4):399-429

Ruimy A, Dedieu G Saugier B (in press) TURC-Terrestrial Uptake and Release of Carbon, a diagnostic model of continental gross primary productivity and net primary productivity. Global Biogeochemical Cycles

Running SW Hunt ER Jr (1993) Generalization of a forest ecosystem process model for other biomes, Biome-BGC, and an application for global-scale models. Scaling processes between leaf and landscape levels. *in* Ehleringer JR Field C (ed) Scaling Processes between leaf and landscape levels,141-158. Academic Press, New York

Sellers PJ, Los SO, Tucker CJ, Justice CO, Dazlich DA, Collatz GJ Randall DA (submitted) A revised land surface parameterization (SiB2) for atmospheric GCMs. Part 2: The generation of global fields of terrestrial biophysical parameters from satellite data. J. Climate

Sellers PJ, Tucker CJ, Collatz CJ, Los SO, Justice CO, Dazlich DA, Randall DA (1994) A global 1 by 1 NDVI data set for climate studies. Part 2: The generation of global fields of terrestrial biophysical parameters from the NDVI. International Journal of Remote Sensing 15(17):3519-3545

Warnant P, Francois L, Strivay D Gerard JC (1994) CARAIB: A global model of terrestrial biological productivity. Global Biogeochemical Cycles 8(3):255-270

Wilson MF Henderson-Sellers A (1985) A global archive of land cover and soils data for use in general circulation climate models. J. Climate 5:119-143

Woodward FI, Smith TM, Emanuel WR (1995). A global land primary productivity and phytogeography model. *Global Biogeochemical Cycles* 9(4):471-490

Zobler L (1986) *A World Soil File for Global Climate Modeling.* Goddard Institute for Space Studies.

General discussion and workshop conclusions

Predicting the response of terrestrial biota to future environmental changes

Brian Huntley
Wolfgang Cramer
Alan V. Morgan
Honor C. Prentice
Judy R.M. Allen

The papers presented above have considered various aspects of the workshop theme, and together enable an assessment of the relative importance of spatial and evolutionary responses to rapid environmental changes. Many of the papers provide indications as to the threshold rates and magnitudes of environmental change to which organisms may be able to respond by means of one or other, or a combination of both, mechanisms. A series of questions and issues raised by these papers were debated at the workshop either in plenary sessions or in smaller groups. The discussion that follows draws upon the reports from the rapporteurs for the separate discussion groups (Margaret B. Davis, Russell W. Graham, Adrian M. Lister, Herman H. Shugart and Elisabeth S. Vrba) and summarises the key points of debate and the conclusions that emerged from the workshop.

Although many questions can be posed in relation to the workshop theme, three may be viewed as being of particular importance:

1. How good is the past as the 'key to the future' – especially a future with extensive human disturbance?
2. Can the rates of climate change during those past episodes of rapid global environmental change to which ecosystems have been exposed be quantified without using biological proxies in a circular manner?
3. What, if any, rôle will evolution play in the response to future climate change?

An important issue when addressing these questions is the extent to which the answers may differ according to which component of the terrestrial biosphere is being considered. In particular, are there fundamental differences between higher plants and animals with respect to these questions?

NATO ASI Series, Vol. I 47
Past and Future Rapid Environmental Changes:
The Spatial and Evolutionary Responses of
Terrestrial Biota
Edited by Brian Huntley et al.
© Springer-Verlag Berlin Heidelberg 1997

The past as the 'key to the future' ?

Although it is clear that the response of ecosystems to future climate change will be strongly influenced by human effects upon the landscape, the continental-scale anthropogenic impacts of the late twentieth century being without precedent, the palaeoecological record nonetheless provides the only basis for understanding some of the processes involved in ecosystem response to global environmental change.

The Quaternary fossil record provides a much longer timescale than ever can be provided by studies of contemporary systems. In consequence the fossil record can help identify processes that are important in modelling responses to climate change, as well as providing data that may be used to test the predictions of such models. Such data–model comparisons are especially valuable as a means both to understand the shortcomings of, and subsequently to improve the performance of, the models used to predict both future environmental changes and the biospheric response to these changes.

The Quaternary record also may be useful in identifying potential refugial areas; areas that may have been relatively less sensitive to large-scale environmental changes and that thus have remained relatively stable in terms of their biota during the alternating glacial and interglacial stages of the Quaternary. However, even if this is possible, the analogy with the future may fail for at least two reasons. Firstly, the trace-gas induced climate changes of the future are likely to differ, in their regional patterns as well as in their magnitude, from the past climate changes that resulted principally from orbital variations. Secondly, the altered composition of the atmosphere itself, particularly the concentration of carbon dioxide, has direct impacts upon plant function.

In additon to these two limitations upon the use of rapid Quaternary climate changes as analogues for the future, two further points must be borne in mind. Firstly, Quaternary fossil data suffer from inherent limitations arising from their taphonomic and taxonomic biases and the temporal resolution that they offer. Secondly, the already substantial, and likely increasing, extent of human disturbance of the global land surface, as a consequence of agricultural and other land uses, renders many areas relatively inhospitable to the wild plants and animals that naturally would occupy them.

These limitations notwithstanding, we conclude that Quaternary palaeoecological and palaeoenvironmental data have a vital rôle in underpinning our predictions of the impacts of forecast future climate changes. Although past conditions provide poor analogues for the future, the responses of organisms to past and future rapid environmental changes will be by means of the same principal mechanisms and will be subject to the same limitations that are inherent in these mechanisms.

Establishing past rates of climate change

The establishment of past rates of climate change using only data that are independent of the biotic response to these changes remains one of the most thorny problems in the study of Quaternary palaeoenvironments. However, a number of such independent data sources are available and deserve greater attention and exploitation than many hitherto have received. The best known of such sources in recent years have been the stable isotope and other records from polar ice cores; these, however, have the limitation that they may not reflect truly *global* rates and magnitudes of climate change and that, in addition, the link to climate is often, as in the case of the $\delta^{18}O$ record from polar ice, indirect or at best complex.

Highest priority must be given to obtaining further records from the principal continental areas where they may be more directly related to the fossil record of the response of terrestrial biota. Laminated lake sediments may offer one important but as yet under-exploited source of palaeoclimatic information; lamination thickness and micro-sedimentology both offer scope to provide palaeoclimatic proxy data. Stable isotope records from lower-latitude ice sheets, from speleothems and from organic remains all also can provide data relating to palaeoclimate conditions.

An alternative strategy that sometimes has been proposed is that of examining multiple biological indicators from a single sedimentary sequence. Although this approach may enable the underlying climatic signal to be extracted from the biotic response, it strictly does not break the circle of cause and effect. Where, however, physical, chemical or sedimentological data can be obtained from the same sequence as is used to extract the fossil evidence of past biota, then a truly 'multi-proxy' approach is possible and holds the greatest potential for separating the underlying environmental changes from the biotic response to these changes.

There is an urgent need for a series of such 'multi proxy' studies if we are to critically assess both the rates and magnitudes of past climate changes over the continents and to establish the rates — and hence gain insight into the mechanisms — of biotic response to these changes. Although laminated lake sediments offer great scope in this respect, they are likely only to provide palaeovegetation data and perhaps evidence of some invertebrate groups; in order to investigate the response even of small terrestrial vertebrates it will be necessary also to undertake linked studies of other sedimentary environments more suited to the preservation of vertebrate remains.

The rôle of evolution in the response to future climate change

In addition to any adaptive evolutionary responses that may be elicited by a changing environment, populations of organisms also are continually subject to non-adaptive stochastic genetic processes that may affect their capacity to respond to environmental change. Factors that will influence the relative importance of these two general classes of genetic process in different taxa include:

- the breeding system and overall reproductive biology of the taxon;
- the dispersal ecology and gene-flow characteristics of the taxon;
- the life-history traits of the taxon, e.g. length of life cycle, life form, functional type, etc.;
- the available genetic variation ('adaptive potential') within the taxon and the distribution of this variation within and between the populations of the taxon; and
- the type and extent of the taxon's present geographical distribution, including its population and/or meta-population structure.

The relative importance of stochastic and adaptive evolutionary change, nevertheless, is extremely hard to predict. Even outbreeding taxa that at present are geographically widespread, occur in large populations and show extensive gene flow will be subjected to stochastic loss of genetic diversity if, in the future, their population size is substantially reduced and this reduction persists over many generations.

In general, long-lived organisms such as trees may exhibit considerable individual resilience in the face of environmental change, whereas short-lived, out-breeding or early-successional taxa are more likely to show a genetic response to changing environments. Despite this, only micro-evolutionary changes of frequency and combinations of existing alleles are likely to occur even in short-lived taxa over the time scales relevant to considerations of forecast future climate change (i.e. the next 1 – 2 centuries); significant morphological and/or physiological evolution will be extremely rare over this time scale.

Given the importance of the various characteristics of organisms outlined above in determining their ability to exhibit adaptive genetic responses and their susceptibility to stochastic events, it is possible to predict which general classes of organism are likely to be particularly susceptible to the type of stochastic genetic events that will reduce their capacity to respond to environmental change. These more susceptible groups include taxa that have low overall levels of genetic variability, a limited migration capacity, limited geographical ranges and/or small total populations. Late-successional taxa often also will be more susceptible as will 'specialists'. It is important to note that some taxa with large overall geographical ranges will exhibit differential sensitivity to environmental change in different parts of their overall range.

The Quaternary fossil record provides general support for these conclusions; it also leads us to expect that rapid extinction of taxa will occur as a consequence of climate change. Which particular taxa will become extinct and the subsequent consequences of these extinctions are beyond our capacity to predict at this time, and may well remain so.

In addition to these three general questions, many more specific issues are raised by the evidence presented at the workshop. Three of these topics that were judged to be of particular importance were addressed by discussion groups during the workshop and their conclusions will be outlined below. These topics relate to three quite distinct areas upon which the workshop subject impinged and are as follows:

1. The extent to which the paradigm of co-evolution of ecosystem components conflicts with the evidence for individualistic responses to past environmental changes. This topic has implications that are central to the whole subject of ecosystem organisation, as well as to the impacts of climate change upon ecosystems and their function.

2. The extent to which enhanced extinction rates, or even mass extinctions, are associated with periods of past rapid environmental change, in particular during the Quaternary. The conclusions here may have fundamental implications with respect to the likely impact of future anthropogenic climate changes in terms of loss of global biodiversity.

3. The extent to which, given the benefit of hindsight, effective conservation measures could have been designed to minimise the impacts upon the terrestrial biota of the rapid global warming during the last deglaciation. Although the latter period may not be a particularly close analogue for the near future, nonetheless, if successful conservation measures cannot now be designed, given a knowledge of precisely what did happen to the environment and the biota, then this has profound implications with respect to contemporary conservation problems.

Co-evolution vs. individualism – consequences for community assembly

One of the key conclusions that emerges from the palaeoecological record is that terrestrial taxa generally exhibit individualistic responses to environmental changes. As a consequence, no-analogue assemblages are frequent in the Quaternary fossil record, indicating the occurrence of palaeocommunities unlike any extensive modern community; in many cases the taxa co-occurring as fossils have non-overlapping contemporary geographical ranges. It seems certain that new assemblages of taxa also will arise in the future in response to global climate change. The individualistic response of taxa renders more difficult the prediction of the precise composition of these future communities.

To-date, palaeocommunity reconstructions principally have been made only for individual taxonomic groups (e.g. higher plants, beetles, terrestrial molluscs, birds, mammals, etc.). Little has been done towards reconstructing past ecosystems and their trophic and other functional relationships. The same in general also is true of attempts to predict the biotic response to future climate change.

Whereas ecologists who study the structure and organisation of ecosystems frequently identify and focus upon a range of inter-specific interactions that apparently indicate co-evolution, it is apparent that many of these interactions may be susceptible to uncoupling as a result of environmental change. Such uncoupling will be especially likely where the interacting organisms respond to different cues with respect to critical stages of their annual cycle of activity (e.g. photoperiod vs. temperature). Parasite–prey interactions may be considered some of the most closely coupled of inter-specific interactions and thus perhaps the least susceptible to disruption as a result of environmental change. Whereas some such interactions indeed may be minimally disrupted by a changing environment (e.g. endoparasites that have a resting transfer stage), others may be more susceptible (e.g. ectoparasites and/or endoparasites that have an active transfer stage); the most susceptible may be those cases where the parasite has a complex life cycle involving alternate/intermediate hosts. In addition, individualistic reorganisation of ecosystems may expose species to 'new' parasites leading to the development of 'new' interactions. Given the rôle that some parasites play in controlling the abundance and even the range of their hosts, predicting geographical range changes by individual species will be rendered more difficult if these changes alter the species' potential interactions with parasites.

In terms of the paradigm of co-evolution, the evidence of individualism and of no-analogue palaeocommunities must cause us to question the extent to which such apparent co-evolution has occurred. It is clear that during the Quaternary the global environment has continually changed and that the present global environment is exceptional in a Quaternary context. Thus contemporary ecosystems have been assembled in their present form only during the Holocene, and in many cases never previously have existed in the same form. Although those exceptional pairs of species that participate in obligate and specific interactions, in most cases of a mutualistic character, may indeed be reciprocally constrained by their combined environmental tolerances, and thus may respond in tandem to a changing environment, most species that interact do so facultatively and/or non-specifically. We should expect them to continue to do so in the future and thus to display individualistic responses to future environmental changes.

The consequences of species' individualistic responses to their environment with respect to community assembly were commented upon more than thirty years ago by West (1964) when he wrote that "our present plant communities have no long history ... but are merely temporary aggregations under given conditions of climate, other environmental factors, and historical factors". As for community assembly rules, it seems that only the most general of rules derived from the functional characteristics of ecosystems are likely to apply; these may be so general as to be almost trivial in the majority of cases.

Rapid global environmental changes and extinctions

Viewed against a long-term geological perspective, the Quaternary as a whole has been a period marked by extinctions. These extinctions have not, however, occurred as a consequence of a single catastrophic event, as seems to be the case for at least some earlier mass extinctions, nor have they taken place steadily as a chronic loss of biodiversity throughout the last 2·4 Ma. Instead, a number of major extinction events can be distinguished within the Quaternary. Some of these events appear to have affected a diverse range of groups of organisms (e.g. at ca. 780, 450 – 400 and 10 ka BP), whereas others have impacted principally only upon one or two major groups (e.g. the major phase of extinction for terrestrial molluscs in Europe that occurred between 400 and 300 ka BP). All of these events, however, appear to relate to times of rapid global environmental change; the fact that a series of such events may be recognised relates both to the repeated rapid climate changes of large magnitude that characterise the glacial–interglacial cycles of the Quaternary and to the differential impacts of each such event between different geographical regions and also to some extent between taxonomic groups that have different principal environmental constraints. Furthermore, although in some cases extinctions in one group may only indirectly result from an environmental change, being principally the consequence of the effect of that change upon other ecosystem components (e.g. larger mammals generally are vulnerable only indirectly to temperature changes through the effects of these changes upon vegetation), these taxa remain vulnerable to extinction as a consequence of such environmental changes.

Examination of the Quaternary palaeontological record allows a number of generalisations to be made as to the taxa at greatest risk of extinction as a consequence of a rapid change in the global environment. Amongst the characteristics of these taxa at high risk are the following:

- large-body size – associated with a low intrinsic rate of population increase, generally with a small total population size and often also with requirements for a large range or territory;
- low levels both of intra- and inter-population genetic variation;

- spatially-limited geographical distribution;
- habitat specialisation – especially requirements for a specific type often of spatially-limited, but sometimes of temporally limited, habitat – e.g. some terrestrial molluscs that require moist, deeply-shaded situations, or the vernal herbs of temperate deciduous woodlands that require a minimum period of high light intensity and warmth in spring followed by shade that restricts other potentially competing herbaceous taxa;
- resource specialisation – e.g. a specialised phytophage feeding upon only one or at most a very few host plants;
- a relatively low position in the trophic pyramid – i.e. herbivores generally seem at greater risk than carnivores;
- absence of behavioural flexibility – e.g. inability to adjust hibernation;
- adaptations to cold conditions – cold-adapted taxa have been especially vulnerable, perhaps because the glacial terminations have been the most rapid large-magnitude climatic changes of the Quaternary, allowing least time for a response whether by migration or adaptation.

It is likely that these same characteristics will render taxa vulnerable to potential extinction as a consequence of future anthropogenic global environmental changes.

It also is evident from the Quaternary record that global climate changes of lesser or, at most, comparable magnitude and rate to that forecast for the next 1 – 2 centuries have resulted in marked extinction events; we should expect the same to be true of the forecast anthropogenic change.

The Quaternary record, as well as the contemporary biogeographic and ecological distributions of many taxa, provide evidence that different geographical areas or types of region are differentially sensitive in terms of the likelihood of extinction of components of their biota as a consequence particularly of global climate warming. Amongst the high risk areas that may be identified are the following:

- high altitude areas of insufficient elevation to enable taxa to respond by moving to higher altitudes – e.g. the Scandes Mountains of Norway and Sweden and peaks in the northern Appalachians where at least some plants and beetles already are restricted to the highest available altitudes, the Pyrenees where at least some land snails are found only on the highest peaks, the high altitude savannas of

Africa whose mammal faunas have no higher altitude areas to which they may retreat;

- coastal areas and areas near sea-level, especially the managed and engineered coasts of the developed world, where sea-level rise is likely to lead to habitat losses;

- isolated oceanic islands often with high levels of endemism — although their endemic floras and faunas have survived Quaternary climatic changes, some species at least are likely rapidly to be affected by the forecast warming of the global climate to mean global temperatures that soon may be warmer than at any time since the Tertiary;

- continental 'cul-de-sacs' – e.g. the tundra areas of northernmost Fennoscandia and sub-Arctic eastern Canada that may be transformed to boreal forest, leaving nowhere to which the tundra taxa may migrate.

Once again it seems inescapable that if global warming of the forecast magnitude and rate does take place then at least regional and in some cases global extinctions will result, on a scale comparable with major extinction events of the past. The resulting loss of global biodiversity will be restored only though evolution over timescales of $10^5 - 10^7$ yr.

Conserving biodiversity in the face of rapid global environmental change

Given the conclusions emerging from the workshop, it is relevant to ask whether successful conservation measures may be possible in the face of any rapid large-magnitude change in the global environment. The most recent such event was the last glacial termination; this event falls within the range of ^{14}C dating and has been well-studied at locations throughout the world. Not only do we have independent evidence of the rate and magnitude of at least the regional climate changes, but we also have excellent palaeovegetation records as well as palaeontological data for a variety of animal groups, vertebrate and invertebrate. These records indicate the mass extinction of larger vertebrates, especially in the northern hemisphere; particularly vulnerable was the fauna associated with the so-called 'tundra–steppe' biome. Examination of this period may provide lessons for those attempting to design conservation strategies for the coming centuries. It also may reveal the extent to which their efforts potentially may meet with success or be rendered futile.

The most extreme expressions of the last glacial termination are those climate records that show warming of several degrees Celsius over a period of between a few decades and a century around 10 ka BP. However, the climate changes at that time were not uniform, rather they exhibited considerable spatial variability. In the British Isles, for example, there was a rapid and very substantial warming that had a profound impact upon ecosystems; this impact included turnover of a large proportion of the flora and extirpation of ca. 90% of the beetle fauna. Although none of the plants or beetles that were lost from the British Isles at this time appears to have become extinct, numerous extinctions did occur elsewhere and in other major groups. In North America there also was a substantial warming, although it was of lesser magnitude than that in north-west Europe; this warming, however, occurred at varying times and was completed markedly before 10 ka BP in some areas. In southern South America the most striking temperature rise also occurred earlier, at ca. 14·5 ka BP. Both in Europe and in North America a variety of larger vertebrates became extinct as a direct or indirect consequence of the climate change at the end of the last glaciation. These extinctions include that of the giant deer *Megaceros giganteus*, the woolly mammoth *Mammuthus primigenius* and the cave bear *Ursus spelaeus* of Europe, and the ground sloths *Nothrotheriops* and *Megalonyx*, the mastodont *Mammut americanum*, the Columbian mammoth *Mammuthus columbi*, the sabre-tooth cat *Smilodon fatalis* and the dire wolf *Canis dirus* of North America. In the Arctic, melting of the sea ice and consequent opening of the sea surface led to a substantially altered climate near to the Arctic coast; summers became cooler whereas winters warmed. As a result the tundra–steppe biome vegetation that had supported populations of large grazing mammals disappeared and the modern tundra developed as perhaps an essentially new biome.

Thus, although the global warming at the end of the last glacial resulted in profound climate changes everywhere, these changes were regionally variable and in some instances locally differed in direction from the global trend. Biodiversity was reshaped in various ways by these climate changes according to their regional character. It seems certain that future climate change similarly will vary greatly from one region to another; furthermore, it may not always be possible to predict the changes that will occur in any given region. It also seems certain that these future climate changes will have substantial impacts upon regional and global biodiversity.

Predicting the regional impacts upon biodiversity also is difficult because the nature of the response of organisms to climate change may differ between different parts of the world. For example, higher latitude continental regions of the northern hemisphere that have floras of relatively low diversity, including many species with extensive geographical ranges, have been characterised by shifts of species' range margins by thousands of kilometres during the Quaternary. Species' glacial and interglacial ranges also often have differed in area by several orders of magnitude. In contrast, in regions where the flora is highly diverse, especially tropical and lower latitude mountainous areas, many species have spatially-limited ranges that do not appear either to have shifted or changed in extent in this dramatic manner in response to Quaternary climate changes. In such high diversity regions plant species may be more specialised in their habitat requirements. The response of such geographically-restricted species to future climate change may not easily be predicted, although it seems unlikely that they will undergo the rapid migrations and large range shifts that are predicted for the temperate forest species of north-west Europe and eastern North America. For at least some geographically-restricted species the response to climate change more probably may be fragmentation and shrinking of their ranges and populations, although not necessarily extinction. However, it also is important to note that in the same way that many species that presently are abundant were rare in the past, some currently rare species may in the future become more abundant; once again the challenge is to predict which these may be and what the effects of their increased abundance might be.

Given the knowledge of what did take place, if faced with the challenge of attempting to conserve global biodiversity in the face of the global environmental changes of the last deglaciation, what conservation strategies might we expect to succeed? Based upon this, what conservation strategy would we recommend be adopted in relation to the global environmental change that, if not already underway, is expected to occur in the near future?

Because the response of organisms was in the past, and is likely in future to be, qualitatively different in different regions, the appropriate conservation strategy also will differ between regions. In regions such as north-west Europe and eastern North

America, where the expected response is one of large-scale range displacement, the conservation of corridors, or better of a network of habitats, that might facilitate such migration will be the preferred strategy to minimise the loss of biodiversity. Elsewhere, in regions of the world where diversity is higher and many species have local distributions and narrower habitat tolerances, the expected response to environmental change may be dominated by more local shifts of range and of relative population size. In these regions the preferred strategy will be to set aside numerous and diverse protected areas that individually and together include the widest possible variety of habitats, thus providing opportunities for species' to achieve their expected responses to environmental change, as well as enabling fragmented populations to survive.

A large part of the biodiversity lost during the last deglaciation was that of the large mammals lost as a result of the loss of the tundra–steppe biome. Because of their size, many of these animals can be expected to have utilised large home ranges; their conservation thus would have required the establishment of very large protected areas. In the world today, establishing such very large protected areas presents an increasing problem as the human population continues to grow, making greater demands upon the environment and causing the habitat available for such protection to shrink in extent. Even areas designated for protection often succumb to the pressures exerted by the growing human population.

The prospect may not be entirely bleak, however, because it is evident that many animal species underwent repeated and substantial fluctuations in population size during the Pleistocene, yet survived. Perhaps many species can survive at relatively low population levels if free from predators and other pressures. Their likelihood of survival might be maximised by establishing protected areas in regions of high relief that may offer the greatest resilience to change and especially to the more stochastic elements of environmental change. However, in the case of the tundra–steppe fauna, it is questionable whether many species would have survived — even given intensive conservation management — because their habitat essentially disappeared. Thus, although appropriate conservation strategies might have facilitated the response of many organisms, including large mammals, to their changing environment and minimised the loss of biodiversity, unless more far-reaching measures had been possible that would have enabled the maintenance of a sufficient area of the tundra–steppe biome, the characteristic fauna of this biome inevitably would have been lost.

As we turn to consider the forecast future climate changes, high latitude and high altitude habitats and biomes once again are at great risk; the tundra biome may face a threat equivalent to that faced by the tundra–steppe ten thousand years ago – and may be equally unlikely to survive without far-reaching measures that limit the extent of the global climate change that eventually takes place. However, other areas that have not faced severe threat during Quaternary global warming events also are at risk. During the Quaternary the poleward range limits of many temperate taxa were shifted far towards the equator during glacial stages, returning polewards during interglacials, whereas their equatorward range limits often suffered much less displacement or were hardly displaced at all. However, future global warming threatens to cause marked poleward displacement of the equatorward limits of many taxa, as well as shifting their poleward limits even further and more quickly than did glacial terminations. The extent to which the poleward limit, especially of trees, will be able to respond may be limited by the rate of permafrost melting and/or by the need for soil development on newly available substrates. Together with the rapid poleward shift of species' equatorward limits, such constraints may lead to severe population reductions. The altitudinal range extension of some trees in mountainous areas also is likely to be limited by the need for soil development on newly available substrates, with the same consequence.

Perhaps one of the areas of greatest uncertainty arises from the lack of consistent scenarios for future changes in precipitation and water availability, and especially for changes in their regional patterns. These factors, although often overlooked, nevertheless have a potential to change the range limits of species and ecosystems similar to that of temperature changes. Soil constraints also may affect the fauna; for example, the peat soils of higher latitudes often will be unsuitable for colonisation by the fauna found today in adjacent lower latitude areas of brown forest soils. Further problems will arise from new combinations of environmental conditions that are without an analogue not only today but also during the Quaternary; for example, new combinations of climatic conditions and seasonal insolation régimes will arise leading to requirements for new combinations of climatic and photoperiodic tolerances. Indeed, the forecast levels of the principal greenhouse gas (CO_2) in themselves represent a no-analogue condition that is outside the range of conditions during the Quaternary as far as we are able to determine them. Predicting the response of ecosystems to this new situation is very difficult, demanding much more than simply a series of short-term measurements of the physiological response of selected short-lived plants to exposure to suddenly elevated CO_2 levels.

A recurrent theme that arose in both the discussion groups and the plenary sessions concerned the availability of the basic skills and information required to research many of the areas essential to understanding and predicting the response of biodiversity to, as well as to monitoring the ongoing impacts of, environmental change. Concern was voiced about:

- the limited availability of base-line studies and of floristic, faunistic and biogeographic data that are essential to any assessment of the extent of ongoing impacts;

- the limited amount of taxonomic research and research into the basic biology of organisms that currently is being undertaken — such research is essential if we are to have the knowledge required to undertake many of the other research programmes that are essential to improve our understanding of biodiversity and of the impacts of environmental change, including monitoring and predicting those impacts; and,

- the limited availability of the training required to produce new generations of researchers – the students who will in due course carry out environmental change research need to receive adequate training in identification skills, underpinned by a sufficient knowledge of taxonomy, nomenclature and the basic anatomical and morphological attributes of organisms — in short there is a need to reinstate much of the 'traditional' teaching in biology that has tended to be displaced by more 'modern' areas of teaching — both are essential and should be seen as complementary rather than competing because many traditional biological problems can be addressed most effectively using modern techniques and approaches, although to recognise and understand the problem requires a knowledge of the traditional basics.

Addressing these limitations of current knowledge, research and training must be seen as worldwide priorities within the context of the fast-developing science of 'global environmental research'.

CONCLUSIONS

The papers presented in the six sections of this volume, plus the summary presented above of the discussions that took place during the workshop, lead to a number of conclusions that are summarised below:

1. There is little evidence to suggest that organisms will be able to achieve any major adaptive evolutionary responses to the forecast rapid changes in the global environment over the next 1 – 2 centuries.

2. Fragmentation and reduction of populations, loss of genetic diversity and the extinction of at least some taxa are likely outcomes.

3. Although large-scale spatial responses may dominate in some regions (e.g. Europe and eastern North America), elsewhere the spatial responses may be on a smaller scale and/or may be less important than shifts in relative population size and spatial extent within diverse landscapes (e.g. New Zealand, low latitude mountainous regions).

4. Large-scale spatial responses are unlikely to be realised at the rates necessary to maintain species' distributions in equilibrium with the changing environment. As a consequence, many taxa in regions where such responses predominate may suffer marked population reductions, with associated, and possibly selective, loss of genetic diversity that will enhance the likelihood of their subsequent extinction. This will especially be the case for species whose retreating range limit is responding directly to the changing environment and thus is not exhibiting any lag.

5. Together the variability in regional expression of the forecast global warming and changing water availability, as well as the overwhelmingly individualistic response of taxa, render difficult the prediction of ecosystem responses at a regional scale.

6. The impact upon ecosystem function of the loss of a 'keystone' species is an area in which the limitations of our present knowledge render our ability to make predictions extremely limited.

7. Valuable lessons can be learned with respect to the conservation of biodiversity by considering the strategies that might have been adopted in order to minimise biodiversity loss as a consequence of 'global warming events' during the Quaternary.

8. The appropriate conservation strategy varies regionally according to the character of the biota, the nature of the region and the likely predominant response of the biota to environmental change.

9. The extinction of some taxa cannot be avoided if the global climate change is unavoidable and if that global change leads to the disappearance of an entire biome and/or a substantial range of habitats.

10. The overall characteristics of taxa that may be especially vulnerable to extinction can be identified in general terms. However, given the individualism of their response and uncertainty about the precise nature of future environmental changes at a regional scale, it is not possible to list particular taxa as being candidates for extinction.

11. Because we remain ignorant of the basic biology of the majority of species, and of their interactions and inter-dependencies, the natural world retains a disquieting capacity to surprise us with unpredictable responses to global climate change. Some species whose dispersal ability is artificially enhanced by some form of human activity may, as a result, be able to migrate quickly enough to remain in equilibrium with the changing climate. Other species that today are rare may rapidly increase their populations if they are able to exploit the habitats disturbed by human activities and/or as a consequence of climate change impacts. Yet other species that today are abundant may rapidly be driven to extinction by the novel environmental conditions of the future, perhaps combined with the pressures exerted by the human population. The extinction of the once extremely numerous passenger pigeon in North America since European settlement stands as an example of such unpredictable loss; the extent to which its extinction may in turn affect the migratory response of the trees and other

plants whose propagules it formerly dispersed only will be revealed when these taxa are called upon to respond to a rapidly changing environment.

Our overall conclusion is that the consequences of forecast global environmental changes are likely to include severe loss of biodiversity at every level, from the loss of genetic diversity in virtually all taxa to the extinction of some, perhaps many, taxa within 1 – 2 centuries.

Because of this, and because we cannot expect to be able to foresee and predict all of the potential consequences for ecosystem function of such biodiversity loss, it is our overwhelming recommendation that it would be prudent to limit the extent of global environmental change by commencing at once to make strenuous efforts to reduce emissions of carbon dioxide and other trace gases.

REFERENCES

West RG (1964) Inter-relations of ecology and Quaternary palaeobotany. Journal of Ecology 52(Supplement):47-57.

LIST OF WORKSHOP PARTICIPANTS

Organising committee

Prof. Brian Huntley, Environmental Research Centre, University of Durham, Department of Biological Sciences, South Road, Durham DH1 3LE, U.K.
Tel: +44 (0)191 374 2432; Fax: +44 (0)191 374 2432;
E-mail: Brian.Huntley@durham.ac.uk

Prof. Wolfgang Cramer, Potsdam Institute for Climate Impact Research, Telegrafenberg, P.O. Box 60 12 03, D – 14412 Potsdam, Germany
Tel: +49 331 288 2521; Fax: +49 331 288 2600;
E-mail: Wolfgang.Cramer@pik-potsdam.de

Prof. Alan V. Morgan, Quaternary Sciences Institute, University of Waterloo, Waterloo, Ontario N2L 3G1, Canada
Tel: +1 519 888 4567 ext. 3029/5633; Fax: +1 519 746 0183;
E-mail: avmorgan@sciborg.uwaterloo.ca

Prof. Honor C. Prentice, Department of Systematic Botany,
Östra Vallgatan 18 – 20, S – 223 61 Lund, Sweden
Tel: +46 46 222 8971; Fax: +46 46 222 4234; E-mail: Honor.Prentice@sysbot.lu.se

Prof. Allen M. Solomon, U.S. Environmental Protection Agency,
200 S.W. 35th Street, Corvallis, Oregon 97333, U.S.A.
Tel: +1 503 754 4772; Fax: +1 503 754 4799; E-mail: solomon@heart.cor.epa.gov

Key speakers and other participants

Dr Judy R.M. Allen, Environmental Research Centre, University of Durham, Department of Biological Sciences, South Road, Durham DH1 3LE, U.K.
Tel: +44 (0)191 374 2000 ext. 4049; Fax: +44 (0)191 374 2432;
E-mail: J.R.M.Allen@durham.ac.uk

Prof. Allan Ashworth, Department of Geosciences, North Dakota State University, Stevens Hall, Fargo, North Dakota 58105-5517, U.S.A.
Tel: +1 701 231 8455; Fax: +1 701 231 7149; E-mail: ashworth@vm1.nodak.edu

Prof. John C. Avise, Genetics Department, University of Georgia, Athens, Georgia 30602, U.S.A.
Tel: +1 706 542 1456; Fax: +1 706 542 3910; E-mail: avise@bscr.uga.edu

Prof. Patrick J. Bartlein, Department of Geography, University of Oregon, Eugene, Oregon 97403-1251, U.S.A.
Tel: +1 503 346 4967; Fax: +1 503 346 2067; E-mail: bartlein@oregon.uoregon.edu

Dr Harald Bugmann, Potsdam Institute for Climate Impact Research, Telegrafenberg, P.O. Box 60 12 03, D – 14412 Potsdam, Germany
Tel: +49 331 278 1144; Fax: +49 331 278 1204;
E-mail: Harald.Bugmann@pik-potsdam.de

Dr Jennifer E.L. Butterfield, Department of Biological Sciences,
University of Durham, South Road, Durham DH1 3LE, U.K.
Tel: +44 (0)191 374 3347; Fax: +44 (0)191 374 2417;
E-mail: J.E.L.Butterfield@durham.ac.uk

Dr Yvonne C. Collingham, Environmental Research Centre, University of Durham,
Department of Biological Sciences, South Road, Durham DH1 3LE, U.K.
Tel: +44 (0)191 374 2000 ext. 4049; Fax: +44 (0)191 374 2432;
E-mail: Y.C.Collingham@durham.ac.uk

Dr John C. Coulson, 15 The Links, Belmont, Durham DH1 2AG, U.K.
Tel: +44 (0)191 386 9107; Fax: +44 (0)191 386 9107

Dr Peter Coxon, Department of Geography, Museum Building, Trinity College,
Dublin 2, Ireland
Tel: +353 1 608 1213; Fax: +353 1 671 3387; E-mail: pcoxon@tcd.ie

Mr John R.G. Daniell, Environmental Research Centre, University of Durham,
Department of Biological Sciences, South Road, Durham DH1 3LE, U.K.
Tel: +44 (0)191 374 2000 ext. 4049; Fax: +44 (0)191 374 2432;
E-mail: J.R.G.Daniell@durham.ac.uk

Prof. Margaret B. Davis, Department of Ecology, Evolution and Behavior,
University of Minnesota, 1987 Upper Buford Circle, St Paul, Minnesota 55108-6097,
U.S.A.
Tel: +1 612 625 5717; Fax: +1 612 624 6777;
E-mail: mbdavis@ecology.ecology.umn.edu

Dr Paul M. Dolman, School of Environmental Sciences, University of East Anglia,
Norwich NR4 7TJ, Norfolk, U.K.
Tel: +44 (0)1603 592533; Fax: +44 (0)1603 507719; E-mail: P.Dolman@uea.ac.uk

Prof. Peter R. Evans, Department of Biological Sciences, University of Durham,
South Road, Durham DH1 3LE, U.K.
Tel: +44 (0)191 374 3357; Fax: +44 (0)191 374 2417;
E-mail: P.R.Evans@durham.ac.uk

Dr Alberte Fischer, Centre d'Etudes Spatiales de la Biosphère,
18 Avenue Edouard Belin, bpi 2801, F – 31055 Toulouse Cedex, France
current address: Potsdam Institute for Climate Impact Research, Telegrafenberg,
P.O. Box 60 12 03, D – 14412 Potsdam, Germany
Tel: +49 331 288 2546; Fax: +49 331 288 2600;
E-mail: Alberte.Fischer@pik-potsdam.de

Dr Russell W. Graham, Research and Collections Center, Illinois State Museum,
1011 East Ash, Springfield, Illinois 62703, U.S.A.
current address: Earth Sciences, Denver Museum of Natural History,
2001 Colorado Blvd, Denver, Colorado 80205, U.S.A.
Tel: +1 303 370 6473; Fax: +1 303 331 6492; E-mail: rgraham@dmnh.org

Dr Alan J. Gray, Institute of Terrestrial Ecology, Furzebrook Research Station,
Wareham, Dorset BH20 5AS, U.K.
Tel: +44 (0)1929 551518; Fax: +44 (0)1929 551087; E-mail: A.Gray@ite.ac.uk

Prof. Philip Grime, Unit of Comparative Plant Ecology, The University of Sheffield, Department of Animal and Plant Sciences, Sheffield S10 2TN, U.K.
Tel: +44 (0)114 282 4315; Fax: +44 (0)114 276 0159;
E-mail: S.Hubbard@Sheffield.ac.uk

Dr Rob Hengeveld, Institute for Forest and Nature Research, Arnhem, P.O. Box 23, 6700 AA Wageningen, The Netherlands
Tel: +31 (0)85 546800; Fax: +31 (0)26 4422175; E-mail: Hengeveld@ibn.agro.nl

Dr Annika Hofgaard, NINA, Tungasletta 2, N – 7005 Trondheim, Norway
Tel: +47 73 910500; Fax: +47 73 915433; E-mail: annika.hofgaard@nina.nina.no

Prof. Håkan Hytteborn, Department of Botany, The University of Trondheim, N – 7055 Dragvoll, Trondheim, Norway
Tel: +47 73 596033; Fax: +47 73 596100; E-mail: hakhyt@alfa.avh.unit.no

Dr George A. King, ManTech Environmental Research Services, US Environmental Protection Agency, National Health and Environmental Research Lab., Western Ecology Division, 200 Southwest 35th Street, Corvallis, Oregon 97333, U.S.A.
Tel: +1 503 754 4310; Fax: +1 503 754 4338; E-mail: george@mail.cor.epa.gov

Dr Adrian M. Lister, Department of Biology, University College London, Gower Street, London WC1E 6BT, U.K.
Tel: +44 (0)171 387 7050 ext. 2670; Fax: +44 (0)171 916 2016;
E-mail: a.lister@ucl.ac.uk

Prof. Richard N. Mack, Department of Botany, Washington State University, Pullman, Washington 99164, U.S.A.
Tel: +1 509 335 3316; Fax: +1 509 335 3517; E-mail: rmack@mail.wsu.edu

Prof. Vera Markgraf, INSTAAR, University of Colorado, Box 450, Boulder, Colorado 80309-450, U.S.A.
Fax: +1 303 492 6388; E-mail: markgraf@spot.colorado.edu

Prof. Csaba Mátyás, Department of Plant Sciences, Faculty of Science, University of Sopron, P.O.B. 132, H – 9401 Sopron, Hungary
Tel: +36 (99) 311 100; Fax: +36 (99) 311 103; E-mail: cm@sun30.efe.hu

Dr Matthew S. McGlone, Landcare Research, P.O. Box 69, Lincoln 8152, New Zealand
Tel: +64 3 325 6701 ext. 3790; Fax: +64 3 325 2418;
E-mail: mcglonem@landcare.cri.nz

Dr Nathalie de Noblet, LMCE Bat 709, DSM, Orme des Merisiers, CEN Saclay, F – 91191 Gif-sur-Yvette Cedex, France
Tel: +33 1 69 08 77 26; Fax: +33 1 69 08 77 16; E-mail: noblet@asterix.saclay.cea.fr

Dr Philippe Ponel, Laboratoire de Botanique historique et Palynologie (Case 451), Faculté des Sciences et Techniques de Saint-Jérôme, Université de Droit d'Économie et des Sciences d'Aix-Marseille, Avenue Escadrille Normandie Niémen, F – 13397 Marseille Cedex 20, France
Tel: +33 91 28 80 13; Fax: +33 91 28 86 68

Dr Richard C. Preece, Department of Zoology, University of Cambridge,
Downing Street, Cambridge CB2 3EJ, U.K.
Tel: +44 (0)1223 336600; Fax: +44 (0)1223 336676;
E-mail: rcp1001@cus.cam.ac.uk

Dr Andrei V. Sher, Severtsov Institute of Ecology and Evolution,
Russian Academy of Science, 33 Leninskiy Prospect, 117071 Moscow, Russia
Tel: +7 095 238 3875; Fax: +7 095 954 5534; E-mail: asher@glas.apc.org

Prof. Herman H. Shugart, Department of Environmental Sciences, Clark Hall,
University of Virginia, Charlottesville, Virginia 22903, U.S.A.
Tel: +1 804 924 7761; Fax: +1 804 982 2137; E-mail: HHS@virginia.edu

Dr Dale S. Solomon, United States Department of Agriculture, Forest Service,
P.O. Box 640, Durham, New Hampshire 03824, U.S.A.
Tel: +1 603 868 7637; Fax: +1 603 868 7604; E-mail: davidh@christa.unh.edu

Dr Martin T. Sykes, Global Systems Group, Department of Ecology,
Lund University, Östra Vallgatan 14, S – 223 61 Lund, Sweden
Tel: +46 46 222 9298; Fax: +46 46 222 3742; E-mail: martin@planteco.lu.se

Mlle Delphine Texier, LMCE Bat 709, DSM, Orme des Merisiers, CEN Saclay,
F – 91191 Gif-sur-Yvette Cedex, France
Tel: +33 1 69 08 31 97; Fax: +33 1 69 08 77 16;
E-mail: dauphin@asterix.saclay.cea.fr

Prof. Elisabeth S. Vrba, Department of Geology, Yale University, New Haven,
Connecticut 06511, U.S.A.
Tel: +1 203 432 5008;Fax: +1 203 458 0918

Prof. Thompson Webb III, Department of Geological Sciences, Brown University,
Providence, Rhode Island 02912-1846, U.S.A.
Tel: +1 401 863 3128; Fax: +1 401 863 2058;
E-mail: Thompson_Webb_III@brown.edu

Guest

Chairman of NATO Committee on the Science of Global Environmental Change

Mr Max A. Beran, TIGER Programme Office, Institute of Hydrology,
Maclean Building, Crowmarsh Gifford, Wallingford, Oxfordshire OX10 8BB, U.K.
Tel: +44 (0)1491 692211; Fax: +44 (0)1491 692313; E-mail: TIGER@IOH.ac.uk

INDEX

The ASI Series Books Published as a Result of
Activities of the Special Programme on Global Environmental Change

This book contains the proceedings of a NATO Advanced Research Workshop held within the activities of the NATO Special Programme on Global Environmental Change, which started in 1991 under the auspices of the NATO Science Committee.

The volumes published as a result of the activities of the Special Programme are:

Springer
and the
environment

At Springer we firmly believe that an international science publisher has a special obligation to the environment, and our corporate policies consistently reflect this conviction.
We also expect our business partners – paper mills, printers, packaging manufacturers, etc. – to commit themselves to using materials and production processes that do not harm the environment. The paper in this book is made from low- or no-chlorine pulp and is acid free, in conformance with international standards for paper permanency.

Springer

CPSIA information can be obtained
at www.ICGtesting.com
Printed in the USA
LVHW081925030520
654913LV00016B/1955